Technology Transfer: From Invention to Innovation

NATO ASI Series

Advanced Science Institute Series

A Series presenting the results of activities sponsored by the NATO Science Committee, which aims at the dissemination of advanced scientific and technological knowledge, with a view to strengthening links between scientific communities.

The Series is published by an international board of publishers in conjunction with the NATO Scientific Affairs Division

A Life Sciences **B Physics**	Plenum Publishing Corporation London and New York
C Mathematical and Physical Sciences **D Behavioural and Social Sciences** **E Applied Sciences**	Kluwer Academic Publishers Dordrecht, Boston and London
F Computer and Systems Sciences **G Ecological Sciences** **H Cell Biology** **I Global Environment Change**	Springer-Verlag Berlin, Heidelberg, New York, London, Paris and Tokyo

PARTNERSHIP SUB-SERIES

1. Disarmament Technologies	Kluwer Academic Publishers
2. Environment	Springer-Verlag / Kluwer Academic Publishers
3. High Technology	Kluwer Academic Publishers
4. Science and Technology Policy	Kluwer Academic Publishers
5. Computer Networking	Kluwer Academic Publishers

The Partnership Sub-Series incorporates activities undertaken in collaboration with NATO's Cooperation Partners, the countries of the CIS and Central and Eastern Europe, in Priority Areas of concern to those countries.

NATO-PCO-DATA BASE

The electronic index to the NATO ASI Series provides full bibliographical references (with keywords and/or abstracts) to about 50,000 contributions from international scientists published in all sections of the NATO ASI Series. Access to the NATO-PCO-DATA BASE is possible via a CD-ROM "NATO Science and Technology Disk" with user-friendly retrieval software in English, French, and German (©WTV GmbH and DATAWARE Technologies, Inc. 1989). The CD-ROM contains the AGARD Aerospace Database.

The CD-ROM can be ordered through any member of the Board of Publishers or through NATO-PCO, Overijse, Belgium.

4. Science and Technology Policy – Vol. 19

Technology Transfer: From Invention to Innovation

edited by

Annamária Inzelt

IKU Innovation Research Centre,
Budapest, Hungary

and

Jan Hilton

Vuman Ltd.,
University of Manchester,
Manchester, Great Britain

Kluwer Academic Publishers

Dordrecht / Boston / London

Published in cooperation with NATO Scientific Affairs Division

Proceedings of the NATO Advanced Research Workshop on
Technology Transfer: From Invention to Innovation
Budapest, Hungary
12–15 November 1998

A C.I.P. Catalogue record for this book is available from the Library of Congress.

ISBN 0-7923-5622-5

Published by Kluwer Academic Publishers,
P.O. Box 17, 3300 AA Dordrecht, The Netherlands.

Sold and distributed in North, Central and South America
by Kluwer Academic Publishers,
101 Philip Drive, Norwell, MA 02061, U.S.A.

In all other countries, sold and distributed
by Kluwer Academic Publishers,
P.O. Box 322, 3300 AH Dordrecht, The Netherlands.

Printed on acid-free paper

Printed in the Netherlands

TABLE OF CONTENTS

PREFACE

Advanced and emerging economies are increasingly based on knowledge and information. The importance of knowledge and technology diffusion requires better understanding of knowledge and technology transfer, and their networks. On 12-15 November 1997 NATO Advance Research Workshop held in Budapest was devoted to these issues. The papers presented in this book were among those contributed by participants from a wide variety of backgrounds at the workshop on "Technology Transfer: from Invention to Innovation".

This NATO ARW was initiated and organised by Vuman Ltd. (University of Manchester) and IKU Innovation Research Centre (Budapest University of Economic Sciences). The workshop was arranged to bring together social scientists, technology transfer managers, specialists in intellectual property rights, primarily from NATO countries and Partner countries.

Discussions focused on the institutional framework of knowledge and technology transfer; intellectual property rights as sources of information and tools for co-operations; international, national and regional aspects of knowledge and technology dissemination, diffusion; networking. The papers selected reflect these topics and show the current situation as it appeared at the workshop.

Annamária Inzelt
Budapest

Eddie Duff and Jan Hilton
Manchester

Co-directors

INTRODUCTION

Transfer of knowledge is as old as civilisation. Tremendous historical examples illustrate well the dangers for society by failing in the transfer of knowledge. It was a burden to people learning about our Planet in medieval ages, that the Catholic Church blocked the dissemination of Galileo Galilee's discovery. In this age of rapid global changes in communication we have to face again the problem of poor social capabilities to transfer knowledge efficiently.

Technology transfer has expanded rapidly over the past twenty years within Western Europe, North America and the Pacific Rim. Industry in these areas recognises that the cost of high quality research is very high and some estimates suggest that some 50% of new products and processes will originate from outside the primary developer. Academic and other research institutions are obvious sources of much of this new technology.

Science and technology may be inspired by science, by technology, by society and disseminate inside and among generations, inside and among organisations and nations. Knowledge dissemination may exist intra-and interdisciplinary, from one or a set of disciplines to technology(ies) and products.

Cross-fertilisation of scientific thinking by dissemination of knowledge is the key factor of scientific development. International mobility, correspondence, any ways of exchanging ideas are important in scientific thinking and development. Knowledge diffusion among disciplines may result in a new track in a science or new sciences (e.g. biophysics, biochemistry, physic-chemistry, cybernetics)

Cross-fertilisation among technologies resulted in a new fundamental type of innovation: *technology fusion.* A new technology can be created through the organic fusing of several technical breakthroughs in a number of different fields. Typical cases of fusion include : mechatronics, biotechnology, new ceramics, and optoelectronics. Collaborative work, continuous interaction, feedback have a growing importance in this process. These changes add other dimension to the innovation model.

We tend to place ownership of knowledge into two categories :- *public and private*. *Codified knowledge* can be both, but *tacit knowledge* is always private. Intellectual property rights as codified knowledge are important tools of knowledge dissemination.

We know that there are tremendous problems with linkages on international, national and enterprise levels. In NATO Co-operating countries, technology transfer is in its infancy; it is crucial for wealth creation and improvement in the quality of life to develop this mechanism . Therefore the aim of this book entitled "Technology Transfer: from Inventions to Innovation" (supported by NATO ARW Science Programme and Co-operation Partners) was to discuss these issues, to identify crucial research issues for science and technology policy researchers and, as a conclusion, to make some policy recommendations. The authors of this book are from NATO member countries (14), from Partner countries (7), from other parts of the world (1) and from international organisations (1).

This book contains selected papers and some recommendation for further co-operation. The first chapter discusses some key issues of knowledge transfer and transmission. The second chapter presents the best practices of knowledge transfer (national and enterprise level). The third chapter goes into detail on the role of codified knowledge (patents) and other intellectual property rights. The fourth chapter monitors Central and East European experiences of knowledge transfer during the transition period, identifying critical issues. The fifth chapter summarises the idea on further co-operation, such as setting up of new networks, collaborative research groups, etc.

Acknowledgements

The co-directors of the workshop are grateful to the NATO Science Committee for its grant to organise a workshop for 61 participants and to the Hungarian government organisation, the OMFB (National Committee for Technological Development) for its in-kind contribution.

We would like to express our thanks to the rapporteurs of the workshop: Duro Kutlaca and Frantisek Bernadic, assistants to the rapporteurs: Noémi Gál and Péter Cserna and language correctors: Elaine Antalffy, Peter Welsh and Marcus Hopwood for their careful and diligent work. Our grateful thanks also go to Beryl Mooney for her patience and attention to detail in the work of formatting and compiling the contents of the book. As well as participants, other distinguished scientists helped the editors as referees. We also would like to thank their generosity: Alexandreu Balaban (Academia Románá) and János Gács (IIASA). Our grateful thanks also go to Larry Rausch, Keith Hodkinson and Jacques de Bandt for their helpful feedback to Chapter V.

CHAPTER I

KNOWLEDGE TRANSFER AND TRANSMISSION

ARE TRANSITION COUNTRIES 'INSIDERS' OR 'OUTSIDERS' OF THE KNOWLEDGE-BASED ECONOMIES?

DR. ANNAMÁRIA INZELT

IKU Innovation Research Centre
Budapest VIII.
Múzeum u. 18.
Hungary 1088 Pf.
tel 36-1-318-5674
fax: 36-1-338-2438
ainzelt@iku.omikk.hu

"For those cities that were great in earlier times must have now become small, and those that were great in my time were small in the time before... Man's good fortune never abides in the same place."
Herodotus (cited by Michael Ondaatje 1993)

Introduction

Man's march from the cave to the information society has depended on the accumulation of knowledge and its transmission between generations and communities. A distinction must be made between knowledge accumulation and utilisation. Knowledge accumulation and its diffusion in time and space is imminent in humankind. It is the very essence of progress, improving living conditions and quality of life. The role of knowledge has changed significantly in this century. This paper concentrates on a narrow issue, namely, which models of innovation help to exploit science and technology for human well-being, improved competitiveness, the wealth of nations, protection of environment, etc. It touches on the transmission of knowledge and concentrates on flows of knowledge, and on the processes and players of transfer. It does not investigate the stock or the production of knowledge, inspirations of science, etc. The first section summarises the shift in the economic role of knowledge, the second investigates the different models of innovation and then goes into details of the tools and capabilities of transfers when feedback loops are important. The third section discusses the organisational issue and the fourth some questions beyond national boundaries. The fifth touches on the measurement of the flow of knowledge and the transfer of technology. The sixth makes some concluding remarks.

A. Inzelt and J. Hilton (eds.), Technology Transfer: From Invention to Innovation, 3–21.

I. The shift in the economic role of knowledge

The recent changes in the world economy have turned the focus of investigation to *the knowledge-based economy*. Knowledge-based economies are those economies which are directly based on the production, distribution and use of knowledge and information. The linkage between science and technology is increasing. As Francis Narin *et al's* recent study highlighted, technology and science are getting closer and closer, have specialised linkages, and are virtually contemporaneous. It shows that a national science base contributes significantly to national technology leadership. (Narin *et al.* 1997) Inversely, more and more analyses show that developments in technology continue to feed into the research process. (Patel and Pavitt 1994) The science system is becoming an integral part of a larger network and system – the *knowledge-based economy*.

The main challenges are the following:

➲Techno-paradigm shift from modern technologies to high technologies
➲Changing configuration of national innovation system (collaborative networks, organisational changes)
➲ Shifts in the Socio-Economic paradigm (Keywords: restructuring employment, highly skilled workers, service sectors, distance work, social relationships, welfare state, role of trade unions, non-unionised sectors, new inequalities, on-line-off-line society, etc.)
➲Globalisation
➲Economies in Transition② (post socialist economies)

Because of these developments *science systems* are in a period of **change** not only in the so-called economies in **transition②** but also in advanced economies. Many experts use the category of "transition economy" even for **G7** (the seven giant countries are: the USA, Great Britain, France, Germany, Japan, Italy and Canada), but the post-socialist economies have to manage two different types of transition: (1) following the changing techno-economic paradigm, and (2) transformation from socialism into democratic society and market economy. I call this latter group of countries with double transformation tasks, "economies in transition on square②".

Here, knowledge covers both *scientific and technological knowledge*. The distinction between them is relevant for science and technology (S&T) policy because of different views on the appropriate role of government (including international organisations) in funding the production of various types of knowledge. Transmission and transfer depend on how the product of knowledge is considered. If knowledge is a "public good", then everybody may share.

According to an OECD description (OECD 1996, p. 23) the key functions of a knowledge-based economy are

i. knowledge **production** – developing and providing new knowledge
ii. knowledge **transmission** – educating and developing human capital resources, and
iii. knowledge **transfer** – disseminating knowledge and providing inputs to problem solving.

ad (i) *Recent changes in knowledge production itself are shifting its transmission and transfer. Interaction between producers and users of knowledge is an important source of production. The crucial transformations in the knowledge system are the following:*

- The rate of growth in science is high and the accumulation of assets of knowledge is fast.
- In some fields the distinction between basic and applied research has become somewhat blurred.
- As a consequence of the **growing** number of scientists and the **increasing** proportion of educated people in the labour force, with technological advances, each of them can now produce more knowledge than ever before.
- The scale of critical mass in personnel and financial **resources,** and in technological equipments for R&D is increasing in an unprecedented way.

The marginal costs of scientific progress are increasing in certain disciplines.

ad (ii) *Knowledge transmission has to fulfil two tasks in the knowledge-based economy:*

- continuous learning of codified information and tacit knowledge,
- unbroken development of competencies to use this information.

ad (iii) Learning may take many different forms:

formal education, formal R&D, training in non-formal settings, learning-by-doing, on-the-job learning, interactive learning in networks, engaging in trade, imitation, reverse engineering, etc.[1] It is increasingly important to acquire a range of skills or types of knowledge. Computer literacy and access to network facilities are becoming more important than literacy in the traditional sense. Availability of on-line systems is a key factor for individuals and organisations. Isolation can be avoided by joining such

[1]As an EIMS (1994) report emphasised, firms themselves face the need to become learning organisations, continuously adapting management, organisation and skills to accommodate new technologies. They also join in networks, where interactive learning, involving producers and users, experimentation and the exchange of information is the motor of innovation.

networks. A new gap may threaten if some of the nations and generations belong to the on-line societies, and the remainder to the off-line societies.

II. Models of Innovation

The economic success of a society increasingly depends on its power of knowledge distribution. This power is based on transmission and transfer capabilities and on the mutual influence of science and technology. In this context, it must be emphasised that technology transfer, from invention to innovation, is not a one-way model. Interactions and networks are of great importance and contain many feedback loops. As David and Foray (1995) emphasised, the economy becomes a hierarchy of networks, driven by the acceleration in the rate of change and the rate of learning. What is created is a network society, where the opportunity and the capability of getting access to and joining knowledge- and learning-intensive relations determine the socio-economic position of individuals and firms.

The contemporary innovation model differs from earlier ones. According to Rothwell's description (1994), we can distinguish five generations of innovation models. The first generation, the technology-push model, dominated from about 1950 to the 1960s. The second generation, the market-pull model, was dominant between the late 1960s and early 1970s. The third generation, the interactive or coupling model, marked the period from the mid-1970s to the early 1980s. The fourth generation, the so-called integrated model, was dominating between the mid 1980s and the early 1990s. Since then, we can speak about a fifth generation: that of systems integration and networking. In the following, those innovation models which occurred in some form in socialist countries will be presented, together with the science-based innovation model, as a challenge for the transition economies②.

II. 1. THE LEGACY OF THE SOCIALIST MODEL[2]

If we apply the terminology of the innovation model to relationships of socialism we can observe the appearance only the first and third generation of these models in command economies. In these countries we can speak about a *deformed version of the first generation* [3] which dominated until the mid-1970s. The scheme of the *chain-link, the technology-push and one-way linear model of planned economies* was much longer than this model in market economies. (Figure 1)

In this model, all development issues, not only the strategic ones, were decided by bureaucracy. All linkages between the players were indirect and co-operation went

[2] This part is taken from Inzelt 1997

[3] In socialist economies the enterprise was not a venture. The organisation did not have any competencies in many functions. The basic deformation was the enterprise itself. The main corporate functions and their organisations were extramural (Inzelt 1988). The market was only a simulated one even in reformed socialist economies.

through the hierarchy. Integration was vertical with some horizontal elements. Inter-organisational relationships were much more important than intra-organisational ones because **most functions of firms were separated from enterprises.**[4] Hungary, as a reformed socialist country, had modified the strict Soviet model in several waves since 1956. In a given framework, the authorities had allowed some limited direct linkages among universities and enterprises and between enterprises. **Enterprises** also had regained some functions of **firms** (Inzelt 1988). Many players of system of innovations were absent from the economic arena and interests in co-operation were limited by orders for goods and services. In 1968 the Hungarian Economic Reform had introduced some market instruments and some first steps towards a *second generation, market-pull innovation model.*

But the size structure of organisations was further deformed during the seventies. It was an inverted pyramid, where micro and small enterprises were missing. Joint ventures with foreign partners hardly existed. Multinationals usually were absent in these economies, on the one hand, their entry was blocked, and, on the other, they usually had very limited interests in doing business with the command economies.)

Figure 1

First Generation Innovation Model by Economic Systems (1950-1965)

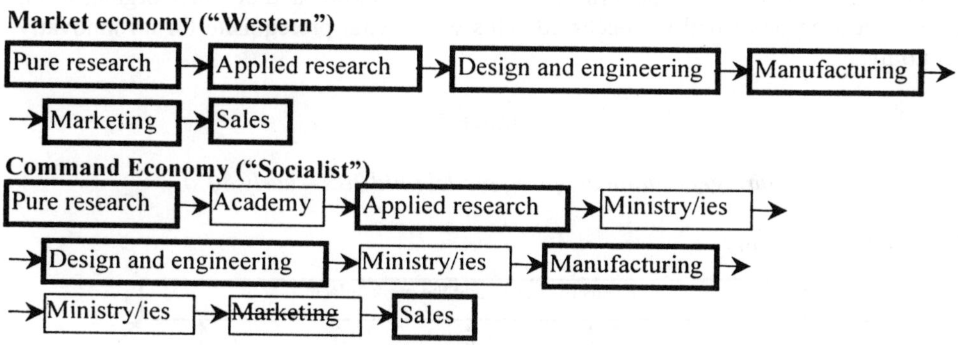

Hungary attempted to overcome the main feature of the socialist system: the lack of innovativeness arising from the fact that the economic players had little or no interest in commercialising R&D results. In the field of S&T policy formulation, the most important attempt was to re-link R&D and business activities and to shorten the path from invention to the practical application of new technologies in the production process. This process of rethinking S&T policy resulted in the Large-Scale Development Programmes that were launched in the 1970s. These programmes

[4] In the context of command economies the literature use the catogory „enterprise" to describe command-led micro-economic organisations which differs from profit-oriented micro-economic organisations of market economies which called „firm".

represented a modified bureaucratic approach to R&D programming and were able to encourage a shift away from the former interrupted innovation approach to the one-way linear innovation model. But this transformation was not enough to reach the feedback loop model; the chain of innovation still remained interrupted. During the late seventies there were some attempts of introducing the *torso of a third generation*, the interactive or coupling model of innovations. The contrasts between the model and its torso are illustrated in figures 1 and 2.

The 'complex' Large-Scale Development Programmes were supposed to provide the background for the change of models. These programmes included joint research, education, production and market orientation agendas in the field of high tech industries, in particular information technologies, computers and biotechnology. Different players in the innovation process were encouraged to work together. The reformed socialist model was able to achieve that different stages of innovation were considered in a dynamic context and not separately. Nevertheless, this reform helped to establish direct horizontal linkages and direct co-operation between institutes/ universities and business, slowly replacing the previous, vertically organised bureaucratic connections. Business enterprises and production areas were less isolated from the domestic R&D sector, and the gap between the technologies used and the state-of-the-art technology diminished somewhat. But the incentives for commercialisation and innovation remained weak, and development and testing activities inadequate. No pattern or network of intra- and/or inter-organisational communication paths had yet occurred. This was a vital prerequisite for an innovative economy.

Figure 2

Third generation coupling model (mid 70's – early 80's)

2.1 Market Economy

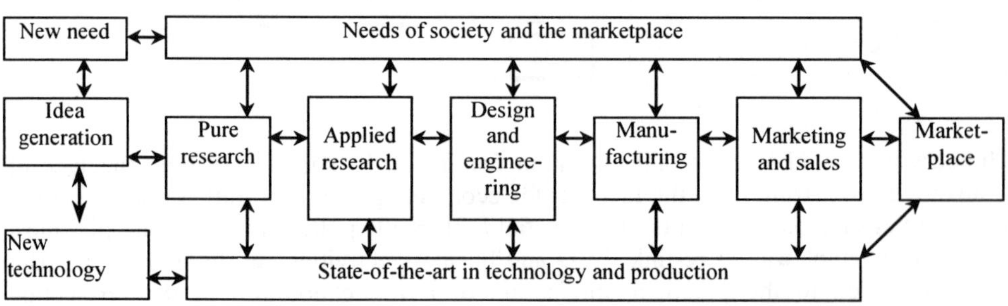

Source: adapted from Rothwell, R. 1993 "Systems Integration and Networking: Towards the Fifth Generation Innovation Process", SPRU University of Sussex, Brighton, mimeo

2.2. Command Economy

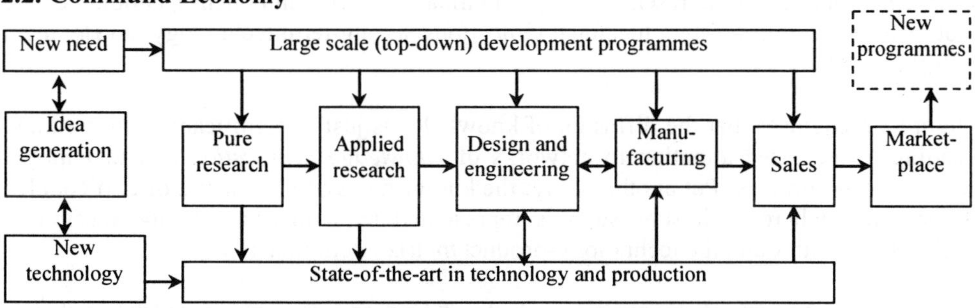

Source: author's own figure

Because the fundamentals of the system remained untouched in Hungary, there were gaps in the net of communication paths. Feedback loops occurred only in a few segments of the economy. The intra-organisation of firms was transformed; to a certain extent they could react extramural challenges. These transformations allowed them to introduce modified innovations but they were not able to adapt themselves to incremental changes.

The enterprise system of socialism gave microeconomic players little possibility to manage innovations that required large-scale reorganisation. However, while former socialist countries had relatively strong scientific capabilities, they were weak in technological development, in distribution of knowledge, innovation and efficiency of technology transfer. One of the least successful areas was the commercialisation of R&D results. There was little pressure for commercialisation in the old socialist model.[5]

As well known from literature, the above-mentioned generation of the innovation model (Kodoma 1992, Rothwell 1994, Nelson 1984) does not fit the high-tech industries of the post-World War II era. The growing awareness of the strategic importance of emerging generic technologies, with increased emphasis on technology and manufacturing, led to the *fourth and fifth generations of models of innovation incorporating integration and networking*. This model describes how forerunner firms work all over the world.

II.2. INNOVATION MODEL OF A KNOWLEDGE-BASED ECONOMY

The innovation model of a knowledge-based economy was described by S. J. Kline and N. Rosenberg (1986) as shown in Figure 3.

In this model, the demand and the supply sides both have equal importance. Demand articulation can be interpreted as the search and selection process among technical options. Information for innovation may be drawn from the *shelf of existing pool of scientific knowledge* and/or from the *shelf of existing **technical knowledge***, where

[5] The ideological structure was based on the types of organisations and linkages among organisations of the 1930's.

marketing and economic research are predominant. When component technologies are not available within existing technical collections, a long-term technology development effort is needed.

Figure 3 recognises that the diffusion of knowledge is just as significant as its creation, leading to increased attention to networks of *knowledge distribution* and to *national systems of innovation.* Put another way: the knowledge system consists of two coupled subsystems, where each subsystem is represented by a matrix, i.e., the discipline-technology matrix and the technology-product matrix (Berkhout 1997).

The linkage pattern of innovations of the Kline and Rosenberg model makes a distinction between two different series of feedback loops: the „short loop" (or activities, such as design, commerce, manufacturing) and the „long loop" (what the information needs to transit by research laboratories). Contrary to the traditional one-way linear model, this new model is conceived as a *system of loops*, established from systematically constructed interactions within the central chain of innovation. The transformation task is different for the „short" and the „long" feedback loops. In some periods we have to manage more „short" and fewer „long" feedback loops, and vice versa.

In the 1980s, when the fourth and the fifth generation models were spreading in market economies, some socialist countries were able to produce good knowledge in new generic technologies and create some islands of knowledge to produce for these technologies. But the generic technologies only slowly penetrated into the socialist economy. In the age of these high-tech industries, command economies were clearly in a technological deadlock.[6] To emerge from this deadlock they had to create a proper economic environment for commercialisation and innovation.

[6]The discrepancy between R&D output and technological applications can be observed in market economies as well. It is worth paying attention to this problem to avoid slipping from faults of the socialist system into failures of the market economy.

Figure 3

Chain-link model of innovation

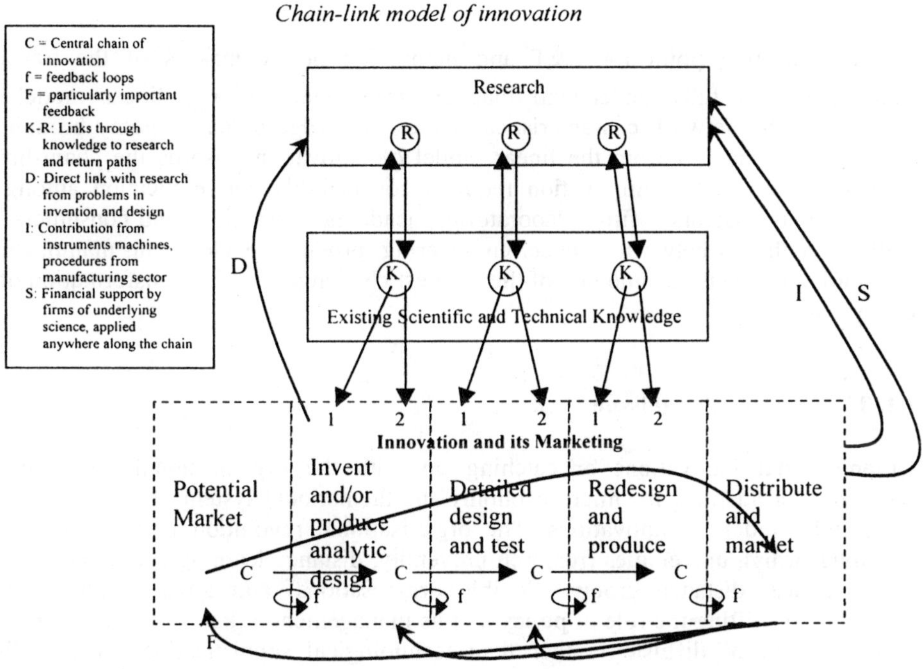

Modified version of Kline, S. J. and N. Rosenberg (1986) "An Overview of Innovation", in R. Landau and N. Rosenberg (eds.) *The Positive Sum Strategy: Harnessing Technology for Economic Growth*, National Academy Press, Washington, DC.

In order to transform the R&D resources into economic success, the former socialist countries have had to redefine their institutional and behavioural operations. The socialist system was very weak in diffusing knowledge and commercialisation.

According to Davies (1979) a distinction can be made between two types of new technology by pay-off period. „Group A" innovations are relatively cheap and simple, whereas „Group B" innovations are expensive and technically complex. The diffusion path is different between the groups and between the type of members in each group. Another distinction is between two types of innovations: the first concerns the development of already existing products or those which derive from an effort of analytical conception using existing procedures to create new products (product improvement and product development). The second concerns the inventions themselves, which imply operations with no links to the past (new products). Both types are based on „short" and „long" feedback loops but they do not use the same circuits for the exchange of information. The division and the co-ordination of tasks for these innovations are different but the common feature is that none of them belongs to the linear model.[7] Recognition of these differences offers some options for strategy formulation.

It was a turning point for S&T and innovation policy makers of the transition economies② to fully understand that the traditional one-way linear model was outmoded. The network characteristics of the knowledge-based economy brought the beginning of changes from the linear model of innovation towards the chain-link or feedback loop models. Innovation requires considerable communication among the different playersactors – firms, laboratories, academic institutions and consumers – as well as feedbacks between science, engineering, product development, manufacturing and marketing. This became one of the greatest challenges for the transformation of the old system.

III. The organisational issue

Organisational innovations or catching up with the organisational revolution are important for adapting the micro-economy and the national system of innovation to the chain-link model of innovations. The organisational revolution touches upon firms (learning firms), universities (research university, distance learning), research institutes (megascience, distance groups, flexible organisations) and governmental agencies (referring to shifting the role of policy and its management tasks). Institutional inertia is a determinant of diffusion rather than technological sophistication. Besides firms, institutions – financial, educational and management institutions –also have to acquire new competencies.

Dissemination and integration of knowledge are the critical processes in each economy. Distribution power depends on learning, adaptation and adjustment capabilities. From this point of view, the role of knowledge production and dissemination organisations is to develop adaptation capabilities, to be open to the world.

In this model of the knowledge-based economy "firms search for linkages to promote inter-firm interactive learning and for outside partners and networks to provide complementary assets. These relationships help firms to spread the costs and risk associated with innovation among a greater number of organisations, to gain access to new research results, to acquire key technological components to new products and processes. Firms determine which activities they will undertake individually, in

[7]It has to be emphasised that technical simplicity is not necessarily correlated to organisational simplicity; product improvement may also complicate the co-ordination task. Substantial organisational changes may be needed to exploit new technology even if it is technically less sophisticated. From the point of view of diffusion, the first group of innovations can be implemented without substantial organisational change though not necessarily technically simple. The second group of innovations can be implemented only with organisational change of a far-reaching character but may be unsophisticated technically.

collaboration with other firms, in collaboration with universities or research institutions, and with the support of government." (OECD 1996, p. 18)

Organisational differences are important factors when comparing the efficiency of advanced economies. One of the greatest challenges for European Union (EU) member states is to promote organisational innovations at micro and macro level with a view to filling in the missing link in transforming large investments of research, technology and productive capacity into economic welfare and the revival of employment growth. (Andreasen *et al,* 1995). This requires changes in governance, institutions, regulations, regionalisations, among and inside organisations to move towards thinking organisations. "Decentralised corporate management systems based on financial controls breed managers in the same mould, whose competencies and systems of command and control are not adequate for the funding of continuous and complex technical change. Firms managed by these systems therefore tend to move out (or are forced out) of sectors requiring such technical change." (Pavitt 1996. p. 16.)

Organisational innovations and changes in organisational patterns have crucial importance for those countries which are less developed than the EU member states. From those countries, which were backward in the middle of this century, we observe a cluster of newly emerging economies. Many others, including CEECs (Central and Eastern European Countries) are „clamping-on" economies. Without going into details of their different systems, it is worth pointing out that the knowledge and technology transfer systems of the newly emerging economies were significantly different from those of economies that are still „clamping-on" and are still far behind. (Bart Verspagen categories) Not only do the organisational structures and government system of newly emerging economies differ from the rest of the world, but so do their technology-supporting institutions (TSIs). The TSIs are competent organisations for technical applications and reverse engineering.

During this century, Japan moved from being an under-developed to a leading economy. It has been successful in mastering the short loops of the innovation chain thanks to horizontal communication between adjacent units, and to the fact that there is a flow of human capital between the units. (Aoki, 1988, p. 269) The Japanese system strongly encourages the circulation of information within the „short loops" of the innovation chain. The weakness of the Japanese model lies in the relation to „invention" for „radically new" products. In such a situation, innovation depends on the good management of „long feedback loops" and multi-disciplinary knowledge. The Japanese method performs less well when it has to acquire knowledge outside the „*Keiretsu*" (for instance from universities).[8]

[8] It is a common misbelief that Japan is weak in the production of important inventions. This misbelief is contradicted by Takabumi Hayashi's analyses (see chapter I). His study illustrates how the Japanese society can produce breakthrough inventions, strong patents. Another problem to be emphasised is the macroeconomic organisational weakness. In contrast to the strong linkages inside the Keiretsu, there are weak links between industry-university.

Organisational revolution, to adapt to the chain-link innovation model, is a much greater challenge for the economies in transition②. Previously, the degree of decision-taking was very centralized and economies semi-closed. The vertical institutional frameworks for technology policies were overwhelmed and horizontal institutional frameworks hardly existed. Hungary has taken some steps in the direction of organisational innovations. The role of bridging organisations, such as technology transfer offices, innovation relay centres, professional associations, is increasing. The structure of transfer organisations and tools for the transfer of knowledge are changing and new types of organisations are emerging.

IV. Beyond National Boundaries

Innovation systems are extending beyond national boundaries to become international. This *globalisation* process has a significant influence on the flows of knwoledge and relationships which exist among industry, academia and government in the development and transfer of science and technology. The transborder flow of scientific information and the internationalisation of science and technology has grown steadily, with post-World War II travel and fellowship support, increasing opportunities for academic exchanges, and more recently, electronic information exchange and the end of the Cold War. Financial support for science, especially by governments, has been crucial in enabling this international interchange.

Investors in R&D allocate their expenditures world-wide and the importance of "soul-deep" co-operation is increasing (Inzelt 1997). This means that foreign business may like to tap national stocks of knowledge. A large volume of FDI (Foreign Direct Investment) in R&D is moving toward regions with large markets, a large pool of talents and a friendly environment (Patel and Pavitt 1994). *Globalisation has once again raised the question: is knowledge (results of basic research) a public good, and – if so – to what extent?* The way different schools of economists take the usefulness of codified information and tacit knowledge into account has important policy implications. Government support for R&D strongly depends on the assumptions made about the nature and usefulness of knowledge. Every Government has to answer tax-payers' questions about the benefit of the nation from national public investment in S&T when scientific mobility is so high (with coexisting brain drain, brain gain, brain waste) and companies more and more internationalised. *If the investment in production and transmission of knowledge is national but its exploitation international, how should nations share costs and benefits?* The globalisation of S&T activities, and the transmission and transfer of knowledge give growing importance to answering the question: If knowledge is a public good, for whom and where? Who should pay the investment in knowledge and for how long? *How can we make a distinction between „free-riders" and partners?*

Intellectual property rights (IPRs) provide tools for the reimbursement of part of the costs of knowledge and technology creation. The extension of IPRs and the Trade Related Agreement on IPRs (TRIPs) are attempts to regulate international diffusion, with a view to reducing the number of "free-riders". But there are still many areas in where the idea of IPRs cannot be employed because of the character of the knowledge concerned or due to such as business secrecy or because of the speed of development.

IPRs aim at covering newly emerging fields, such as biotechnology, information technology. The penetration of information technology in all S&T developments raises a new question: should the exchange of scientific data, especially across national boundaries, be made free of charge or against payment? Restriction of international diffusion would be quite anachronistic in this age of globalisation. This question is important not only for those countries which are the largest investors in R&D but for all countries investing in the production of knowledge. The task in coming years is to solve the problems raised by globalisation and by rapid diffusion of new technologies in space. National benefits from domestic efforts on R&D strongly depend on the capability of firms to combine knowledge assets (research) with other firm-specific assets. If the latter are missing, the national economy can hardly benefit from its stocks of knowledge. Missing firm-specific capacity of exploiting knowledge assets are the legacy of the innovation model of socialism.[9] Therefore the main problem for many CEECs is that they are „non-riders" (or „donkey riders") whereas their science may be an easy prey to „free-riders". To diminish the negative influence of such free-riders, these countries need an open and flexible interface between research and its applications. Improving competitiveness and innovativeness strongly depends on the transferring capabilities of all the players. The redeployment of organisations and the establishment of new ones are crucial for the operation of „short" and „long" feedback loops. Local and international networks also have to be strengthened to benefit from all this tacit know-how.

V. The measurement of flows of knowledge and transfers of technology

Another important issue is the information on the diffusion of knowledge. There are a number of important avenues of disembodied knowledge diffusion inside the country and between countries. The current measurement system is weak in quantifying and mapping the paths of knowledge and innovation diffusion in an economy and in global terms.

Measuring the flows of knowledge and transfers of technology is more complicated than measuring inputs into knowledge. We need combinations of old and new indicators to measure the role of knowledge diffusion, technology transfer, alliances, networks, etc. in the dissemination process. Some of them are already factored into existing indicators, others have not yet been factored. A distinction can be made between the traditional and

[9]We have to keep in mind Pavitt's statement (1996 p. 19.) "The main economic value of basic research is not in the provision of codified information, but of the capacity to solve complex technological problems, involving tacit research skills, techniques and instrumentation, and membership of national and international research networks."

the knowledge-based economies more by the characteristics of flows than by the size and the structure of the stocks of knowledge.

Two different types of flows of knowledge have to be measured: (1) *embodied diffusion* which insists of the introduction into production processes of machinery, equipment, components and intermediate goods incorporating new technology, and (2) *disembodied diffusion,* i.e. the transmission of knowledge via technical expertise, technology in the form of patents, licences or know how. It is much easier to measure the first type of flows than the second.

OECD Member countries have adopted a number of internationally harmonised indicators which are applied by most of them. These are the following:

- the *Technology Balance of Payments* is one of the most traditional measures of international flows of S&T which captures international movements of technical knowledge through payments of licensing fees, patents and other direct purchases of knowledge. (OECD, 1992)
- the *Innovation surveys* capture information about firms or technologies by object or subject approaches and map the distribution of knowledge. (OECD, 1996a)

Some other methodologies are applied by a few OECD Member countries to measure the national and international dissemination and diffusion of knowledge and technology. These methods which have not yet been harmonised include:

- **the *Science Citation Index*, and its analysis provide an analytical tool to explore intra- and inter-disciplinary flows of knowledge in the realm of basic research. Knowledge illustrated in patent citations also shows the flow of information inside and among industries, between university and industry, and across borders.**
-

- *Information Technology (IT) –related transfer of knowledge.* A great amount of E-mail traffic and data transfer today represent the knowledge dissemination among researchers world-wide and between university and industry. Information technology is likely to offer *new measurement techniques* at lower costs to describe the development and the distribution of knowledge. Not only new indicators but also non-traditional measurement methods are needed to capture the transfer of knowledge among key players and institutions in the economy. If electronic data collection techniques are employed, this will be a first minor step towards the renewal of our information capabilities. If IT can be used as a file data gathering method this will be a breakthrough in our information system. It may offer a much cheaper and less time-consuming information basis. The mush-rooming of www and intra-, extra- and internets indicates that this possibility is on our doorstep.

- The globalisation of R&D is making it increasingly important to obtain detailed information on the phenomena of *inward and outward FDI (Foreign Direct Investment) in R&D, co-operative research agreements*, etc. Some attempts have been made to measure these flows, by numbers and types of agreements, co-patenting, co-publications, etc.
- The mobility of scientists and engineers shows the capability of adaptation on the one hand, and the knowledge transfer on the other. (OECD 1996c) The acquisition and the diffusion of knowledge is also realised by mobility. Several countries have good information on how mobility of scientists and engineers supports the diffusion of knowledge. What are the losses and profits originating from these flows?

In Hungary there are no up-to-date statistics on mobility. The mobility of researchers may be illustrated in the following stylised input-output matrix (Figure 4).

Figure 4

Mobility of Researchers

From/To	Research institutes	Higher education candidates	Higher education researchers	Other research institutes	Public sector	Business sector	Abroad
Research institutes		😐	😐	😐	😐	😐😐😐	😐😐😐
Higher education candidates			😐			😐	😐😐😐
Higher education researchers				😐	😐	😐😐😐	😐😐😐
Other research institutes						😐	😐
Public sector					😐	😐	😐
Business sector					😐	😐😐😐	😐😐😐
Abroad	😐	😐😐	😐	😐		😐😐	

Flow rate: 😐=low; 😐😐=medium, 😐😐😐=high

Despite the fact that brain drain, brain gain and brain waste issues are widely discussed in Hungary, there is only sketchy information on this phenomenon. An input-output table of this type allows us to illustrate both intra- and international flow from a bird's-eye view. Employment of foreign personnel means (incoming) international transfers of knowledge. It is much broader than the brain drain - brain gain issue. The flow of skilled workers is also an important part of the dissemination of tacit knowledge. The mobility of personnel may occur in institutionalised forms, e.g. international transfer of consulting services, intra-firm transfers, joint ventures. In these cases not only the personnel but also organisational knowledge is transferred. Experiences of knowledge transfer – organisational competencies – by foreigners in Hungary during the first years of privatisation offer good examples (Inzelt 1994).

If the decision-makers wish to create a proper environment for the potential players they need adequate information on a large number of variables, such as: expenditures on R&D, efficiency of knowledge- and technology transfers, innovation activities, international comparisons of R&D inputs and outputs, mobility of sciences and engineers, costs of transfers to judge the efficiency of R&D investments, and to determine how well these investments contribute to the production capabilities and to the competitiveness of the country and, by this means, to the improvement of living standards and quality of life.

Concluding remarks

The mutual transfer of knowledge and technology is indispensable for successful integration into the increasingly globalised world economy. One of the key issues is that the economic environment should become innovation-friendly. Business organisations should demand technological innovations and be capable of implementing and disseminating them. A further prerequisite is that countries should continue to generate new scientific results. There should be an adequate intellectual base able to implement, disseminate and even develop new scientific-technological achievements domestically. Moreover, it is important that the country should be capable of becoming actively connected to international co-operation in all fields of R&D, science and technology, both as recipients and as donors. All this should be promoted by the general economic policy, the science policy and the government budget.

Public expectations for a knowledge-based economy are high. People hope social welfare is not too far away. Scientific knowledge may help them to protect the environment for future generations. Knowledge and its dissemination will contribute to job creation, to the improving of the economic and social wealth of the nation.

Transition economies are moving from an outsider position of the knowledge-based economy. Their knowledge distribution power is improving. From the critical domains of knowledge distribution we can observe improvements in the role of bridging institutions. The capability of organisations to absorb new technology (privatised mainly by foreigners) is improving. Technology-oriented start-ups are using science results and they usually have a good absorptive capacity. The other side of the coin is that a large majority of firms still do not use scientific results, many domestically owned medium-size firms are weak in learning and do not have an absorptive capacity. The role of the public sector is still confused, many bridging institutions (for example technology transfer oganisations of the universities) are still missing and others are in their infancy. Transition economies may join more and more international networks, however these are only partially opened to them. Summing up this observation we cannot state that transition economies are insiders of knowledge-based economies, but they are taking off from an outsider position. Our response to the question in the title therefore has to be : these countries are midway between outsider and insider positions and there are still a lot of traps on this avenue.

My motto cited Herodotus, who was right: fortune never abides in the same place but humankind has the capability to create fortune by the production, transmission and transfer of knowledge and technology. These can support the emergence of sleeping economies and preserve the prosperity of advanced economies. Examples of technologically strong regions – Southeast Asian economies - demonstrate that any country at any time may become a prosperous economy with hard work – though not without risks. The peripheries can catch up to the frontiers. Scientific research, science and technology policies, transmission and transfer organisations have to help all nations to join the sunny part of the world. Every cloud has a sunny side.

References

1. Andreasen L. E, Coriat, B, den Hertog, F., and Kaplinsky, R. (eds.) 1995 *Europe's Next Step. Organisational Innovation, Competition and Employment*, Frank Cass Co. Ltd, Great Britain
2. Aoki, M. 1988 *Information, Incentives and Bargaining Structure in the Japanese Economy*, Cambridge University Press, Cambridge
3. Balázs, K. and Török, Á. 1996 "Tudás- és technológiatranszfer-szerveztek és mechanizmusok a fejlett országokban és az átalakuló Magyarországon. Összefoglaló tanulmány", (Science and technology transfer organisations in advanced economies and in Hungary. Summary report) OMFB, Budapest
4. Berkhout, A. J. 1997 "The Knowledge System for Product(ion) Innovation." Paper presented to an OECD workshop, September 1997, Budapest
5. Chandler, A. 1992 "Corporate Strategy, Structure and Control Methods in the United States in the 20th Century", *Industrial and Corporate Change*, No 1. pp. 263-284.
6. David, P. and Foray, D. 1995 "Accessing and Expanding the Science and Technology Base" *STI Review*, No 16, OECD, Paris
7. Davies, S. 1979 *The Diffusion of Process Innovations*, Cambridge University Press
8. EIMS (European Innovation Monitoring System) 1994 *Public Policies to Support Tacit Knowledge Transfer*, Proceedings of SPRINT/EIMS Policy Workshop, May, 1993
9. Hayashi, T. 1997 "R&D Capabilities of Eastern Asian Countries - centring on Asian NIEs" in: *Technology Transfer: from Invention to Innovation* (eds.: Inzelt és Hilton), Kluwer, 1998 forthcoming
10. Inzelt, A. 1988 *Rendellenességek az ipar szervezetében* (Abnormalities in the Organisation of Industry) Közgazdasági és Jogi Könyvkiadó, Budapest.
11. Inzelt, A. 1994 "Restructuring of the Hungarian Manufacturing Industry" *Technology in Society,* 16, 1, 1994, pp. 35-63.
12. Inzelt, A. 1996 OECD "The Main Results of Hungarian Pilot Innovation Survey, (Application of First OECD/EC Questionnaire)" Conference proceeding "Conference on Implementation of OECD Methodologies for R&D/S&T Statistics in Central and Eastern European Economies", OECD Paris
13. Inzelt, A. 1997 "To foster a Society in a Fellow Traveller Country" in: *The 21st Century Opportunities and the Challenges for the Asian Economies: Science, Technology and Innovation Policy* (eds. Heitor M, Gann D, Gibson D, Martins, R. Shariq S, Xuelim C) forthcoming
14. Kline , S. J. and N. Rosenberg 1986 „an Overview of Innovation” in R. Landau and N. Rosenberg (eds.) *The Positive Sum Strategy: Harnessing Technology for Economic Growth,* National Academy Press, Washington D.C.
15. Morita, Akio 1989 *Made in Japan, Akio Morita and Sony* (with Edwin M. Reingold and Mitsuko Shimomura), Hungarian translation, Árkádia, Budapest
16. Narin, F. Hamilton K. S. and D. Olivastro 1997 "The Increasing Linkage between US Technology and Public Science" *Research Policy*, October 1997, pp 317-330.

17. Neumann, John von, 1958 *The computer and the brain*, New Haven, Yale University Press
18. OECD 1992 Technology Balance of Payments, Paris
19. OECD 1996a Oslo Manual, OECD, Paris
20. OECD 1996b The Knowledge-based Economy, OECD Paris
21. OECD 1996c Canberra Manual, OECD, Paris
22. Ondaatje, Michael 1993 *The English Patient*, Vintage International, New York, 1993, cited page 142
23. Patkós, A. 1997 "Excellence and Social Relevance in Hungarian Higher Education" in: *Technology Transfer: from Invention to Innovation* eds..: Inzelt, A. and J. Hilton, Kluwer, 1998, forthcoming
24. Pavitt, Keith 1996 "National Policies for Technical Change: Where are the Increasing Returns to Economic Research?" (Paper prepared for publication in the *Proceedings of the National Academy of Sciences*, Washington, D.C.)
25. Patel, P. and K. Pavitt 1994 "Uneven (and Divergent) Technological Accumulation among Advanced Countries" *Industrial and Corporate Change*, No. 3. pp 759- 787.
26. Rosenberg, Nathan 1983 *Inside the Black Box: Technology and economics*, Cambridge University Press
27. Rothwell, Roy 1994 "Industrial Innovation: Success, Strategy, Trends" in Dodgson, Mark and Rothwell Roy, *The Handbook of Industrial Innovation*, Edward Elgar, Cheltenham, UK pp 33-53
28. Utterback, James M. 1994 *Mastering the Dynamics of Innovation*, Harvard Business School Press, Boston, Massachusetts
29. Verspagen, Bart 1997 "A Global Perspective on Technology and Economic Performance" paper presented on NATO ARW Quantitive Studies for S&T Policy in Economies in Transition, 23-25 October 1997, Moscow

KNOWLEDGE AND TECHNOLOGY TRANSFER

Jacques De Bandt
Directeur de Recherche CNRS
LATAPSES/IDEFI, 250 avenue A.Einstein, Sophia Antipolis, 06560 Valbonne, France.

This paper is about the implications (at different levels, for different types of actors) of the « learning economy » This is essentially an economy in which values are produced on the basis of knowledge production processes.

"Technology transfer" is an ambiguous term. It suggests, wrongly, that technology is a well defined finished product which can thus be transferred and used as such. The transfer of an equipment, in which some technology is embodied, leads (eventually on the basis of some training) to use the equipment, but is of course not the transfer of the technology, which is embodied. In order to use, to maintain, to repair it, some know how has to be transferred, but learning by using and by doing will do the rest. In order to undo it (reverse engineering), and to reproduce it identically, still much more knowledge has to be acquired, through learning (with some external inputs and assistance). In order to produce new equipment or a new product, similar but more adapted to particular circumstances and needs, the importance of autonomous new knowledge production processes is increasing, even of course if one can benefit from external inputs and assistance. And for any new product, process or equipment, the autonomous knowledge production process becomes decisive.

Of course, any such process is necessarily based or existing knowledge - existing knowledge need not to be reinvented - but the way to use or apply it for specific applications has to be invented - and fed by external inputs. But whatever the inputs, the required knowledge has to be specifically produced.

For this reason, we concentrate on knowledge production.

1. The emergent learning economy .

Three preliminary remarks may be useful :
i. I use this term as meaning (this will be further explained below) that knowledge is becoming a key variable/resource/asset for the production of value and wealth. Different terms are being used : "service economy ", "quaternary economy ", "knowledge, information or learning economy" these terms are not really equivalent, but they are all referring to the (decisive) importance of the interactive creation of information or knowledge for producing value.
ii. I do not intend to enter the discussion about the actual (full) existence of the learning economy. We can of course speak of the learning economy either as the existing model or economy, or as something in the future, which is merely emergent, or even as

A. Inzelt and J. Hilton (eds.), Technology Transfer: From Invention to Innovation, 23–35.

something which lies in the future. The three pictures may be said to be true, dependent on how we look at it. The learning economy is already apparent through various manifestations or phenomena, is driven by strong tendencies, but is of course not yet a full-fledged comprehensive system or model. This means that the learning economy is building up, that manifestations are more and more widespread, and that much more of it is still to come.

iii. It is probably true that, like in many other domains, the perception or awareness of the players is somewhat lagging. This may be partially due to the fact that the actual manifestations of the learning economy are only partial, and are more underlying trends than clear facts. Some type of analysis is needed in order to come to the understanding of this reality in its complexity. This is certainly also due to the fact that most players, accustomed to their environment and the ways things are, are not very keen to accept the idea of fundamental changes and the uncertainties which go with them.

To be honest, one has to recognise that this is not only true for public opinion or for most players at large, but even for scientists : economic theory has visibly quite a lot of trouble in recognising the importance of and integrating the new realities of service relations, information and knowledge production.([i])

I would suggest the following are typical examples or manifestations of the "learning economy " :

- the existence of quite profitable fabless companies : able to appropriate a fair share of the value of the goods produced, without producing them.
- poorly profitable subcontracted production of standardised products : for example, T-shirts are still produced and consumed, in still larger quantities, even while their cost/price tend asymptotically towards zero.
- the increasing share of publications in so-called scientific journals published by enterprises (not by "scientists")
- the squeeze of non skilled labour in developed countries
- the new skill and training requirements
- the new importance of project management
- the integration in management of the design and management of information systems, organisational learning, knowledge engineering, and the like.
- the network and networking partnerships phenomena
- the importance of R&D alliances, licences
- the importance of local innovation systems, tacit information and knowledge..
- the changing nature of investments and assets (less material, more non material)
- the changing nature of productivity (industrial productivity is not relevant any more)
- the importance of reputation, quality norms, certification.... and so on.

What then is the "learning economy "?

Negatively, it is of course not an « industrial economy » any more : it is a significant shift away from the industrial or energetic economy, in which the central activities were the direct transformation of materials into goods and which was, for that reason, focusing on the rationalisation of the material and energetic processes involved, on the basis of both standardisation and the use of roundabout methods and huge volumes of physical capital. Technical progress resulted from rationalising the design of new processes and products : design was typically an important separate upstream activity. Typical components were : (more or less) specialised capital goods, economies of scale, inventories, huge transport costs (notwithstanding progress in transport and communication), degrees of capacity utilisation, ... and soon, all aspects of "energetic" processes.

Positively, the learning economy is based on new modes of production, i.e. new methods of performance in the field of the production of products and thus of values and wealth. To be clear, not only is the mode of production entirely renewed, but the same can be said, at least to a certain extent, of the modes of consumption, organisation, financing... Central are the new ways of producing and organising.

This means of course that the visible aspects of the activities change, with the rapidly changing relative importance of the non material dimensions as compared with the material ones. Much more important however are the ways the players see the relations between them, on the basis of which values are created and exchanged, and more particularly the content and the conditions of such relations. The things they take for granted, the representation they have of the activities and roles of the various players and of their interests relative to the others, the criteria and norms to which the players refer and the rules to which they accept to obey are all likely to be, in some essential way, different.

Of course a significant shift away from the old model doesn't mean that all the components of the old system are formally disappearing. The specialised, rationalised material production of standardised goods or pieces of course remain important, certainly quantitatively speaking. Goods are and remain in any case a central and decisive part of the economy, as they are in our daily life. There is no doubt about that. However,

i. the material and energetic basis of the economy (in terms for example of raw materials and energy resources consumed per unit of GDP) has been steadily decreasing in recent decades.
ii. the relative importance of direct transformation labour and costs, within the value chain, is constantly being reduced.
iii. the relative importance (share of GDP) of manufacturing (or industry at large) is decreasing, while the share of tertiary activities (to which should be added the tertiary part or manufacturing activity proper) is continuously increasing. Even in quantitative terms, industrial activities cannot be said to be the driving force of the economy any more.([ii])

iv. the same is true, and even more so, in terms of employment ([iii])

v relative productivity ([iv]) is highest in services, particularly in informational business services, which are more and more in a position which enables them to appropriate an increasing share of the value added to which they are contributing. This means essentially that the terms of trade between industry and informational business services turn out to be less and less favourable for the first ([v]) and more and more favourable for the last.

2. Knowledge production processes

What is at hand ?

All situations in which the former existing solution (all possible specifications - weight, size, colour, durability, energy consumption, flexibility...of a product, process, organisation, system ..) is not or not thought to be up to (existing, differentiated, or new) requirements. Such a situation can correspond to quite different cases :

- the competitive process, meaning that monopoly or innovation margins are continuously eroded, imposing the search for a continuous stream of new innovations.
- the environment, making that some of the basic data are changing : population, climate, geo-politics...., thus creating uncertainties
- the technologies, making existing ones obsolete and opening up a series of new opportunities
- new ways of living, tastes, psycho-sociological elements, demand for variety...
- increasing complexities of realities and of the involved systems

The spectrum of such cases is very large indeed : going from severe constraints which have to be lifted to more artificial fashion problems, from new problems asking for radically new solutions to demand for variety asking for marginal or more or less superficial adaptation. In all cases, the specific solution means value and money.

This all means that there is a strong need or demand for a new solution (a new product, the solution of a new problem, a wider variety of products, the adaptation of the product to particular circumstances or particular needs or tastes....) and that, because the client/consumer/user is ready to pay what is needed in order to get this (customised) solution, there is both a strong incentive to supply this solution, while the margins on standardised repetitive products are under strong competitive pressures, and very large technical opportunities exist, because of a the actual spectrum of technological advances. The number and proportion of cases or situations in which the supplier and the client are exchanging something which is not completely the same as before or as compared with what they are exchanging with others, for whatever reasons, are increasing steadily. In order to do so they have had to inform each other and to develop a common knowledge of what is needed, before assembling information and knowledge in order to find the required solution. The client is asking for such a new solution, because he needs it and he knows better solutions are likely to be possible. The supplier proposes it, because he knows that a better solution would indeed seem to be possible and this is a way for him to secure his income.

There is no use in trying to understand what is the driving force behind this. This process would seem to be neither demand-driven nor supply driven, but both at the same time, interactively. Clients/customers/users need better solutions, in some cases cannot do without these(because of negative consequences of the old solution, strong constraints or even survival problems), consider that it may be worth while to pay more for a more appropriate or specific solution, have more or less the idea, message or experience that such better solutions are indeed more or less possible.... Suppliers are undergoing more or less the same evolution, but in some complementary way. They have competencies and capabilities, out of which they want to raise their income, they need to produce and sell, they have difficulty in making their earnings within the context of strong competitive pressures, they know they can sell more specific solutions and they know that it should be possible to adopt such new solutions. And the new information technologies can be very helpful.

Somewhat schematically, it can be said that there are two main stages in this evolution towards customisation In the first stage, which we can call an intermediate stage, the solution consists in defining *a priori* whole series of components, parts and products according to different specifications, allowing the client/customer/user to choose from still larger series of alternatives, whose possibility has been introduced at the level of design. We are still in the previous system in which design is preceding production. On the basis of some a priori ideas (even if they are based on past experiences in related domains) products are designed, rationalised, produced. The client can choose between alternatives. There are two special cases here : one in which the producer is designing different products but standardises as much as possible as many as possible of the parts and components; the other in which the client is asked to choose among various alternatives for the various parts and components, the product being thus only a combination of the parts and components as chosen. In such cases (they are not quite different but for the fact that the parts and components the producer tries to standardise are normally not the same as those among which the consumer is choosing) the logic is combinatory : variety or differentiation is obtained mainly through a combination of elements. The elements themselves are defined and designed beforehand. Information technologies are to a large extent a precondition for introducing combinations within production processes, on a large scale.

The second stage is really of a different nature. Neither the solution, nor the elements of the solution can be defined beforehand. It is not possible merely to combine previously designed elements. At this stage, the relations between producer/supplier and client/consumer are fundamentally changed, to the extent that the client/consumer must somehow be integrated in the design and production phases of the process. The whole process can, besides the material parts of it, which are increasingly only of secondary importance, be described in terms of knowledge production : the production process consists mainly in both producing information/knowledge on the problem (the need for another solution) and producing information /knowledge on the solution, which has then to be worked out (including normally some material production operations). The new solution has indeed to be both adapted or appropriate and to be efficient or cost-effective.

Two remarks may be useful as concerns this production of knowledge on the problem or question :

i. What is called the problem here is a specific problem : the problem as raised by specific players, with their specific needs and constraints, in a specific context and in specific circumstances, ... In order to be able to solve this problem, this has to be "known". The production of knowledge here means that, starting from some rather general problem or question, broadly defined, it is necessary to produce precise and relevant information and knowledge as concerns the various dimensions of the problem at hand. There is of course no possibility to solve the problem or answer the question if this problem or question is not thoroughly analysed. One obviously needs a circumstantial analysis or diagnosis. This defined, the issue would seem to be self evident. But in real life, the precise identification of the problem is really central.

ii The two knowledge production processes thus distinguished are not to be seen as independent. In the same way the solution cannot be an absolute or ideal solution, but must be adapted to the specific case at hand, as there is no use in trying to analyse the problem in some absolute sense, independently of any possible solution. While the problem has to be thoroughly analysed, the analysis must necessarily take into account impossible and possible lines of solution. Conversely the characteristics of the problem, in terms of players and circumstances, will also give indications about feasible or unfeasible solutions (for example, in terms of time, costs...).

3. The nature of knowledge and of knowledge production

In order to understand better what is at stake when speaking of the production process in terms of knowledge production, we have to be more explicit as concerns both the nature of the knowledge to be produced, the nature of competencies and knowledge/information which are required and mobilised and the nature of the knowledge production process involved.

As concerns the nature of knowledge to be produced, the production of knowledge has been described in terms of understanding and solving a problem or question. It is thus necessary to take account of the fact that due to differences as concerns the questions raised, the nature of knowledge produced is similarly of different nature. L.Dibiaggio (1998) has usefully suggested a distinction between three types of information or knowledge. The first category includes all types of elementary or "simple" information for which, the relation between question and response being predefined or pre-codified, the meaning is self-evident : the term data is often used in this sense ([vi]). The second category – "problem" information or knowledge - includes knowledge in conditions in which the solution (or the answer to the question) is unknown and has to be "produced", but in which one knows that a solution is available. The solution is not known, but can be found. The third category – "complex" information or knowledge - includes knowledge which is produced in order to answer a question or solve a problem, in conditions in which it is not known from the start whether there exist an appropriate solution or not. Of course, there doesn't necessarily exist a strict frontier between these categories, be it only because things are not static and may be changing in due course. In any case, for example, as concerns the second category, the idea about the existence of a solution may appear to be false or the solution or response may prove to be not really adequate or appropriate, while as concerns the third category, it is not impossible that a satisfactory approximation of the expected solution can be found by replicating an existing solution.

We can draw a parallel between the distinction just made between the second and the third category and what has been said above about the combinatory stage in the process of differentiation of solutions and products. In the combinatory stage, the required elements exist, are thus predefined, but have to be assembled or combined. There is no uncertainty as concerns the possibility of defining an adequate solution.

The production of knowledge requires both competencies and procedures. It of course requires competencies, i.e. information/knowledge which have already been acquired, on the basis of previous learning (training, experience), and which can thus be mobilised in order to produce further knowledge([vii]). We know that, in a learning economy - even if this is usually not recognised as a distinct feature of the learning economy - competencies are what matters, from the standpoint of competitiveness. Of course this is too general to be meaningful. Without being able here to go in the details, let us quickly point out the following :

- competencies are of course knowledge, but oriented towards doing things (in our case, in the field of production)
- the "volume" of competencies or specific non material assets can only be evaluated in terms of the capabilities to solve new production problems([viii]).
- competencies are not only about doing things technically speaking, but also about being cost-effective and selling.
- in order to do things (in the field of production), different types of knowledge are required (natural sciences, technologies, environment, organisation, people...), which are complementary from the standpoint of production requirements.
- more and more important, among the required competencies, are those competencies (including collective or organisational competencies) for handling and managing both information and knowledge as such (managing information systems) and interactive learning processes.

- the (competitive) enterprise has « core » competencies, i.e. a specific set of acquired competencies corresponding to her strategic lines of action and giving her some strategic advantage in her domain of activities. These competencies constitute her specific non material assets.
- competencies need to be neither (too) general, nor (too) specialised, but both.
- competencies are partly codified knowledge (easily appropriated, transmitted or diffused) and partly tacit (which can be mobilised, but not easily communicated)
- competencies can be partly collective, as the result of past collective learning processes
- competencies are not static, they are spontaneously undergoing processes of deterioration and obsolescence, they need the systematic management of maintenance and accumulation.

Because different, complementary competencies are required, the knowledge production process has necessarily collective dimensions, the question being then how to select and to bring the necessary complementary resources together and how to organise their contribution within this collective learning process. It has been said above that the competitive or dynamic enterprise has « core » competencies. These constitute a necessary condition for being able to launch dynamic learning and accumulation processes, but not a sufficient condition. The competencies of the firm or the competencies which can effectively be mobilised within the firm are always (of course more or less) insufficient for entering into complex knowledge production processes. While this may be to a certain extent evident in the case of SME's, which, being small, have only limited resources and always lack different types of competencies, this is also true, and more and more so, in the case of large firms. This has already been indicated above, but it is important to stress that, because of the complexity of the knowledge to be produced and thus of the complexity of the interactive (cognitive) processes involved, central among the competencies are the competencies needed for being able to develop such organisational learning processes. It may be quite difficult, or even impossible, and quite costly, to organise the required multidimensional interactive processes (see below) within big organisations (because of organisational rules and procedures, decision and control processes...)([ix]). But, going further, one could say that the firm is neither a self-contained system including a priori all the necessary competencies required for answering new questions or questions lying at the frontier of actual activities ([x]), nor a system within which, even if a rich variety of competencies exists, co-operations are spontaneously emerging or developing ([xi]).

What is thus needed, is to « assemble » the required complementary resources and competencies, and to design and implement some kind of contractual and organisational structure allowing for the effective co-operation of those different actors and competencies. Series of problems arise, object of a rapidly expanding literature, concentrating on the various aspects and problems of networks and networking, alliances, partnerships, R&D agreements, consortia, co-operation...., i.e. various forms, degrees and duration of co-operative schemes. Let us just mention some of the problems involved :

- identification of partners (various types of players willing to enter into the game and various motives for doing so) and competencies (competencies being to a certain (large) extent based on tacit knowledge, some aspects of competencies are difficult to identify and may be becoming apparent only in due course)
- contracts defining contributions, some guarantees, ownership, distribution of results, with all what is known as concerns the imperfection of such contracts and the danger of opportunistic behaviour. Such contracts need to be somehow evolutionary, in order for example, to enlarge the consortium.
- the organisational set-up and project management (and monitoring) : leadership, decision processes, routines, location, concurrent activities, conflicts,....

This brings us to the knowledge production process itself. I want only to stress some of the major dimensions and requirements :

<u>The process is targeted</u> : as indicated, knowledge is produced in order to, and to the extent that, a (new) solution is found for a (new) problem. By new solution, we mean of course an innovation, with the understanding that the innovation is not necessarily an innovation in the absolute sense, but at least for the players involved, that the innovation can be a major or a minor one, that innovations can concern products, processes, organisations, markets...

Speaking of <u>new solutions to new specific problems</u>, this means, negatively, that we are not (but by chance) in any repetition or replication of existing solutions. This means, positively, that the analysis of the exact specific dimensions of the question/problem is as important as the technical dimensions of the answer. The target can of course be more or less well defined and clear. It can of course happen by chance that the problem or question is very clearly defined form the start, but in most cases, the problem or question can only be defined in rather general terms, and the definition has to be made more precise progressively (partially, on the basis of first indications as concerns possible solutions). In particular cases, it may appear (or seem) that the question or problem is very clear and some solution applied elsewhere (in another country or place, in another sector or domain..) can be readily applied. If so, we can of course speak of " transfer of technology " in the strict sense of the word.

Two important reservations here :

* there has always been (see the debates on « appropriate technologies ») and there always is some systematic bias leading to the underestimation of the importance of the specific dimensions of the problem or question at hand. And while it may of course be true that a similar standardised

production problem can adequately receive the same technical solution, what the learning economy is about, as indicated above, is the increasing importance of such specificities. The deep and circumstantial analysis of the specific dimensions of the problem or question is an absolute necessity, without which there is no use in applying an existing solution.

* even when, because of existing solutions to similar problems, it seems rather clear that a solution exist, the players have to go through the « problem » knowledge production process implying also, even if at some lower level, several of the interactions which are required in order to be sure that the relevant solution has been found and mastered.

<u>The required interactions are multidimensional</u> : schematically speaking, interactions have to take place both between three orders or spaces of information/knowledge and within those different spaces. : having complementary competencies, the associated actors or competencies have to contribute collectively to pieces of knowledge at their level or in their domain of expertise, but they have also to confront those pieces of knowledge with those which are developed in the other spaces.

Three types of interactive processes have to be organised simultaneously and iteratively :

i. on the one hand, among and with different types of scientific and technological competencies : they have to select and combine the scientific and more so technological elements of the possible and feasible solution.

ii. on the other hand, among and with different design, engineering and production competencies : they have, taking account of technical, resource and cost constraints, to select and combine the technical engineering elements of the possible solution;

iii. finally, with the client/user, and thus also with economic and market competencies : some way or another, user requirements, constraints and possibilities have to be confronted, so as to "customise" the possible solution.

And in none of those spaces can the elements of solution be made definitive, before being confronted with the elements of solution in the other spaces. This requires several iterations and, in many cases, some kind of continuous iteration.

The <u>requirements of the interactive (cognitive) processes are very demanding</u>. Three major potential dangers exist here :

i. For many small-scale applications, because of the organisational set-up (for example, large organisations), the costs appear to be rapidly quite excessive : there is a kind of essential imbalance between the high costs of such a specific solution and the small scale or limited production run which is concerned. It is possible that, in such case, this solution is not even looked for : such processes being too expensive within large firms, and out of reach of small firms. This can be considered as a case of "organisational failure ".

ii. Because of the complexity of the interactive processes involved and of the lack of the competencies and /or organisation, which are required for organising and mastering such processes, these don't succeed in heading towards their expected result or target. This is thus a pure failure.

iii. Because of the complexity of the interactive processes involved and, in some cases, the role of some intermediaries or service providers whose interests lie in the process itself more than in the result, several simplifications, approximations or reductions are introduced within the process, which thus tends to be more or less systematically impoverished, as is necessarily the outcome. A solution is thus found - the simplification increases of course the probability of a solution - but only a "poor" solution, i.e. a solution which, from the technical and/or economic standpoint, is only of limited interest. Poor solutions, in a strongly competitive world, bring no benefit.

What has been said above about replicated solutions, based on the underestimation of the specific dimensions of the question or problem, can be considered as belonging to this category.

4. Some implications

If these are the new realities of production, there are necessarily enormous implications. I list some :

i. the increased competitive importance of competencies, the firm being more and more defined with reference to her « core » competencies, which are built up within specific strategies. Competencies cannot be defined per se : they can only be defined with reference to strategic productive objectives or targets, i.e. for doing things in a particular domain. Of course some of the pieces of competencies are transferable.

ii. new competence requirements : I have suggested the idea of the "tripod " (J.De Bandt, 1995, 1997) : meaning that in order to participate in such knowledge production processes (which will be the case of the large majority of workers), three types of training and competencies need to be combined : disciplinary knowledge of some realities (nature, engineering, economy, firms and people) with pluridisciplinary openings; competencies in information (hard or software) and information system (how to design and manage information systems); interactive competencies (communication, cognitive processes, language and codification).

iii. the strongly increased importance of tacit knowledge. Contrary to those who consider that tacit knowledge had progressively been reduced by more and more systematic codification efforts, the idea is that knowledge production processes are made of successive phases of production of both tacit and codified knowledge.

iv. new training processes and loci : rapidly increasing importance of training (learning) within targeted organisational (collective) learning processes within firms. From that standpoint, the current trend towards "professionalisation " of training is only a kind of transition.

v. new research processes, institutions and loci : the idea being that while the " new production of knowledge" (Gibbons et al.1994) is developing, in response to problems or questions and outside the disciplinary boundaries, the specificity of science itself or the scientific systems is to come to an end (J.De Bandt, 1997).

vi. the new forms of the production process : information/knowledge contents, non material investments, non separation of design, production and consumption, cognitive interactions, process maturing over time,...

vii. the emphasis on networking partnerships, involving series of players, with different statutes, competencies, interests, logic....

viii. the renewed and much increased importance of co-operation and non-market relations among actors : within the conflict-concourse (or competition-co-operation) relations between decentralised actors, of which economic theory has traditionally been willing to consider only the competition part, the co-operative dimensions are increasingly important and decisive (in the sense that co-operation becomes a decisive dimension of competitiveness).

ix. the "nature" of the firm : while the boundaries of the firm tend to more or less disappear, the new major organisational requirements have to be met at the level of sub-systems (of actors) or projects.

x. the new importance of the collective dimensions of production activities : it is well known that, the same way technologies have been shown to be social products, knowledge production processes have deeply embedded collective or social dimensions.

xi. the necessity to define new rules for « industrial » property, taking account of the non-material dimensions and the absence of technical specifications, of the importance of knowledge as such, of the nature of knowledge, of the collective dimensions of knowledge production processes...

xii. the new reference for the evaluation of the value (and wealth) produced : the user value of the specific dimensions of the answer or solution.

References.

1. W.J.Baumol, S.A.B.Blackman, E.N.Wolff (1989), Productivity and American Leadership : the Long View, MIT Press.
2. J.De Bandt (1997), De la science à la connaissance : changement de paradigme ? *Revue d'Economie Industrielle*, N°79, 1°trim.
3. J.De Bandt, (1997), Peut-on et comment adapter la théorie économique aux activités et relations de services ? Is it possible and how to adapt economic theory to service activities and relations (preliminary title), VIIth Annual Conference of RESER, Roskilde, september .
4. J.De Bandt (1996), Business Services : Markets and Transactions, *Review of Industrial Organization*, Volume 11, N°1, February.
5. J.De Bandt (1996), Coopération, accords interentreprises, concurrence, in J.Ravix, ed. *Coopération entre les entreprises et oganisation industrielle*, CNRS Editions.
6. J.De Bandt (1996), *Services aux entreprises : informations, produits, richesses*, Economica.
7. J.De Bandt (1989), Can we measure Productivity in Service Activities ? in A.Bressand and
8. L.Dibiaggio (1998), Information, Connaissance et Organisation, Thèse de doctorat, Université de Nice Sophia Antipolis (janvier)
9. K.Nicolaïdis, eds., *Strategic Trends in Services : an Inquiry into the Global Service Economy*, Harper and Row.
10. M.Gibbons, C.Limoges, H.Nowotny, S.Schwartzman, P.Scott and M.Trow (1994), *The new Production of Knowledge: the Dynamics of Science and Research in Contemporary Societies,* SAGE Publications.

TECHNOLOGY-KNOWLEDGE DIFFUSION PATTERNS IN THE UNITED STATES

LAWRENCE RAUSCH and JENNIFER SUE BOND
National Science Foundation
United States

This paper will discuss methods for identifying and tracking transfers of technology between economic sectors and across national borders. It begins by first outlining recent developments in U.S. science and technology policy and then follows with selected indicators of intersectoral co-operation in the United States. Evidence of the importance of partnering was also uncovered in a 1994 pilot innovation study and the paper will briefly describe those findings. Finally, indicators of technology diffusion between nations will be examined using U.S. data on trade in advanced technology products and in intellectual property.

1. Overview of U.S. Science &Technology Policy

In an April 1997 report to the U.S. Congress entitled Science and Technology Shaping the 21st Century, President Clinton noted, "We live in an age of possibilities. A hundred years ago we moved from farm to factory. Now we move to an age of technology, information and global competition. These changes have opened vast new opportunities for our people, but they have also presented them with stiff challenges." The 1997 report in effect reaffirmed Administration priorities first announced in August 1994 in a report entitled, Science in the National Interest. That report summarised the Administration's research portfolio and pointed to scientific and technical opportunities deserving greater emphasis in the coming years. It also noted that the U.S. Government's science and technology (S&T) portfolio must evolve in response to changes in the global-based economy – a borderless, competitive economy driven by information and technology.

Science in the National Interest was the first major post-Cold War review of national science policy. It outlined five broad goals: (1) maintain leadership across the frontiers of scientific knowledge; (2) enhance connections between fundamental research and national goals; (3) stimulate government, industry, and academic partnerships that promote investment in fundamental science and engineering and effective use of physical, human, and financial resources; (4) produce the finest scientists and engineers for the 21st century; and (5) raise the scientific and technical literacy of all Americans.

A. Inzelt and J. Hilton (eds.), Technology Transfer: From Invention to Innovation, 37–55.

The Clinton Adiministration's science and technology policy is fostered through investments in research and development (R&D), tax incentives, export controls, the encouragement of technology transfer, investments in worker skills, defence conversion, and the promotion of a modern information infrastructure. Partnerships between government, industry, and academia are also being used to implement this S&T policy.

2. Indicators of Partnerships and Collaborations

One of the strengths of the U.S. innovation system is the interaction and collaboration between sectors. Evidence that such partnerships are increasing in the United States can be seen in various indicators including the following:

- Industrial R&D support to academic institutions has grown more rapidly than support from all other sources in recent years. In constant dollars, industry-financed academic R&D increased by an estimated average annual rate of 8.1 percent between 1980 and 1997. Industry's share grew from 3.9 percent to an estimated 7.1 percent during this period.

- Industrial interactions with academia extend beyond financial arrangements. Co-authorship by industrial researchers has grown since the 1980s, and in particular with academia. Co-authorship between industrial researchers and academic colleagues increased from about 30 percent of all cross-sectoral articles co-authored in 1981 to 57 percent in 1995.

- Increasingly, scientific collaboration in the U.S. involves scientists and engineers from different employment sectors. In 1995, just under one-quarter of all academic papers involved such cross-sectoral collaboration—6 percent with industry, 8 percent with the Federal government and not-for-profit sectors, 3 percent with FFRDCs, and 2 percent with other sectors. In the other sectors, over half of their cross-sector collaborations involved academic authors.

- Industrial firms are utilising academic research in their patent applications. The number of science article citations on U.S. patents increased from 8,600 in 1987 to 47,000 in 1996. Citations to biomedical research and clinical medicine grew the fastest.

- Academic patenting has increased rapidly. The number of academic patents, while small, rose more than sevenfold in just over two decades, from 250 annually in the early 1970s to more than 1,800 in 1995. Patenting by academics in biomedical fields has been especially strong.

- Co-operative R&D is now an important tool in the development and leveraging of S&T resources. There has been a major upswing in the number of inter-and intra-sectoral partnerships and international S&T

partnerships since the early 1980s. The formation of new research joint ventures continued to grow in the 1990s with the largest increases in the two most recent years (1995 and 1996).

- Technology transfer activities became an important mission component of Federal laboratories in the late 1980s and more than 3,500 new co-operative research and development agreements were developed between 1992 and 1995.

3. Survey-Based Indicators Of Innovation And Technology Transfer

The need for better information about the innovative activities of U.S. firms, the innovative process, and the factors that affect it, all led to the National Science Foundation's 1994 study of innovation activities in U.S. industry. The pilot study of 1,000 U.S. firms involved in manufacturing industries or in software development was designed to produce national-level estimates of innovative activities.[10]

CHARACTERISTICSOF U.S. INNOVATORS

One-third of respondents answered positively to either having introduced a new product or process during the reference period (1990-92) or had plans to introduce a new product shortly (during 1993-95). This one-third figure is an estimated national average for the United States (coverage spanned manufacturing industries and the U.S. software industry.)

Certain industries reported above average levels of innovation in particular U.S. industries manufacturing computer hardware (84 percent of companies were innovators), precision instruments and equipment (74 percent), pharmaceuticals (69 percent), and chemicals (68 percent).

Process innovation appears as prevalent as product innovation with nearly equal numbers of companies introducing new innovative processes and products during the 1990-92 period. Almost all (97 percent) of these companies now identified as innovators plan to introduce new innovations (product or process) during 1993-95.

Innovating firms were more likely to export than were non-innovators: 50 percent of companies that reported introducing an innovation during 1990-92 also reported export sales in 1992, compared with only 38 percent of non-innovating firms.

The Pilot Study provided insights to several other important questions:

[10] The universe consisted of all manufacturing companies and one service-sector industry (Computer Programming, Data Processing and Other Computer-Related Services) with 20 or more employees. The 1,000 companies included in the pilot study were segmented by employment size into 4 categories: 1,000 or more employees, 500-999 employees, 100-499 employees, and under 100 employees.

- **Where do U.S. innovators get information that leads to the development and introduction of new products?** The three most important sources identified (answers indicating sources as being either "very significant" or "crucial") were internal sources, clients and customers, suppliers of materials and components. The least important (combining answers of "slightly significant" and "insignificant") were government labs, technical institutes, and consulting firms.

- **What are the key factors involved in the decision to innovate?** The three most important factors (answers of "very significant" or "crucial") were a desire to improve product quality, increase or maintain market share, and extend product range within main product field.

- **What channels did U.S. innovators most often use to gain access to new technology?** The three channels mentioned most often were hiring skilled employees, purchasing equipment, and using consultants. It is interesting to note that two of these three involve people.

- **What channels did U.S. innovators most often use to transfer new technologies out of the enterprise?** The three channels for transferring technology most often mentioned by innovators was communication with other companies, mobility of skilled employees, and R&D performed for others

- **What methods did U.S. innovators employ to appropriate the benefits of their new innovations?** The three methods or practices most often mentioned by innovators were having a lead time advantage over competitors, maintaining trade secrets, and obtaining patents.

- **How important is R&D to the innovation process?** According to respondents, 84 percent of all innovators performed R&D in 1992 and 91 percent of all innovators plan to undertake R&D during 1993-95. Innovators reported R&D activity in a wide spectrum of technology areas.[11] The top three areas were software, materials synthesis and processing, and flexible integrated manufacturing.

- **Did R&D or other innovative activities involve external partners?** More than half (52 percent) of all innovating companies in the survey had co-operative R&D or innovation-related arrangements with other enterprises or institutions. Customers and suppliers were the most common partners in

[11] The 1993 National Critical Technologies Report identified nine technologies important to the long-term security or economic prosperity of the United States. The report noted that more information was needed about current levels of R&D activity in these areas. The pilot study included this question in response to the request. This report was prepared by the U.S. Office of Science and Technology Policy and the National Critical Technologies Review Group for the President.

these co-operative activities; competitors, industry-operated labs, and government labs were the least common.

International cooperation and the sharing of lessons learned from national surveys have greatly improved the survey instruments used to develop national statistics on industrial innovation. The revised Oslo Manual[12] in large part reflects this progress and will certainly guide the National Science Foundation on its future survey activity.

4. Indicators of International Technology Diffusion

Existing national statistics can offer many insights into technology transfers between countries. This paper discusses two: statistics on product trade and trade in intellectual property.

U.S. TRADE IN ADVANCED TECHNOLOGY PRODUCTS

In order to track and better understand trade in new technology products, the U.S. Bureau of the Census has developed a classification system for exports and imports of products that embody new or leading-edge technologies. This classification system allows trade to be examined in 10 major technology areas. These 10 advanced technology areas are:

- **biotechnology**—the medical and industrial application of advanced genetic research toward the creation of new drugs, hormones, and other therapeutic items for both agricultural and human uses;

- **life science technologies**—application of scientific advances (other than biological) to medical science (for example, medical technology advances such as nuclear resonance imaging, echocardiography, and novel chemistry, coupled with new production techniques for the manufacture of drugs, have led to new products that allow for the control or eradication of disease);

- **opto-electronics**—development of electronic products and components that involve emission or detection of light, including optical scanners, optical disc players, solar cells, photosensitive semiconductors, and laser printers;

- **computers and telecommunications**—development of products that process increasing volumes of information in shorter periods of time, including facsimile machines, telephonic switching apparatus, radar apparatus, communications satellites, central processing units, computers,

[12] OECD/Eurostat 1997.

and peripheral units such as disk drives, control units, modems, and computer software;

- **electronics**—development of electronic components (except opto-electronic components), including integrated circuits, multilayer printed circuit boards, and surface-mounted components, such as capacitors and resistors, that result in improved performance and capacity and, in many cases, reduced size;

- **computer-integrated manufacturing**—development of products for industrial automation, including robots, numerically controlled machine tools, and automated guided vehicles that allow for greater flexibility in the manufacturing process and reduces the amount of human intervention;

- **material design**—development of materials, including semiconductor materials, optical fibre cable, and videodiscs, that enhance application of other advanced technologies;

- **aerospace**—development of technologies, such as most new military and civil airplanes, helicopters, and spacecraft (with the exception of communication satellites), turbojet aircraft engines, flight simulators, and automatic pilots;

- **weapons**—development of technologies with military applications, including guided missiles, bombs, torpedoes, mines, missile and rocket launchers, and some firearms; and

- **nuclear technology**—development of nuclear production apparatus, including nuclear reactors and parts, isotopic separation equipment, and fuel cartridges (nuclear medical apparatus is included in life science rather than this category).

Industry analysts who are expert in foreign trade data, knowledgeable about product manufacturers, and aware of how products are manufactured identify whether internationally traded products fall into any of these categories. To be included in a category, a product must contain a significant amount of one of the leading-edge technologies, and the technology must account for a significant portion of the product's value.[13] Since the characteristics of products the United States exports are likely to be

[13] This classification system relies on the judgments of knowledgeable government experts for product identification. These judgments are reviewed by other experts to minimize the impact of subjectivity. There is no single preferred methodology for identifying high-technology products or industries. The identification of those industries considered to be high-tech has generally relied on some calculation comparing R&D intensities. R&D intensity, in turn, has typically been determined by comparing industry R&D expenditures and/or numbers of technical people employed (i.e., scientists, engineers, and technicians) with industry value added or the total value of its shipments. The information produced by these R&D-intensity-based

different from the products the nation imports, experts evaluated exports and imports separately.

THE IMPORTANCE OF TECHNOLOGY PRODUCT TRADE TO OVERALL U.S. TRADE

U.S. trade in technology products accounted for 17 to 20 percent of all U.S. trade (exports plus imports) in merchandise between 1990 and 1996. (See text table 1.) Total U.S. trade exceeded $1.4 trillion in 1996; $285 billion involved trade in technology products. Trade in technology products accounts for a much larger share of U.S. exports than imports (25 percent versus nearly 16 percent in 1996) and makes a positive contribution to the overall balance of trade. After several years in which the surplus generated by trade in technology products declined, preliminary data for 1996 show a larger surplus than in 1995.

EXPORTS OF U.S. TECHNOLOGY PRODUCTS

Japan and Canada are U.S. industry's largest nation customers; each country is the destination for about 12 percent of total U.S. technology exports. European and other Organization for Economic Cooperation and Development (OECD) countries are also important consumers of U.S. technology products. New markets have developed in several newly industrialized and developing economies, especially in Asia. Technology purchases by these economies now rival levels sold to many of the advanced European countries.

Japan and Canada are among the top three customers across the range of U.S. technology products
(Japan ranks in the top three 10 of 11 technology areas; Canada in eight). (See figure 1.) Germany is a leading consumer of U.S. products in three technology areas: life science products, opto-electronics, and nuclear technologies. While several other advanced nations are also important customers for particular U.S. technologies, notably the United Kingdom (telecommunications and aerospace), France (aerospace), and Belgium (biotechnology), several of the newly industrialized and emerging Asian economies now rank among the largest customers for U.S. technology products.

U.S. Exports to Developing and Transitioning Nations. Trends in a nation's purchases of foreign-made products that contain cutting-edge technologies give some indication of that economy's ability to incorporate new technologies, and may suggest a national direction regarding technology development. Data on U.S. exports of technology products to Asia, Latin America, and Africa provide a measure of these trends. The United States is just one of several suppliers of technology products to developing

classification systems is often distorted by the inclusion of all products produced by the selected high-tech industries, regardless of the level of technology embodied in the product. The advanced technology product system of trade data allows for a highly disaggregated, more focused examination of technology embodied in traded goods compared with that possible with any industry-based classification system.

nations. Japan, Germany, and other OECD countries are major suppliers of technology products and may be more dominant than the United States in particular regions. Nevertheless, the United States is the leading supplier of technology products to the international marketplace and, therefore, U.S. exports can serve as an indicator of other nations' technological activity.

Asia. Asia is an important customer for U.S. technology products. Japan, one of the largest consumers, has been joined by other Asian economies that have emerged as eager customers of U.S.-made advanced technologies. In 1994, the newly industrialized economies (NIEs) of Asia - Hong Kong, Singapore, South Korea, and Taiwan - together were the recipients of 17 percent of all U.S. technology exports; in 1990, they accounted for 12 percent. Singapore, the smallest of these Asian economies, has become the second largest Asian market after Japan, ahead of the much larger economies of South Korea and Taiwan.5 Electronics, computers, and telecommunications products account for approximately 66 percent of all U.S. technology products exports to Singapore. Bringing in these information-based advanced technology products from the United States helps Singapore move even more quickly toward its national goal of developing an information-based economy. Aerospace technology accounts for another 28 percent of the United States' technology exports to Singapore. U.S.-made commercial aircraft and parts have helped to make Singapore's national airline an important provider of air transportation in the region.

The two larger NIEs, South Korea and Taiwan, are on a similar level of economic development as Singapore; yet, on a per capita basis, these economies purchase far fewer technology products from the United States. Either they are more technologically self-sufficient or they have sources of advanced technologies elsewhere. Other indicators would suggest that they are more technologically self-sufficient than the other NIEs. Still, they continue to be important customers for U.S. technologies. South Korea purchased 5.1 percent of U.S. technology exports in 1994 and Taiwan 4.5 percent. As with Singapore, electronics, computers and telecommunications, and aerospace products are the primary U.S. technologies shipped to these Asian NIEs.

Many of the fastest growing export markets for U.S. technology products are also found in Asia. U.S. technology exports to four emerging Asian economies (EAEs) - China, India, Indonesia, and Malaysia - approached $9 billion in 1994, which represents a doubling over the past 5 years. Malaysia, the smallest of the four in terms of landmass and population, was the biggest consumer of U.S. technology products in 1994; purchases totaled nearly $4.6 billion. By comparison, U.S. technology exports to China exceeded $3 billion in 1994, two-and-one-half times the value exported to China in 1990. Aerospace products are the primary U.S. technology exported to China and Indonesia (63 percent and 76 percent, respectively, of each country's technology imports from the United States). India's imports of U.S. technology products were concentrated in two areas: aerospace products (41 percent of India's technology imports from the United States in 1994) and information technologies (28 percent). U.S. electronics were the primary technology export to Malaysia in 1994.

China, India, Indonesia, and Malaysia together have over 2 billion people and represent many of the world's fastest growing economies. They are also committed to technology-driven, economic development. As the recent trends in U.S. exports suggest, these market dynamics have not escaped the attention of U.S. technology producers.

South America. U.S. exports of advanced technology to four of the larger South American countries (Argentina, Brazil, Chile, and Peru) grew from $2.2 billion in 1990 to $3.2 billion in 1994. These dollar volumes represented 2.3 percent (1990) and 2.7 percent (1994) of all U.S. technology exports, about one-sixth the amount exported to the Asian NIEs.

U.S. information technologies accounted for 68 percent of all technology exports to the four-country South American region in 1994; in 1990, aerospace products dominated the region's technology imports from the United States. Brazil is the region's largest consumer of U.S. technology products, but since 1990 its relative share has declined. Conversely, Argentina's imports of U.S. technology products tripled in dollar value from 1990 to 1994, although it still purchased less than 60 percent of the amount purchased by Brazil in 1994.

The U.S. technology area that experienced the most growth in exports to South America is computer software technology, which grew 600 percent in just 5 years ($129 million in 1994, up from $14 million in 1990). U.S. life science technologies (pharmaceuticals and medical equipment) and biotechnology exports to South America also increased sharply from 1990 to 1994.

Central and Eastern Europe. U.S. exports of technology products to three Central and Eastern European countries - Hungary, Poland, and Russia - are small in comparison with those exported to the Asian NIEs, and only one-third the amount exported to the four South American countries. Together they accounted for just 0.8 percent of all U.S. technology exports in 1994. Obstacles' preventing greater exports to these three countries include the lack of foreign exchange and suitable infrastructure to fully utilize many advanced technologies.

Two product areas account for over 80 percent of U.S. technology exports to this area of Europe: aerospace and information technologies. Hungary and Poland together import less than Russia, and that gap has widened during the 1990s. Russia spent nearly $344 million on U.S. aerospace products in 1994, after spending less than $40 million in the previous 2 years. Exports of U.S. information technologies to Russia have also increased; shipments doubled from 1992 to 1993 to reach $185 million, with another $177 million purchased in 1994.

U.S. technology exports to Hungary and Poland, like those to Russia, are primarily aerospace and information technologies. Consistent with a need to retool manufacturing facilities, the fastest growing technology area for U.S. exports to Eastern Europe is robotics technology.

Africa. U.S. exports of technology products to the three top importing African countries (Kenya, Nigeria, and South Africa) totaled just $485 million in 1994, representing 0.4 percent of all U.S. technology exports that year and about one-half what was exported to the three Eastern European countries. Exports of U.S. technology to South Africa are 10 times exports to Nigeria and Kenya combined, and will likely increase in the near future, with the lifting of the international boycott on investment and trade. Well over 80 percent of South Africa's technology imports from the United States are concentrated in information technologies and aerospace. Software products are clearly the fastest growing area for U.S. technology exports to South Africa. In 1990, U.S. software sales totaled just $490,000, but they rose dramatically each year thereafter and reached $51 million by 1994.

As observed for other developing countries, U.S. technology exports to Nigeria and Kenya are made up primarily of information and aerospace technologies. U.S. sales of life science technologies to both nations were also strong, in particular to Nigeria in 1992 and 1993. During 1990-92, weapons accounted for 10 to 16 percent of Kenya's technology imports from the United States.

U.S. Imports of Technology Products

The United States is not only an important exporter of technologies to the world, but also it is a major consumer of foreign-made technologies. Imported technologies enhance productivity of U.S. firms and workers, improve health care for U.S. residents, and offer U.S. consumers more choice.

Technology products represent a significant and growing share of all goods and services imported by the United States. U.S. imports of technology products grew steadily during the 1990s.

The three technology areas that account for the bulk of technology products shipped from the United States account for a slightly larger portion of technology products shipped to the United States. In 1994, 3 of the 10 technology categories accounted for 89 percent of total U.S. technology product imports: information technologies (50.8 percent of technology imports in 1994), electronics (26.3 percent), and aerospace (11.6 percent). Electronics was the fastest growing technology area for both U.S. imports and U.S. exports.

Top Suppliers, by Technology Area. The leading economies in Asia and Europe are important suppliers to the U.S. market in each of 10 technology areas. (See figure 2.) Japan is a major supplier in seven advanced technology categories, Germany in four. Consistent with their status as major industrialized nations, Canada and the United Kingdom also supply a wide variety of technology products to the United States and are among the top three in several technology areas.

A large volume of technology products comes from the newly developed and developing Asian economies, in particular, Malaysia, Singapore, and South Korea. Growing technology product imports from these Asian countries and from other regions into one of the most demanding markets in the world suggest a widening of technological capabilities globally.

U.S. Technology Imports from Developing and Transitioning Economies. The market competitiveness of technology products from developing and transitioning economies provides an important test for their science and technology enterprises: whether or not they are sufficiently developed to generate advanced technology products that meet the demands of the international marketplace. A nation's ability to export cutting-edge technology products to technologically advanced markets, like that of the United States, provide such an assessment. Trade data also provide an indication of a developing economy's focus with respect to technology development.

Asia. In 1994, the Asian NIEs - Hong Kong, Singapore, South Korea, and Taiwan - together supplied nearly 26 percent of all U.S. technology imports. Five years earlier, they accounted for 23 percent .Singapore ranks second in the region, behind only Japan and ahead of the much larger economies of South Korea and Taiwan. Information products and electronics together account for over 95 percent of all technology product imports by the United States from Singapore.6 These two technology areas also dominated U.S. technology imports from South Korea and Taiwan.

Other Asian economies are developing into important suppliers of technology products to the U.S. market. U.S. technology imports from four EAEs - China, India, Indonesia, and Malaysia - approached $10 billion in 1994, up from just $2.1 billion in 1990. Once again, Malaysia stands out, with exports to the United States twice that of the others' exports combined in 1994. Electronics has been the major U.S. technology import from Malaysia, although U.S. imports of information technology products grew quickly and, by 1994, ranked a close second. U.S. technology imports from China, India, and Indonesia are primarily information products. In 1994, information products accounted for 82 percent of all U.S. technology imports from China, 78 percent from India, and 88 percent from Indonesia. Indian scientists and engineers have gained worldwide recognition for their skills in software engineering. Those skills have helped spur India's exports of software products. In 1994, software products accounted for 13 percent of India's information technology exports to the United States.

The growing bilateral trade activity evident from the preceding discussion cannot be explained solely as shipments between U.S. companies and affiliates in those countries. Other S&T indicators in particular, patenting and bibliometric trends and numbers of scientists and engineers in the working population also point to the expanding technological capacities developing across Asia.7

South America. Trade in technology products between the United States and South America appears to be very one sided. While U.S. exports of advanced technology products to four of the larger South American countries (Argentina, Brazil, Chile, and Peru) grew from $2.2 billion in 1990 to $3.2 billion in 1994, U.S. technology imports were valued at just $363 million in 1990 and declined to $152 million in 1994. This is contrary to a generally rising trend in U.S. imports from these four South American countries in all other goods and services. Information products and aerospace technologies accounted for over 90 percent of U.S. technology imports from these four countries during the 1990s.

Figure 6-3.
Three major export markets for U.S. technology products: 1994

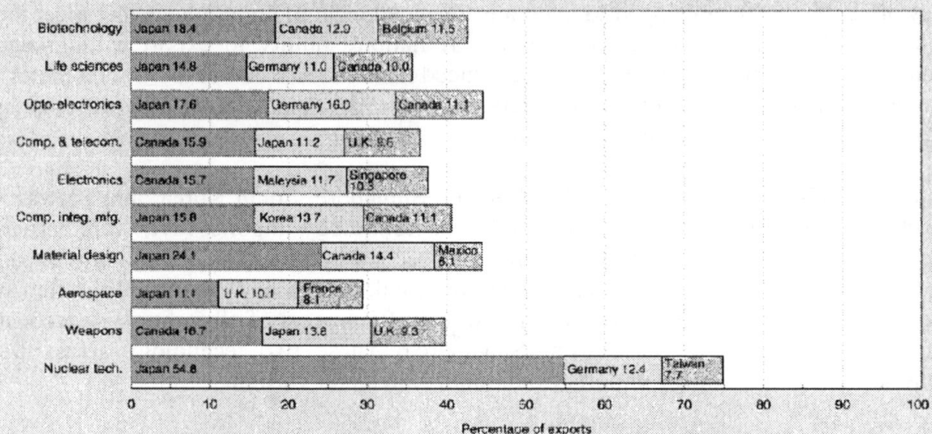

See appendix table 6-2.

Science & Engineering Indicators – 1996

Central and Eastern Europe. U.S. technology imports from the three former Eastern bloc countries, Hungary, Poland, and Russia, were larger than those imported from the four Latin American countries in 1994 and appear to be on an upward trend. Still, together they accounted for just 0.2 percent of all U.S. technology imports in 1994. Life science technologies (pharmaceuticals and medical equipment) make up a major portion of U.S. technology imports from Russia (76 percent in 1994) and Hungary (42 percent).[14] Aerospace technologies accounted for nearly half of all technology imports from Poland. Information products and electronics were two growth areas for Hungary and Poland. Imports of nuclear technologies from Russia were quite large (9 percent of the total) in 1994.

Africa. The United States exports more technology products to Africa than it imports from Africa. This trade surplus is less than the surplus generated from bilateral trade with Latin America during the 1990s; it is similar to the surplus observed for Eastern Europe. U.S. imports of technology products from three African countries (Kenya, Nigeria, and South Africa) totaled just $14 million in 1994, representing 0.1 percent of all U.S. technology imports that year. Imports of technology products from South Africa

[14] In 1994, The Republic of Slovenia was the leading foreign supplier of biotech products to the United States; in 1993, it was the ninth leading supplier. Biotechnology research in Slovenia focuses on cloning and gene expression in *E. coli,* studies of metabolic regulation, and development of processes for new fungal metabolites according to reports from its Ministry of Science and Technology.

are 10 times imports from Nigeria and Kenya combined. In 1994, 57 percent of U.S. technology imports from South Africa and 33 percent from Nigeria were life science technologies. Information products dominate U.S. bilateral trade with Kenya. Life science technology appears to be the fastest growing technology area for African exports to the United States.

5. U.S. Royalties and Fees Generated from Intellectual Property

The United States has traditionally maintained a large surplus in international trade of intellectual property. Firms trade intellectual property when they license or franchise proprietary technologies, trademarks, and entertainment products to entities in other countries. These transactions generate net revenues for U.S. firms in the form of royalties and licensing fees.

U.S. ROYALTIES AND FEES FROM ALL TRANSACTIONS

U.S. receipts from all trade in intellectual property reached $26.9 billion in 1995, a 21 percent increase over 1994. The 1995 surplus continued the steady upward trend that has resulted in a doubling of U.S. receipts in just six years. During the 1987-95 period, U.S. receipts were generally four to five times as large as U.S. payments to foreign firms for intellectual property. Most (about 75 percent) of the transactions involved exchanges of intellectual property between U.S. firms and their foreign affiliates. (See text table 2.)

Figure 6-5.
Three major foreign suppliers of technology products to the United States: 1994

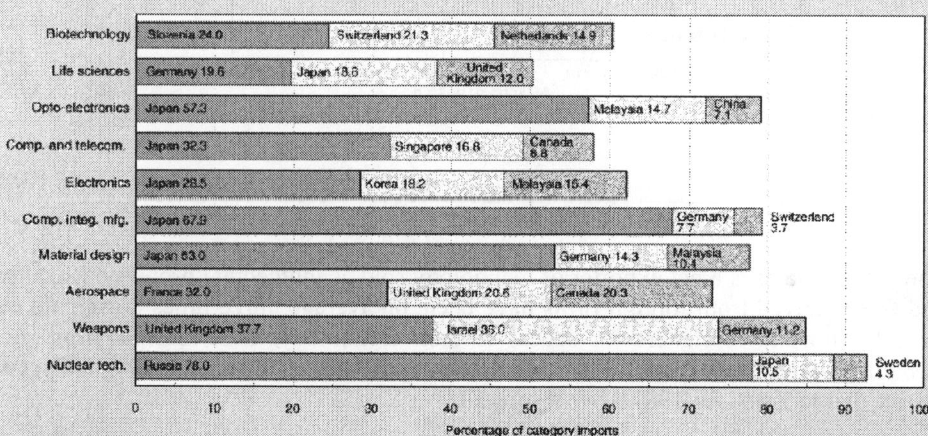

See appendix table 6-2.

Science & Engineering Indicators – 1996

Exchanges of intellectual property among affiliates continue to grow faster than those among unaffiliated firms. This trend suggests a growing internationalisation of U.S.

business and a desire to retain a high level of control on any intellectual property leased overseas.

U.S. ROYALTIES AND FEES FROM TRADE IN TECHNICAL KNOWLEDGE

Data on royalties and fees generated by trade in intellectual property can be further disaggregated to reveal U.S. trade in technical know-how. These data describe transactions between unaffiliated firms where prices are set through a market-based negotiation. Therefore these data better reflect the exchange of technical know-how and its market value at a given point in time than do data on exchanges among affiliated firms. When receipts (sales of technical know-how) consistently exceed payments (purchases), these data may indicate a comparative advantage in the creation of industrial technology. Examining the record of resulting receipts and payments also provides an indicator of the production and diffusion of technical knowledge.

Figure 3.
U.S. royalties and fees generated from the exchange of industrial processes between unaffiliated companies: 1995

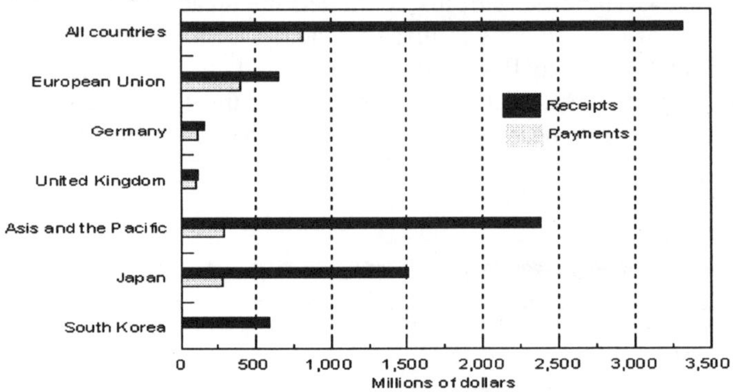

Science & Engineering Indicators - 1998

The United States is a net exporter of technology sold as intellectual property. Royalties and fees received from foreign firms have been, on average, three times those paid out by U.S. firms to foreigners for access to their technology. U.S. receipts from such technology sales exceeded $3.3 billion in 1995, up from $3.0 billion in 1994, and nearly double that reported for 1987. (See figure 3.)

Japan is the largest consumer of U.S. technology sold as intellectual property. In 1995, Japan accounted for over 45 percent of all such receipts, while the European Union (EU) countries together represented about 20 percent. Another Asian country, South Korea, is the second largest consumer of U.S. technology sold as intellectual property and has maintained that position since 1988. In 1988, South Korea accounted for 5.5

percent of U.S. receipts, its share rose to 10.7 percent in 1990, and the latest data show South Korea's share rose to 17.6 percent in 1995.

To a large extent, the U.S. surplus in the exchange of intellectual property is driven by trade with Asia. In 1995, U.S. receipts (exports) from technology licensing transactions were eight times U.S. firm payments (imports) to Asia. As previously noted, Japan and South Korea were the biggest customers for U.S. technology sold as intellectual property—together they accounted for over 50 percent of total receipts in 1995.

In comparison with Asia, the U.S. experience with Europe has been very different. Over the years, U.S. trade with Europe in intellectual property bounced back and forth, showing either a small surplus or deficit each year. In 1995, U.S.-Europe trade produced the largest surplus in the nine years examined, the result of a sharp decline in U.S. purchases of technical know-how from the smaller European countries.

Foreign sources for U.S. firm purchases of technical know-how have changed somewhat over the years, with increasing amounts coming from Japan. Europe still accounts for nearly 60 percent of the foreign technical know-how purchased by U.S. firms, with France, Germany, and the United Kingdom being the principal European suppliers. But since 1990, Japan has been the single largest foreign supplier of technical know-how to U.S. firms.

SUMMARY

These indicators serve to highlight the trend toward greater intersectoral and international collaborations in U.S. science and the growth in technology being transferred across national borders. Also apparent from the current literature is the science and technological activity by an even wider set of nations. All this suggests new markets for a nation's technology exports, additional foreign sources for advanced technologies, and new opportunities for scientific collaboration. The need for better indicators that can inform policymakers on current trends and future directions has also been an outcome of the increased globalisation of science and technology.

Text Table 1. U.S. International Trade in Merchandise

	1990	1991	1992	1993	1994	1995	1996
Total exports (billions of U.S. dollars)….. …..	393.0	421.9	447.5	464.8	512.4	575.9	611.5
Technology products (percent) ……..	24.1	24.1	23..9	23.3	23.6	24.0	25.3
Other merchandise (percent) ……..	75.9	75.9	76.1	76.7	76.4	76.0	74.7
Total imports (billions of U.S. dollars)………..	495.3	488.1	532.4	580.5	663.8	749.4	799.3
Technology products (percent) ……..	12.0	13.0	13.5	14.0	14.8	16.7	16.3
Other merchandise (percent) ……..	88.0	87.0	86.5	86.0	85.2	83.3	83.7
Total trade (billions of U.S. dollars)………..	888.3	910.0	979.9	1045.3	1176.2	1353.3	1410.8
Technology products (percent) ……..	17.3	18.1	18.3	18.1	18.6	19.9	20.2
Other merchandise (percent) ……..	82.7	81.9	81.7	81.9	81.4	80.1	79.8

Note. Total trade is the sum of total exports and total imports

Source. U.S. Bureau of the Census, World Wide Web Site

Text Table 2. U.S. receipts and payments of royalties and fees associated with affiliated and unaffiliated foreign residents 1987-98			
	Foreign Residents		
	Total	**Affiliated**	**Unaffiliated**
	Millions of Dollars		
Receipts			
1987	9,914	7,629	2.285
1988	11,802	9,158	2,646
1989	13,064	10,207	2,857
1990	16,634	13.251	3,384
1991	18,107	14,395	3,712
1992	19,715	15,718	3,997
1993	20,333	15,707	4,616
1994	22,274	17,422	4,949
1995	26,863	21,618	5,333
Payments			
1987	1,844	1,296	547
1988	2,585	1,410	1,175
1989	2,602	1,778	824
1990	3.135	2,206	929
1991	4,076	2,996	1,080
1992	5.074	3,381	1,684
1993	4,765	3,364	1,401
1994	5.518	3,810	1,708
1995	6,312	5,148	1,163
Balance			
1987	8,070	6,333	1,738
1988	9,217	7.746	1,471
1989	10,482	8,429	2,033
1990	13,439	11,045	2,455
1991	14,031	11,399	2,632
1992	14,641	12,337	2,303
1993	16,558	12,363	32,215
1994	16,766	13,612	3,141
1995	20,641	16,471	4,170

Note. Due to rounding, totals will not always compute from detailed data

Source. Bureau of Economic Analysis, *Survey of Current Business,* Vol. 76 No. 11 (Nov. 1996)

References

1. Abbott, T.A., III. 1991. "Measuring High Technology Trade: Contrasting International Trade Administration and Bureau of Census Methodologies and Results." Journal of Economic and Social Measurement 17: 17-44.

2. Alic, J.A. 1993. "Technical Knowledge and Technology Diffusion: New Issues for U.S. Government Policy." Technology Analysis and Strategic Management 5(4): 369-83.

3. ———. 1994. Science in the National Interest. Washington, DC. Corporate Technology
4. Information Services, Inc. 1995. Corporate Technology Directory 1995. Rev. 10.1. Woburn, MA.

5. Kim, L. 1997. *Imitation to Innovation: The Dynamics of Korea's Technological Learning.* Harvard Business School Press, Boston, Massachusetts.

6. Mowery, D.C., and N. Rosenberg. 1993. "Chapter 2." In National Innovation Systems, edited by Richard R. Nelson. New York: Oxford University Press, pp. 29-75.

7. National Research Council, Hamburg Institute for Economic Research, and the Kiel Institute for World Economics, 1996. *Conflict and Cooperation in National Competition for High-Technology Industry.* National Academy Press, Washington, DC.

8. Office of Science and Technology Policy (1997), *Science & Technology Shaping the Twenty-First Century,* Executive Office of the President, Washington, DC

9. Office of Science and Technology Policy (1995), *National Critical Technologies Report,* National Critical Technologies Panel, Washington, DC

10. Organisation for Economic Co-operation and Development (OECD). 1996. Structural Analysis Database for Industrial Analysis, Analytical Business Enterprise R&D file
11. (STAN/ANBERD). Paris.

12. Romer, P.M. 1996. "Why, Indeed in America? Theory, History, and the Origins of Modern Economic Growth." *American Economic Review* Vol. 86, No 2 (May).

13. Science Resources Studies Division, National Science Foundation. 1993a. Human Resources for Science and Technology: The Asian Region. NSF 93-303. Washington, DC: NSF.

14. ———. 1995. Asia's New High-Tech Competitors. NSF 95-309. Arlington, VA: NSF.

15. Souder, W.E. 1995. "The Evolution and Transfer of National Technologies: A Conceptual Model With Comparative Studies." Journal of Scientific & Industrial Research 54 (April 1995): 231-242.

TECHNOLOGY TRANSFER FROM RTOs

Definition / Setting the scene

G.G. HUYLEBROECK
Bergbosstraat 217, B-9820 Merelbeke, BELGIUM

Abstract. In a global market high wage economies tend to be ever more knowledge based. Innovation has become a keyword for the western economies, and although innovation is not always technology based it often is. Through technology transfer knowledge providers can contribute to Schumpeter's creative destruction that alters the behaviour, competitiveness and performance of an economy. This partly explains the growing interest in continental Europe for technology transfer from research institutions to industry as a hitherto poorly utilised way to convert research results into products, processes or services.
Before venturing on technology transfer knowledge providers and their staff should be fully aware of the potential benefits and threats of this complex activity. Therefore at least a basic understanding of the process, of the way it creates wealth and for whom wealth can be created is needed.

DEFINITIONS

INNOVATION

Information technology, the combination of computers and telecommunication, has drastically altered the way in which companies and markets function during the last three decades. Huge amounts of information are now available to anyone anywhere, and they can be processed much faster than ever before. This has led to some remarkable changes in our society : people and capital have become increasingly mobile, consumer lifestyles have been harmonised across the world, and the knowledge content of products has increased dramatically, leading to ever shorter product life cycles. In order to remain competitive companies are compelled to continually add new elements to products and services, or to the methods of producing them. This addition of new elements is called *innovation*

.

Applying functional changes to an existing product is only one example of innovative behaviour. Others include the development of a new design for a product or its packaging, an alteration of a production or a management process, or - rather rarely - the introduction of a conceptually new product. This means that not all innovation has a technological component. Examples of innovative companies whose success is not

A. Inzelt and J. Hilton (eds.), Technology Transfer: From Invention to Innovation, 57–67.

primarily determined by the technological content of their products abound (e.g. IKEA, McDonalds, Bang & Olufsen, Club Mediterranean, Dell, ...).

Usually innovation comes about after careful market research : an innovation that does not satisfy a customer need is bound to be unsuccessful. If innovation is handled this way one speaks of a *market pull* approach, as opposed to *technology push*. The latter happens when an innovation is driven by the mere availability of a newly developed technology. This of course doesn't eliminate the necessity of customer approval - in general innovative products will have to comply with the simple rule that they have to be better *and* cheaper than existing solutions.

TECHNOLOGY TRANSFER

Throughout this contribution innovation will be limited to *technology based innovation*, i.e. the kind of innovation involving two different phases : invention and exploitation. The scope will be further limited to those cases where the invention has been developed by one organisation (A), while the exploitation of the invention is taken care of by another organisation (B). This necessitates a transfer of the technology from A to B, usually by means of a *license*. This process, called *technology transfer*, is not as simple as it looks. Therefore the following definition used by Technology Access is most appropriate :

> " Technology transfer is the set of business relationships by which technology developed in one place or for one purpose by one organisation, is turned into a commercial product or process, by another organisation. We usually restrict our interest to inventions that require further development to be ready for market. "

This definition makes clear what technology transfer is all about : it is about business - not about science - or more precisely about business relationships. Of course all business is people's business, but this is even more so for technology transfer. The business relationship not only includes licensing professionals, but it also extends to the inventors and to the technology's end-users : technology is hard to transfer if the people who master it and those who will apply it are not intensively involved throughout the process.

The definition also shows that technology transfer can be an inter-company activity as well as a transaction between a research institution (universities, private or government laboratories or hospitals - commonly referred to as *Research and Technology Organisations* or RTOs) and a company. Companies can have different strategic reasons to license their technology to possible competitors, including recycling of R&D investments, a lack of funds for further development, circumventing import restrictions or the promotion of their technology as an industry standard. We will not go into those details but focus our attention instead on RTO - industry transactions.

RTOs can transfer their commercially interesting inventions either to existing, well-established companies, or to companies that have been specifically incorporated to exploit the invention. The latter companies are called *spin-off* or *spin out companies* (there is no real difference between these terms : while most countries speak of a spin-off company, people in the UK will use the word spin out). Generally speaking a spin-off company is a company that would not have existed if a specific research activity had not been carried out within a certain knowledge-based organisation. Other definitions of the term are more restrictive, and limit spin-off companies to those companies that are created by employees who leave their former employer in order to exploit know-how that they have acquired while working for that employer.

While it is clear *why* research institutions license their inventions to companies (the production and sales of products are not a part of their mission statement) it is not at all obvious *whether* they have to license their inventions in the first place. After all most RTOs are non profit organisations while technology transfer is a pure business activity. In other words : is technology transfer reconcilable with a research institute's mission ? This matter will be addressed in the fourth chapter of this paper.

TECHNOLOGY TRANSFER COMPARED TO OTHER TRANSFER MECHANISMS

Although science and technology are only one aspect of innovation, and RTOs are only one possible input to the innovation process, research shows that RTOs in general and academia in particular are an important source of science and technology for innovation. Several publications pay attention to this phenomenon. A recent review of the evidence and significance of RTO's contributions to industrial innovation can be found in [1]. Technology transfer is not the only possible way for RTOs to enhance a region's innovation capacity. Other possible transfer mechanisms include (continuing) education, services and joint or contract research. Figure 1 shows how these mechanisms are related. The y-axis can be interpreted as a measure of complexity, of added value and of wealth creation for the RTO.

WEALTH CREATION AND THE TECHNOLOGY TRANSFER VALUE CHAIN

A value chain is a concept introduced by Harvard professor Michael Porter in [2]. It is based on the assumption that the processes in a firm can be broken down into a chain of different activities, where each link in the chain adds value to the product or service produced by the company. When applied to the technology transfer process (figure 2) this concept gives a revealing insight into an important question : how and for whom is wealth created ?

For a correct interpretation of the chain it is important to realise that figure 2 represents a *value* chain. This means that the x-axis is a value axis, not a time axis. The more one proceeds to the right, the more value is added to the initial research results.

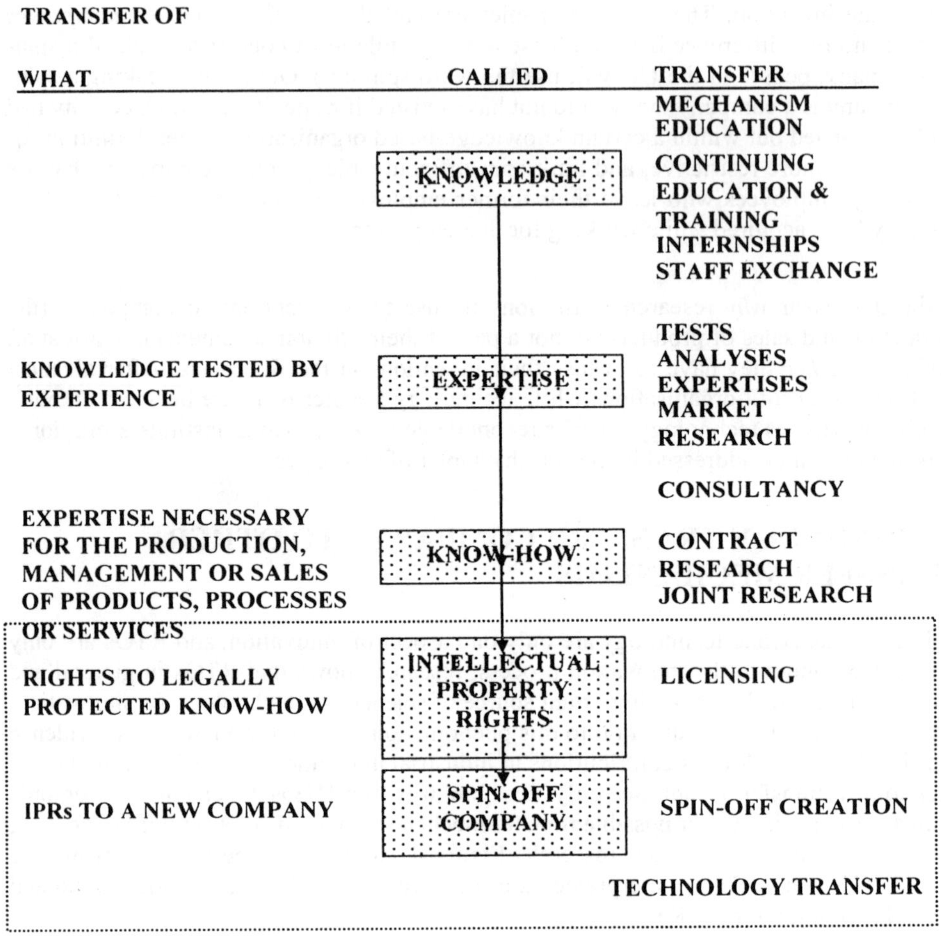

Figure 1. Transfer mechanisms between RTOs and industry

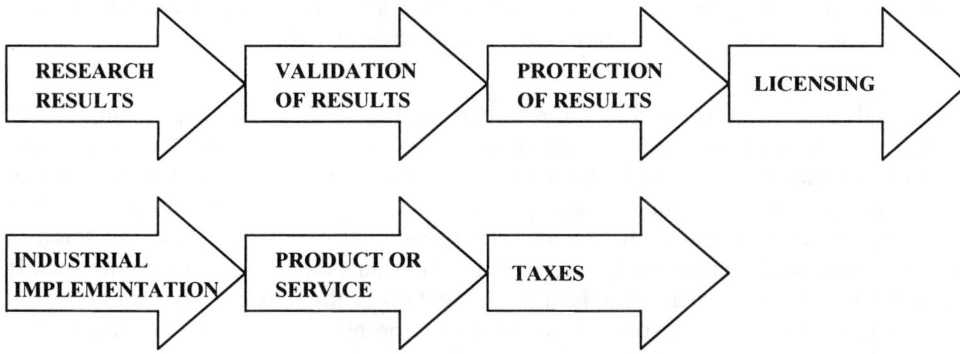

Figure 2. The technology transfer value chain

On a time axis it often happens that certain activities are realised before the links preceding them. Research results are often protected (patented) before they are validated. Sometimes they are even licensed before a patent is applied for. On a time axis the technology transfer often shows feedback loops : during the validation process new insight is gathered, leading to new research to be performed and patent applications to be withdrawn and rewritten. However, these time considerations do not affect the value chain : only validated results can be commercially successful and thus add value to the research results, even if the technology's validation consists solely of the commercialisation process itself.

The first link in the chain, representing the output of the research process, is also the input of the technology transfer process. Wealth has now been created for the researchers and for the suppliers of the research resources, and through overheads also for the institution's administration. Although they may have a tremendous scientific value, the research results have no economic value yet : they are situated at the very origin of the value chain.

The first action that really adds value to the research results is the *validation* of said results. Validation means delivering the proof of the results' potential to be turned into a commercially exploitable product or service. This proof may be obtained in many different ways : by market research, prototype development, β-testing, clinical trials, in vivo or in situ tests, benchmarking etc. Validation adds value (which is a *subjective* measure of worth) to the research results because it reduces the commercial partner's risk, thus making the exploitation of the invention more realistic and attractive. This in turn gives the RTO a bigger leverage during the license negotiations (on the supposition that the validation is completed before the negotiations !). The lack of funds to perform the validation - referred to as the *development gap* - is a well-known problem, often causing the results to be abandoned because of a catch 22 situation : the RTOs cannot fund the validation phase themselves and industry will show no interest to exploit the results as long as they are not validated. Obviously, the validation phase creates wealth for the parties carrying out the validation and for their suppliers.

The (legal) protection of the research results is the next link in the value chain. Mostly protection means filing a patent application, although on a few occasions know-how license agreements are concluded between RTOs and industry. In the latter case secrecy is the only possible protection. Patent protection creates a monopoly for the owner of the patent rights, who can exclude third parties from making or selling the invention as it is described in the patent claims. A license creates an exception to this prohibition. If the license is exclusive then the industrial partner will be the only one allowed to exploit the patent claims in the territories where patent protection has been obtained. This monopoly creates a competitive advantage for the *licensee* (the company that has obtained the license), which adds value to the results. Wealth will be created for the patent agents and for the employees of the patent offices.

Patent applications on validated results are useless to RTOs if they don't get licensed : until a licensee is found the value is negative, because patent applications cost money. Therefore finding a licensee and concluding a good license agreement are essential. During this stage wealth is created for the negotiators and their service suppliers (airlines, hotels, restaurants etc.). Wealth can also be created for the RTO if the license contract stipulates the payment of a *lump sum*. A lump sum is an irrevocable fixed amount of money, payable when the contract is signed or at periodic due dates.

The next link in the value chain is the industrial implementation of the licensed results. Engineering to production and the installation of operational production lines are costly activities, generating wealth for the employees involved and for specialised engineering or consulting agencies. Clearly the centre of gravity has been shifted towards the industrial partner now, although the input of the inventors will still be needed until the stage of full production has been reached.

The sales of products or services is the ultimate goal of the transfer process. If all goes well wealth will now be created for all parties involved : for the industrial partner through profit on sales, for the partner's employees via their salaries, and for the RTO (i.e. depending on the institution's income sharing policy for the inventors, for the inventors' or other research groups, and for the central administration) through *royalties*. Royalties are periodic payments to the licensor (the party that grants a license) as a consideration for the rights granted in the license. Usually royalties are calculated as a percentage of the licensee's net sales.

The economic activity produced by the successfully accomplished transfer process will also create wealth for the government (at a local or a national level), through the different taxes collected (income tax, VAT, tax on the company's net profit etc.). Although this outcome will already be noticeable if the RTO's region only receives a royalty stream (the industrial activity being based abroad) it is clear that a region will only benefit optimally if the licensor is also a local company. In this case secondary effects contributing to the region's wealth are an increase of the GDP, an improvement of the trade balance and a decrease in unemployment.

By separating this government income mechanism from the product/service sales it is clear that the value chain becomes circular : part of the government income received can be reinvested in research. Another way of looking at this last link in the chain is to consider the taxes collected as a return on investment for the government - the RTO's main sponsor.

TECHNOLOGY TRANSFER AND THE RTO'S MISSION STATEMENT

Having looked into the meaning of technology transfer, its relationship to other transfer mechanisms and how it can create wealth, a basic question still needs to be answered : can technology transfer be reconciled with an RTO's mission statement ? Because different RTOs will have different mission statements only a generic answer can be provided by the enumeration of possible benefits arising from technology transfer.
Some RTOs do not have a mission statement. However, the need for an explicit knowledge of the mission cannot be sufficiently emphasised. From the mission statement the organisation's goals and objectives can be deducted. Without a good knowledge of these objectives it is not possible to adequately allocate resources, or set quantitative goals to evaluate the organisation's performances. In other words, doing technology transfer without knowing the organisation's mission is like walking on thin ice.

POSSIBLE BENEFITS OF TECHNOLOGY TRANSFER

Benefits to mankind

Contrary to the popular belief that the publication of interesting research results will automatically lead to their industrial application, a lot of unpatented product technology never gets commercialised. This is particularly true for the pharmaceutical sector, where it can cost over 300 million US dollars to get a new compound past the phase III clinical trials. No company will invest this amount of money without the assurance of a temporary monopoly created by a patent. This is easy to understand : getting a product to the market involves risks and costs. Without patent protection competitors can just sit by, and walk in without taking the risks if the commercialisation proves to be successful. A much quoted example of a good technology that (almost) failed to be exploited by lack of patent protection is penicillin. Although the first penicillin was discovered by Ian Fleming in 1929, it was not until the outbreak of the second world war that large-scale production of penicillin took place.

Sometimes RTOs can decide not to protect their technology with a view to commercialisation, offering it for free instead. This can be a sound choice if the free dissemination of a technology offers exceptional advantages to mankind, e.g. in the case of techniques that will be primarily applied in third world countries (e.g. water purification technology). Other examples of freely disseminated technology include the graphical user interface X-Windows from MIT, and CERN's World Wide Web technology.

Contribution to the regional innovation capacity

RTOs are not isolated islands - they are embedded in a community, and they are almost always paid by this community. Therefore they have a responsibility towards their funding region. In chapter 3 it has been shown that this funding region can obtain a return on investment from a successful technology transfer, especially when the licensee is also a regional company. According to estimates by Dr. Ashley Stevens at Dana Farber Cancer Institute and based upon the 1993 AUTM (Association of University Technology Managers) licensing survey, tax revenue to the US state, local and federal government resulting from non-profit technology transfer programmes was estimated to be over 9 billion US dollars [3]. In the same study it was estimated that non-profit research institutions helped to create or retain over 300,000 industrial jobs in the United States.

The technology transfer process also stimulates the inventors' sense of entrepreneurship. If the latter are academics they can transfer their sense of entrepreneurship to their students. Encouraging entrepreneurial behaviour is essential for a dynamic economy.

RTOs should realise that giving preference to intra-regional technology transfer can have a negative impact on their license income. Foreign companies may be willing to pay a bigger lump sum or higher royalty rates than domestic ones. Spin-off companies create more local jobs, but are usually unable to pay big lump sums.

However, this does not mean that intra-regional technology transfer should be part of an RTO's mission statement exclusively for semi-altruistic reasons. The enhancement of the funding region's innovation capacity through technology transfer is in several ways also beneficial to the RTO.

The circular reasoning. It is clear that high-class research is difficult to uphold in an impoverished region. By contributing to the funding region's wealth the RTO also contributes to its own wealth.

Providing employment for researchers or alumni. In most cases the exploitation of a new technology will create a demand for highly trained people. If the RTO is a university, technology transfer can create more employment for the university's alumni. A special case concerns the inventors themselves. It is well-known that technology transfer is most successful when it is coupled with people transfer, i.e. when at least one of the inventors leaves the RTO to develop the transferred technology with his new employer. Since it often happens that at least one of the inventors is a doctoral student, regional technology transfer can create interesting job opportunities.

Safeguarding the funds for fundamental research. Funds for basic research are under constant pressure. Industry demands that research should be more applied - after all, research is sponsored with the taxpayer's money, and isn't industry one of the biggest taxpayers ? So in their view it would only be fair that funds would preferentially be allocated to these projects that are *directly* beneficial to industry itself.

Intra-regional technology transfer can provide two ways to relieve this pressure. Firstly, successful transfers can increase the regional innovation potential without decreasing the funds for fundamental research. Secondly, the most promising inventions often result from basic research (e.g. biotechnology). A better transfer of these results within the RTO's own region (provided the region has the proper industrial tissue to exploit these new technologies) will be a more convincing justification to maintain the funds for fundamental research than the pertinent argument that applied research is only possible when it is founded on basic research.

Dynamic image. Through its impact on the regional economy an RTO can obtain a dynamic image, especially when it is surrounded by a cluster of spin-off companies. If the RTO is a university this can have a positive effect on the university's attraction for prospective students.

Benefits to the RTO

The most obvious benefit to RTOs is the income stream that is generated by successful technology transfer projects. Generating money has been the driving force for most UK universities to establish professional technology transfer offices when their budgets had been drastically cut back during the Thatcher period. The 1993 AUTM Licensing survey reported a royalty income of 350 million US dollars for US institutions [4]. On average the licensing income received represents 1 to 3 % of the institutions' annual budget. It has to be emphasised though that most income is concentrated in a few licenses. Of all MIT's licenses (205 in 1995) only two yield more than 500 thousand US dollars per year, and yet they comprise 27 % of the total yearly income [5].

Benefits to the inventors

Inventors can benefit from technology transfer in various ways. The knowledge that the results of their work are actually applied by industry, leading to real products or services, can be intellectually satisfying. The part of the license income that will be allocated to the inventors' laboratory can enable them to do research that was otherwise too expensive, or to enhance the research or working conditions. If the RTO decides to share part of the license income with the inventors they can benefit personally from the financial rewards gained from their invention.

POSSIBLE DISADVANTAGES OF TECHNOLOGY TRANSFER

Introducing technology transfer, which is basically a business oriented activity, into an RTO with its own specific culture and values can have drawbacks too. There is of course the financial risk involved : licenses can generate income, but patent applications that do not get licensed out will only cost money.

Even when the transfer programme is successful - and especially then - intra-institutional tensions may arise between the beneficiaries of licensing income and those who know they will never make exploitable inventions. Institutional policies can aim at a partial redistribution of the license income between all research groups, but this will not eliminate the tension : in most case, now the inventors will be dissatisfied because the income that *they* have earned is given to other groups.

Technology transfer activities may put researchers in conflict of interest situations, especially when the transfer involves the creation of a spin-off company. Institutions should be aware of these possible dangers and anticipate them by elaborating a detailed conflict of interest policy.

Although patenting does not prevent publishing (actually a patent is a publication) or the dissemination of the research results, publications may be delayed if the results are patentable - at least until the patent application has been submitted. Even then some care has to be taken concerning the dissemination of know-how related to the patent application in order not to jeopardise the chances of successful commercialisation.

Finally some RTOs can have problems with more fundamental issues, like the nature of patenting itself or the granting of exclusivity to one company through a license.

SUMMARY

In this contribution technology transfer has been defined, particularly in relation to innovation and to other, more conventional, transfer mechanisms used by public research institutions. It has been shown how and for whom technology transfer can create wealth. Finally an overview of possible benefits and disadvantages of technology transfer has been given, that can serve as a guide for institutions who wish to reconcile technology transfer with their mission statement.

References

1. Faulkner, W. and Senker, J. (1995) *Knowledge frontiers*, Clarendon Press.
2. Porter, M.E. (1985) *Competitive advantage : sustaining superior performance*, Free Press, New York.
3. Leahy, H.S. (1994) *Philosophy of licensing for non-profit research institutions*, in *AUTM Manual*, Association of University Technology Managers Inc., II-1 p. 5.
4. Hoffman, D.C. (1994) *AUTM Licensing survey fiscal years 1993, 1992, 1991*, Association of University Technology Managers Inc.
5. Pressman, L., Guterman, S.K., Abrams, I., Geist, D.E. and Nelsen, L.L. (1995), *Pre-production investment and jobs induced by MIT exclusive patent licenses : a preliminary model to measure the economic impact of university licensing*, J. of the AUTM **VII**, 27-47.

ON THE NEED OF NEW MECHANISMS FOR THE PROTECTION OF INTELLECTUAL PROPERTY OF RESEARCH UNIVERSITIES

P. CONCEIÇÃO[1,2], M. V. HEITOR[2] AND PEDRO OLIVEIRA[2]

1) IC[2] Institute, The University of Texas at Austin, USA

2) Instituto Superior Técnico
Av Rovisco Pais, 1096 Lisboa Codex, Portugal

ABSTRACT

This paper discusses the need of reforming the current systems of intellectual property protection, aiming at reflecting the challenges created by the advent of the knowledge-based economy. In a previous paper we argued that the rationale for undertaking intellectual property protection in 'research universities' is the strengthening of the institutional integrity of universities (Conceição et al, 1998). In the present paper, we briefly analyse the economic impact of the American and European systems of intellectual property protection, which were designed to meet the needs of the industrial era. In particular, we observe that the nature of today's inventions is rapidly rendering the current system inadequate and ineffective, in particular in the areas of life sciences and information technologies. Today's technologies and inventions have created new potential forms of intellectual property that cannot be handled using the current system. We focus our analysis in the university sector, and discuss four main challenges research universities are currently facing.

1. INTRODUCTION

In a previous paper we argued that technology transfer, including the protection of intellectual property, should be explicitly acknowledged in the context of the university function (figure 1) as a way to achieve the requirements of preserving the university's institutional integrity (Conceição et al., 1998). By institutional integrity of the university we refer to the idea based on Rosenberg and Nelson (1996), Dasgupta and David (1994), David (1993) and Pavitt (1990), in that universities have developed over the centuries an institutional specialisation by which they perform a unique societal role, by virtue of leaving largely public the results of their research and teaching activities. This specialisation has been accompanied by the emergence of other institutions, such as firms, that have developed their own features, namely the fact that they try to achieve profit by privatising the outcomes of their production processes. Although universities are important in creating technology, they are crucial in creating science, the non-

A. Inzelt and J. Hilton (eds.), Technology Transfer: From Invention to Innovation, 69–85.

excludable portion of software, as discussed in section 3. A threat to the institutional integrity of the university would mean, for example, that there would be fewer incentives to produce non-excludable software, and that there would be a lack of investment of other institutions in producing this type of valuable knowledge.

The above analysis should be considered in terms of the current understanding of innovation, in that the challenge of technology innovation requires the consideration of the entire process from research and development (R&D) in the laboratory to successful commercialisation in the marketplace. Traditionally, successful commercialisation of R&D was the result of an automatic process that began with scientific research and then moved to development, financing, manufacturing, marketing and subsequent internationalisation, without sustaining connections among academic business and government leaders. Today, the relationships between technological innovation and economic wealth generation, markets, and job creation involves more than capital investments. It demands an integrated and interactive approach that blends scientific, technological, socio-economic and cultural aspects in rapidly moving environments.

Kiln and Rosenberg (1986) argued that there are complex links and feed back relations between firms (where the innovation takes place) and the Science and Technology system, and proposed the interactive model of innovation. Myers and Rosenbloom (1996) extended this model to explicitly express organisational capabilities and the special characteristics of innovations. In this model, In this model, organisational capabilities are considered as the foundations of competitive advantage in innovation, and include firm-specific knowledge, communities of practice, and technology platforms. Firm-specific knowledge represents the accumulated learning of the organisation, which is pertinent to the business. This shall be distinguished from the body of generally accessible knowledge. The specific knowledge of a firm is embodied in the firm's personnel and its technology platforms, products, and processes. Communities of practice are ensembles of skilled technical people with expertise on working across the organisation. These communities span organisational divisions and provide both a repository for the firm's expertise and a medium for communication and application of new knowledge. Technology platforms are an output of the design process, which provide a common framework on which families of specific products and services can be created over time. A platform comprises an ensemble of technologies configured in a system or subsystem that creates opportunities for a variety of outputs.

In the context of this model, the question which does arise is how research (namely, at the university level) contributes most effectively to the profitable execution of the chains of innovation. Using the terminology of Myers and Rosenbloom (1996), "effective" research will contribute to the base of general knowledge, but "productive" research requires the corporation to build the organisational capabilities (i.e. firm-specific knowledge, communities of practice, and technology platforms). At the university level, it is clear that it is research, not commercial design and development, which is the field in which academia, are expected to excel. Technology transfer activities are valued in order to secure intellectual property rights, to assess the valuation of technological opportunities and to implement transfer strategies.

Figure 1: A Framework for the Interaction of University R&D and Technology Transfer in the Context of Innovation

(The shadow area includes the priority activities on which a *research university* should concentrate)

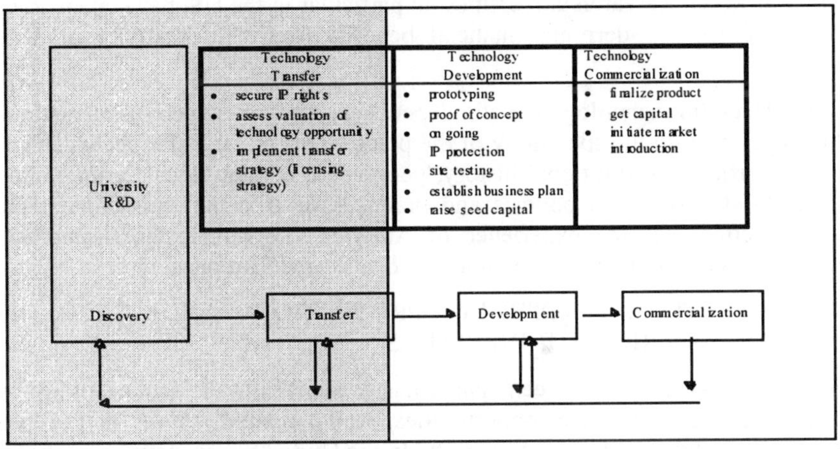

Source: Conceição, P., Heitor, M., Oliveira, P. (1998)

The aim of this paper is to present the motivations and to discuss the challenges, for the development of a new system of intellectual property protection for universities, considering the relation between R&D and economic growth, and the challenges created by the advent of the knowledge based economy.

In this context, the following section presents the reasons that motivate the development of intellectual property policies for universities, and gives experiences of American and European universities. In section 3 we briefly introduce aspects of the economics of knowledge and intellectual capital, namely by classifying knowledge according to the degree of rivalry and exclusion, and discussing some of the elements considered in the decision of protecting intellectual property. Section 4 discusses the main challenges universities are facing in the context of the current socio-economic climate, and discusses the need of reforming the actual systems of intellectual property protection. Finally, a summary of the main findings is presented.

2. THE CONTEXT FOR INTELLECTUAL PROPERTY PROTECTION

The important strategic role that universities can play in helping nations to meet public policy goals has been extensively recognised, including public safety, quality of life, health care, environmental protection and economic competitiveness (e.g. Mowery and Rosenberg, 1989; Readings, 1996; Lucas, 1996; Ehrenberg, 1997). This has been achieved by the creation and distribution of knowledge, improving the competencies and skills of the labour force, and contributing to the development and commercialisation of new technologies. In this context, the protection of the intellectual property is a key institutional mechanism, since it provides incentives for the private production of R&D. Research universities, in particular in the US, have taken advantage of this mechanism by deriving financial benefits from the creativity of academic scientists.

However, successful technology transfer depends on a complex web of interactions, and is highly dependent on the specificity of the place where the transfer occurs (Kim et al, 1997). The particular institutional history, geographic context, legal setting, and other factors, demand customised policies and practices. It is within this context that we present in section 2.1. the experience of American universities in protecting their intellectual property, and then, in section 2.2., discuss the European context.

2.1. INTELLECTUAL PROPERTY PROTECTION AND TECHNOLOGY LICENSING IN AMERICAN UNIVERSITIES

American universities have been particularly successful at contributing to the accomplishment of commercial opportunities, whilst related actions in Europe have been erratic in quality and scarce in quantity. In the United States, new innovations have benefited from a close interaction between universities and the community, as recently discussed extensively by Rosenberg and Nelson (1986). In the context of the complex web of relationships between universities and firms, intellectual property by universities represents a small portion. Nonetheless, it is worthwhile to look in further detail at the impact of intellectual property actions by US universities in order to make two points: first, that the existence of explicit strategies for intellectual property protection in the US has provided the generation of a sizeable aggregate level of income. On the other hand, the impact of the income at the institutional level is negligible on average.

An economic impact model developed by the Association of University Technology Managers (AUTM, 1998; AUTM, 1998) shows that more than $24,8 billion of US economy activity can be attributed to the results of academic licensing. This figure includes both pre-production investments ($4 billion per year) and post-production sales of products by licensees ($17 billion per year). The increasing value of new academic discoveries is illustrated in table 1, which shows that academic institutions, hospitals, and research institutes earned more that $415 million in royalties in 1995.

Table 2 shows the evolution during recent years of the research expenditures in American universities. There is an increasing rate of growth of royalties received in comparison with R&D expenditure, which reflects the growing importance of intellectual property rights. Some American universities have been particularly benefited from R&D income and royalty payments. As a case in point, MIT has received $38 million in licence revenues. At Stanford University since 1969, when the Office of Technology Licensing was founded, royalties have surpassed $111 million, capitalising from inventions such as Recombinant DNA ($53.4 million) and FM Sound ($13.9 million). Other universities have also benefited from significant funding from licensing fees, normally associated with a particular invention. Besides MIT and

Stanford, "big-winners" include University of Wisconsin (with Warfarin and Vitamin D), Michigan State (CIS Platinum) and University of Rochester (Hemphilus Vaccine).

Table 1: Gross royalties received by US Universities, Hospitals and Research Institutes

Royaltie s received	**US Universities ($ million)**	**% chang e**	**US hospitals & research institutes ($ million)**	**% chang e**
1991	$122.9	-	$32.0	-
1992	$159.0	29	$45.4	20
1993	$212.7	34	$62.1	0
1994	$236.7	11	$71.7	9
1995	$270.8	14	$83.0	85

- Source: AUTM Licensing Survey 1991-1995

Table 2: Total research expenditures in US Universities, Hospitals and Research Institutes

Total researc h expend.	**US Universities ($ million)**	**% chang e**	**US hospitals & research institutes ($ million)**	**% chang e**
1991	$10,264.9	-	$776.7	-
1992	$11,033.0	7	$858.9	11
1993	$11,655.6	6	$1,014.0	18
1994	$12,801.4	10	$1,063.0	5
1995	$13,297.4	4	$1,180.9	11

- Source: AUTM Licensing Survey 1991-1995

Despite the impact of patents income at a few American universities and the overall growth illustrated in tables 1 and 2, we should stress that, on average, the share of royalties in the total research expenditures remains small, and below 0,2%. In addition the analysis of table 3, considering the value of the patents listed, shows that the number of universities in which the protection of intellectual property is relevant, is very small.

Table 3: Number of patents in some US universities

University	Patents in 1995
University of California	219
Massachusetts Institute of Technology	107
University of Texas	90
Stanford University	55
University of Wisconsin	47
Cornell University	41
California Institute of Technology	38
Iowa State University	37
University of Florida	33
North Carolina State University	31
State University of New York	31
University of Michigan	29
Virginia Polytechnic Inst. & State University	29
Johns Hopkins University	28
University of Minnesota	27
Duke University	26
University of Pennsylvania	26

- Source: AUTM Licensing Survey 1991-1995

A recent profit/loss analysis of technology transfer programs in U.S. universities, hospitals, and research centres (Trune and Goslin, 1998) concluded that only 40,5% of all institutions generated enough royalties to offset the cost of maintaining the administrative office (technology transfer office costs, patent fees, and legal expenses). This was estimated by taking one-third of the royalties less the cost of maintaining the technology transfer offices. The most profitable offices appeared to be those within technological institutes, universities with medical schools and hospitals/research centres (table 4).

- *Table 4: Profit and Loss Calculations for Technology Transfer Offices*

Loss/Profit	Medical Schools	Technological Institutes	Universities with medical schools	Universities without medical schools	Hospitals and research centres
Loss					
> 200	1	-	12	5	4
100-200	6	2	8	9	6
0-100	3	1	13	24	4
Profit					
0-100	2	1	7	7	4
100-200	-	-	6	1	-
200-300	-	-	2	2	1
300-400	1	1	-	-	-
400-500	-	-	1	1	1
500-1,000	1	2	4	2	2
1-5,000	-	-	9	2	3
>5,000	-	-	2	-	3
Number (%) profitable	4/14 (28.6)	4/7 (57.1)	31/64 (48.4)	15/55 (27.3)	14/28 (50.0)
Mean profit ($000)	1.7	176.8	540.6	40.3	1,1135.1
Range of profit and (loss) ($000)	(297)-808	(179)-650	(619)-11,830	(984)-3,233	(501)-10,583

Source: Trune and Goslin, 1998

Concerning the calculations for overall university program, the study concluded that only 48,8% of these institutions operated at a profit (table 5) This figure was estimated by taking two-thirds of the royalties income less all university expenditures required to operate the technology transfer program.

Table 4: Profit and Loss Calculations for Overall University Programs

Loss/Profit	Medical Schools	Technological Institutes	Universities with medical schools	Universities Without medical schools	Hospitals and Research centres
Loss					
> 200	2	1	14	7	2
100-200	1	2	7	12	7
0-100	5	-	7	18	2
Profit					
0-100	1	1	5	7	1
100-200	1	-	2	1	1
200-300	1	-	4	2	3
300-400	-	-	4	-	1
400-500	1	-	3	-	-
500-1,000	1	2	4	3	-
1-5,000	1	1	11	4	8
>5,000	-	-	4	1	3
Number (%) profitable	6/14 (42.9)	4/7 (57.1)	37/64 (57.8)	18/55 (32.7)	17/28 (60,7)
Mean profit ($000)	198.9	405.9	1,310.9	206.6	2,629.3
Range of profit and (loss) ($000)	(376)-2,019	(310)-2,007	(709)-25,200	(1,263)-6,709	(237)-21,340

Source: Trune and Goslin, 1998

Although the figures of the tables do not represent the specific trends of the leading American research universities mentioned above, the expectation is that this share of royalties in the total research expenditures will remain negligible.

2.2. The European Situation

The above analysis shows that the impact of licensing income is, on average, negligible in the American academic system. This is not a reason not to develop a strategy for intellectual protection in universities. In fact, some specific institutions may, indeed, benefit from generous payoffs from patents. However, we argue that there is a deeper and more fundamental rationale for employing university policies of intellectual property protection. In this section we discuss briefly how the perception that Europe is trailing behind in terms of producing innovations has led to a sense of urgency in terms of pushing universities, and R&D in general, towards more applied type of research.

In recent years, namely since 1992, the relative weakness of European industry has been discussed, particularly in terms of the objectives of the Single European Act and of the provisions of the Maastricht Treaty. Three major indicators have been mentioned, mainly Europe's competitive edge has been blunted; its research potential is being eroded; and, finally, a weak position with regard to future technology. It is clear that the EU has a relatively much lower level of R&D overall than America and Japan. In 1995 the ratios between total R&D expenditure and gross national product were 2.45% in the USA, 2.9% in Japan and only 1.91% in the European Union.

In addition, whilst the demand for research personnel is constantly growing, the supply can hardly keep up, especially in Europe, where the number of technology students and academia is far less than in the other competing parts of the world. Even more important than the absolute number of researchers, are their qualifications, the ability to meet the needs of developing industries and the extent to which the capital they represent is utilised.

Overall, the lower investments in both financial and human terms give cause for concern, especially in a context where intangible assets and intangible investments are the best guarantees for future wealth formation. Nevertheless, besides the weak European figures, analysis has shown that the problem is based on the European weakness in integrating R&D and innovation in an overall strategy, which both exploits and orients the results achieved. This weakness stems from a combination of factors, namely: the still inadequate links between universities and enterprises; the lack of facilities for business start-ups by researchers; the lack of venture capital to help firms through the development phase and the reluctance of private-sector financiers to invest in new activities; the insufficient account of R&D in business strategies and the lack of co-ordinated strategies between businesses, universities and the public authorities; and the targeting on markets which are too small and the weak capacity to foresee future needs and demand on the market (e.g., European Commission, 1996; Archibugi et Pianta, 1996; Wallmark, 1997).

Despite an outstanding scientific performance, Europe is far behind the US and Japan in terms of its technological and commercial performance. The results indicate that one of Europe's major weaknesses lie in its inferiority in terms of transforming the results of scientific research into innovations and competitive advantages. This has recently led to a shift in the European R&D policy towards seeking economic relevance in science and technology. The evolution towards the definition by the European Commission of the

5th Framework Program, as well as the First Action Plan for Innovation in Europe (which was released in early 1997), confirm this perception.

Additionally, empirical evidence shows that the European system for awarding patents to innovators is too expensive and too atomised, since there is no single European patent system (Schmitt, 1998; Ferné, 1998). Therefore, the cost of securing patent protection in every member state is high and discourages companies from exploiting their innovative potential. As a case in point, a typical European patent giving protection in eight countries costs around 20,000 ECU (including fees charged by the European Patent Office and national patent offices, patent attorneys charges, but not including translations), which in the US would cost 1,500 ECU, and in Japan only 1,100 ECU. The evidence demonstrates the need for a truly European system free of institutional and national barriers.

The technological and economic changes are making the current system of intellectual property unworkable and ineffective, since it was designed to meet the needs of the industrial era. Nowadays, new technologies and inventions have created new potential forms of intellectual property that cannot be handled in the same way traditional inventions were, in particular in the fields related with life sciences and information technologies (Thurow, 1997).

It is within this context that we discuss the need for a new system of intellectual property for the European universities, a calling that has been made due to the perception that Europe is lagging far behind in terms of innovation. We argue that intellectual property policies are important for universities and society wide, but that the rationale should not be gathering more financial resources, but rather to preserve the institutional integrity of the university. To make our argument clearer it is important, firstly, to understand the economics of knowledge, namely in terms of the impact of the privatisation of research results.

3. THE ECONOMICS OF KNOWLEDGE AND INTELLECTUAL CAPITAL

This section aims at developing a theoretical framework to analyse the issue of intellectual property. In economic terms, intellectual property awards private rights to knowledge. We discuss the economic features of the knowledge and the economic distinction of the private and public knowledge.

Knowledge has very specific characteristics that make it economically different from objects, (Nelson and Romer, 1996). Using the traditional classification utilised in public finance, economic goods can be classified according to the degree of rivalry and exclusion.

Rivalry is associated with scarcity and expandability of a product, and reflects the idea that, if rivalry exists, a product can only be used by one person at a time. Objects are typically rival goods. However, the knowledge contained in, say, a book, is non-contentious. The fact that I am reading and enjoying a book does not preclude others from reading the exact same book. The same happens with music stored in a CD, or with a software program.

Excludability is associated with the property rights over a product. A product is excludable if the owner has the legal power to prevent others from using it. Knowledge can be made excludable, through intellectual property rights. In the case of a book, the author holds the copyright, and may not wish people to read the book unless they pay a fee for it (buying the book), or that they read at all (taking the book out of print).

Goods with high levels of both excludability and rivalry are designated as private goods. In this case, there are private incentives for production, for the producers can appropriate completely the benefits arising from the use of these goods by others. On the other extreme, goods with low levels of both rivalry and excludability are public goods. For these goods, such national defence and public roads, there are no private incentives for production. Governments normally intervene in the provision of public goods. There are also non-excludable rival goods, such as fisheries. In this case, there is rivalry in consumption, but difficulty in excluding people from using the product. Fisheries are part of a broader class of such type of goods named common pool resources (CPRs).

Romer and Nelson (1996) consider that all objects are classified as hardware, material things that are non-human. Knowledge, on the other hand, is divided into wetware, the knowledge stored in the brain's wet computer, and software, knowledge that is codified and is stored outside the human brain. The relevant distinction between these two types of knowledge is that wetware (more familiarly referred to as human capital) is a rival product, since it is linked to each individual human being. Software, on the other hand, is non-rival, in the sense described above.

Non-rival software has a low marginal cost of reproduction and distribution (making it difficult to exclude people from its use) and is associated with high fixed costs of original production. These properties have substantial impact in terms of generating economies of increasing returns, as Romer (1990) has argued, leading to the new theories of growth. In the context of this paper, we are interested rather in the mechanisms of production non-rival software, for competitive markets will not allocate resources efficiently in order to produce this type of goods. The reason is, as we argued, the absence of private incentives.

David (1993) proposes three types of alternatives to yield the conditions for the production of non-rival software. The first, patronage, consists of giving direct subsidies to producers, on the condition that the goods must be publicly available at virtually zero costs. The second, procurement, is based on the direct production of the goods by the government, awarding specific contracts to private agents whenever necessary. Finally, the third, property, is associated with the privatisation of the non-rival software, awarding the producer monopolistic rights that yield returns large enough to cover the fixed costs of production. Specific legal instruments include patents, copyrights, and trade secrets.

Both patronage and procurement rely on a direct intervention of the government, by which the non-rival software remains non-excluded, and, therefore, effectively public property. Property grants private producers on new knowledge exclusive property rights in the use of their creations. This yields the private incentives in which markets operate efficiently. In terms of the matrix of the Figure 2, the issue then is to opt between making software excludable, or non-excludable, since the non-rivalry characteristic is always present.

In the world of science and technology, there is a tendency to consider science as public property and technology private. Science rests on the publicly available scientific journals and is freely and rapidly disseminated throughout the scientific community and the society at large. Technology is associated with more practical applications exploited by the firms that engaged in its development, and is protected by patents or other instruments of privatising software. Figure 2 illustrates this distinction. Naturally, at the university level there is both science and technology. Indeed, universities often engage in aggressive and effective programs to protect their intellectual property. The motivation is clear: to derive financial benefits from the creativity of academic scientists.

Figure 2: Technology and Science: Two Types of Software

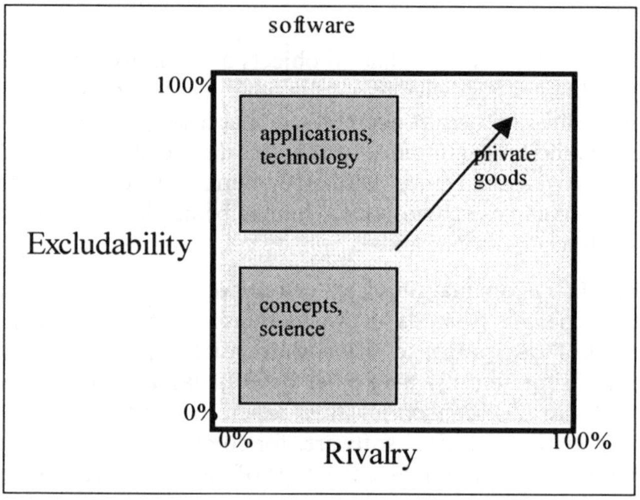

There are several elements that enter into the decision to protect intellectual property, but in this section we should make a note to argue that it is not often that the decision to protect is the most beneficial society wide. In fact, for the common good, it is often more useful to leave the scientific achievements as public goods, especially when they are associated with concepts still in an early phase of development.

4. DISCUSSION: CHALLENGES FOR FUTURE DEVELOPMENTS IN INTELLECTUAL PROPERTY PROTECTION MECHANISMS

The framework established in figure 1, section 1, may theoretically preserve the University's institutional integrity, but brings together a series of questions and challenges for research universities. First, it is clear from the discussion above that the economic impact of protecting the intellectual property is expected to be negligible, at least in average terms. In addition, we may select four main challenges universities are currently facing:

1st Challenge: Balancing innovation and diffusion

Establishing intellectual property rights make software excludable yielding private incentives to production. This strategy is often implemented in commercial computer software programs, books, and music CDs. However, there are two difficulties with this strategy. First, it is sometimes difficult to implement and enforce intellectual property rights, especially at the international level, due to the easiness in copying and reproducing software. Secondly, and most importantly, establishing property rights on software may have perverse effects, since if the benefits are given only to an inventor turned monopolist they will not spread society-wide. In other words, too much emphasis may be being given to innovation at the expense of diffusion, which can slow the overall rate of technological change, or knowledge diffusion and adoption. To illustrate this, Nelson and Romer (1996) ask what would have happened if the concept behind a worksheet, first introduced by Lotus, would have been given exclusive rights. The competition between Lotus, Microsoft, and Borland (with their products Lotus 123, and Excel) might never have happened. Therefore, technology policy in general, and University policy in particular, should not only focus on promoting innovation by restricting access to information, so that innovative firms accrue monopolistic profits temporarily.

2nd Challenge: Beyond the excludable/non-excludable dichotomy of software

We have noticed that establishing intellectual property rights makes software excludable, yielding to private incentives to production. This may be appropriate when the software under analysis is, say, a new formula for Coca-Cola. The new software will benefit only one company. When the software under consideration has a potential society wide impact, like, for example, the cure for cancer would have, then this software production should be induced through patronage or procurement. It is in the public interest that the results be available society wide. This is the dichotomy between making software excludable or non-excludable by the black and white areas in the left-hand rectangle representing non-rival software.. As Soete (1997) has pointed out, some software may not benefit only a firm, nor the entire society. It can benefit an industry, a region, a group of citizens, a number of countries. In this case, the incentives for collective action should be focused on the subjects affected. To subsidise such an effort through general taxation may not be justifiable. Kyriakou (1997) proposes a couple of instances by which focused mechanisms for collective action within the group of subjects that would benefit from the software may be generated. However, the field here is wide open for innovative institutional settings that need to go beyond the pure public/private approach for giving incentives for software production.

3rd Challenge: Integrating intellectual property protection systems

Empirical evidence shows that the impact of intellectual property policies depends on a variety of factors such as history, endowments, market structure, education, openness to trade and investment, and related business regulations, just to mention a few. On the other hand it is known that the variety of systems, for instance in Europe, creates bureaucracy and inefficiency. The European system for awarding patents to innovators is too expensive and too atomised. Since there is no single European patent system, the cost of securing patent protection in every member state is high and discourages companies from exploiting their innovative potential. This situation contrasts with both the American and Japanese situation, which have a single patent system and legal framework allowing protection in the whole territories. As a point in case, a typical European patent giving protection in eight countries costs around 20,000 ECU (including fees charged by the European Patent Office and national patent offices, patent attorneys charges, but not including translations), which in the US would cost 1,500 ECU, and in Japan only 1,100 ECU. The evidence calls for the need of a truly operational system free of institutional and national barriers.

4th Challenge: Facing new knowledge

The technological and economic changes are making the current system of intellectual property unworkable and ineffective, since it was designed to meet the needs of the industrial era. Nowadays, new technologies and inventions have created new potential forms of intellectual property that cannot be handled in the same way traditional inventions were, in particular in the fields related with life sciences and information technologies. Thurow (1997) illustrate this situation by comparing the invention of a new gene and the invention of a gearbox, concluding that such different inventions can not be handled by the same system of intellectual property protection. Additionally, the recent controversies about laws governing biotechnological innovations, such as cloning of human beings and changing human genes, reinforced the need of a new system of intellectual property, clearly stating how far can elements of the human body constitute patentable inventions.

SUMMARY

This paper discusses the need of reforming the current systems of intellectual property protection, adapting it to the challenges created by the advent of the knowledge-based economy. In a previous paper we argued that the rationale for undertaking intellectual property protection in research universities is the strengthening of the institutional integrity of universities. In this present paper, we observe some facts that render the current system inadequate and ineffective, and discuss some challenges research universities are currently facing, namely i) the need to balance innovation and diffusion; ii) the excludable/non-excludable dichotomy of software; iii) the need to integrate intellectual property protection systems; and iv) the challenges created by recent advances in knowledge.

References

1. Archibugi, D., Pianta, M. (1992), *The Technological Specialisation of Advanced Countries*, Kluwer Academic Publishers.

2. Archibugi, D., Pianta, M. (1996), Measuring Technological Change through Patents and Innovation Surveys, *Technovation*, **16** (9) 451-468.

3. Association of University Technology Managers - AUTM (1997), *Licensing Survey, 1991-95: A Five-Year Survey of Technology Licensing (And Related) Performance for U.S. and Canadian Academic and Non-profit institutions, and Patent Management Firms.*

4. Association of University Technology Managers - AUTM (1998), *Licensing Survey, 1996 Survey Summary.*

5. Brooks, H. (1993), 'Research Universities and the Social Contract for Science', in Branscoomb, L. M., Empowering Technology, MIT Press, Cambridge, MA.

6. Conceição, P., Gibson, D., Heitor, M., Shariq, S. (1997), 'Towards a Research Agenda for Knowledge Policies and Management', *Journal of Knowledge Management*, **1** (2), 129-141.

7. Conceição, P., Heitor, M., Oliveira, P. (1998), 'University-Based Technology Licensing in the Knowledge Based Economy', *Technovation.*

8. Conceição, P., Heitor, M., Oliveira, P. (1998), 'Expectations for the University in the Knowledge Economy', *Technological Forecasting and Social Change.*

9. Correa, C. M. (1994), 'Trends in Technology Transfer: Implications for Developing Countries', *Science and Public Policy*, **21** (6), 369-380.

10. David, P. (1993), 'Knowledge, Property, and the System Dynamics of Technological Change', in Summers, L. H., Shah, S. (eds.), *Proceedings of the World Bank Annual Conference on Development Economics 1992.*

11. Daguspta, P., David, P. (1994), 'Towards a New Economics of Science', *Research Policy,* 487-521.

12. Ehrenberg, R. (1997), *The American University - National Treasure or Endangered Species*, Ithaca, Cornell University Press.

13. European Commission (1996), *Green Book on Innovation,* Luxembourg: Official Publication Services of the European Union.

14. Ferné, G. (1998), 'Patents, Innovation and Globalisation', OCDE Observer, **210**, 23-27.

15. Jones-Evans, D., Klofsten, M. (1997), *Technology, Innovation and Enterprise - The European Context,* Macmillan Press.

16. Kim, S.G., Ro, K.K., yu, P.I. (1994), 'Intellectual Property Protection Policy and technology Capability', *Science and public Policy,* **21** (2), 121-130.

17. Kyriakou, D. (1997), "Technology Policy Strategy: Between Research and Development", The IPTS Report, 12, 12-18.

18. Kline, S. J., Rosenberg, N. (1986), 'An Overview of Innovation', in Landau, R., Rosenberg, N. (eds.), *The Positive sum Strategy: Harnessing Technology for Economic Growth*, Washington D.C.: The National Academy Press.

19. Leonard-Barton, D., Doyle, J. (1996), Commercializing Technology: Imaginative Understanding of User Needs, in 'Engines of Innovation', eds R. S. Rosenbloom and W. J. Spencer, Havard Business School Press, 177-207.

20. Lucas, C. (1996), Crisis in the Academy - Rethinking Higher Education in America, New York, St. Martin's Press.

21. Mitra, J., FORMICA, P. (1997), *Innovation and Economic Development*, Dublin, Oak Tree Press.

22. Myers, M., and Rosenbloom, R. (1996), Rethinking the Role of Industrial Research, in 'Engines of Innovation', eds R.S. Rosenbloom and W.J. Spencer, Havard Business School Press, 209-228.

23. Mowery, D. C., Rosenberg, N. (1989), *Technology and the Pursuit of Economic Growth*, Cambridge: Cambridge University Press.

24. Nelson, R. R., (1986), *National Innovation Systems*, Oxford University Press, Oxford, 1993.

25. Nelson, R. R., Romer, P. (1996), 'Science, Economic Growth, and Public Policy', in Smith, B. l. R., Barfield, C.E.; *Technology, R&D, and the Economy,* Brookings, Washington, D.C.

26. OECD (1996), Innovation, Patents and Technological Strategies, Paris.

27. Pavitt, K. (1997), Do Patents Reflect the Useful Research Output of Universities ?, *SPRU electronic working papers series.*

28. Pavitt, K. (1990), What Makes Basic Research Economically Useful', *Research Policy*, 109-119.

29. Readings, B. (1996), *The University in Ruins*, Cambridge, Harvard University Press.

30. Rip, A., Van der Meulen, B. J. R. (1996). 'The Post-Modern Research System', *Science and Public Policy,* **23** (6), 342-352.

31. Roberts, E.B. (1991), *Entrepreneurs in High Technology-Lessons from MIT*, New York, Oxford University Press.

32. Romer, P. (1990). 'Endogenous Technological Change', *Journal of Political Economy,* **98** (5), S71-S102.

33. Rosenberg, N. , Nelson, R. R. (1996). 'The Roles of Universities in the Advance of Industrial Technology', in Rosenbloom, R. S., Spencer, W. J., *Engines of Innovation*, Harvard Business School Press, Cambridge, MA.

34. Salomon, J. J. (1995), 'The 'Uncertain Guest': Mobilizing Science and Technology for Development', *Science and Public Policy,* **22** (1), 9-18.

35. Skoie, H. (1996), 'Basic Research - A New Funding Climate?', *Science and Public Policy,* **23** (2), 66-75.

36. Schmitt, A. (1998), "Patent Law in Europe: Can the Hoped for Benefits be Achieved ?", The IPTS Report, 23, 29-34.

37. Soete, L. (1997), "The Impact of Globalization on European Economic Integration", The IPTS Report, 15, 21-28.

38. Sullivan, N. F. (1995), *Technology Transfer: Making the Most of Your Intellectual Property*, Cambridge University Press.

39. Trott, P., Cordey-Hayes, M., Seaton, R. (1995), 'Inward Technology Transfer as an Interactive Process', *Technovation*, **15** (81), 25-43.

40. Thurow, L. (1997), 'Needed: A New System of intellectual Property Rights', *Harvard Business Review*, 95-103

41. Trune, D., Goslin, L. (1998), 'University Technology Transfer Programs: A Profit/Loss Analysis', *Technological Forecasting & Social Change,* **57**(3), 197-204.

42. Wallmark, J. T. (1997), 'Inventions and Patents at Universities: the Case of Chalmers University of Technology', *Technovation,* **17**, 127-139.

CYCLIC CROSS-BORDER TECHNOLOGY TRANSFER FOR MULTINATIONAL INNOVATIONS

SHERMAN GEE
Office of Naval Research
800 N. Quincy Street, Arlington, VA 22217, USA

Abstract

Technological innovation plays a major role in spurring a nation's economic growth. The technological innovation process has mainly been examined in the context of an industrial enterprise or organisation. Not as widely considered, however, are multinational innovations, where the principals of an innovation belong to public or private organisations in two or more nations. Yet multinational innovations can provide a strong stimulus to cross-border technology transfers and co-operative research and development. Multinational innovations also offer distinct advantages to developing nations where technical capabilities and resources are limited. They offer promising avenues especially for Co-operating Partner (CP) nations in efforts to build market-oriented industrial economies.

This paper examines the multinational innovation process and how it can become the driving force for cross-border technology transfers and co-operative research and development. Multinational innovations will be discussed in the context of a recent co-operative effort involving six NATO nations and the NATO C3 Agency. The evidence clearly shows that multinational collaboration to achieve successful innovations necessarily produces cyclic or recurrent, need-oriented, cross-border technology transfers. The process produces major benefits to the participating nations not only in the form of new technical information but also in the upgrading of technical skills and know how, which is essential for effectively using the technical information for new or improved product development. More importantly, multinational collaboration enables the successful realisation of innovations which otherwise would not be possible or affordable if one nation elected to proceed alone. Hence, the pursuit of innovations in a multinational setting appears especially suited to CP nations. Indeed, the six-nation NATO co-operative effort could serve as a useful model by which CP nations could effectively draw upon complementary technical skills and resources resident elsewhere to develop innovative products that are competitive in world markets. This would be an especially important aspect for CP nations in their movement towards fuller partnership within the NATO community.

A. Inzelt and J. Hilton (eds.), Technology Transfer: From Invention to Innovation, 87–100.

1. Introduction

Technology transfer is an important factor in technological innovations. [1] Technology transfer could well make the difference between a successful innovation or one that languishes. It could also be the ingredient that propels the innovation towards an early successful realisation. Different nations approach technology transfer differently, depending on individual national goals, policies and the associated technical, socioeconomic and industrial infrastructure. [2] Yet technology transfer has a role in both industrialised and developing nations alike.

The importance of technology transfer comes into sharp focus in relation to innovations that involve multiple nations. Multinational technological innovations by their very nature requires close collaboration among the member nations that in turn provides a strong stimulus for recurrent, or cyclic, technology transfers. This paper examines multinational technological innovations in the context of a recently concluded NATO project called Communications Systems Network Interoperability (CSNI).[3] In particular, it focuses on the significant role that cyclic technology transfers had in contributing to the successful conclusion of the multinational innovation.

2. Multinational Technological Innovation

Multinational technological innovation is an innovation involving participants from two or more nations. It provides the framework for international collaborative R&D where member nations can contribute their expertise and at the same time leverage the technical skills found in other nations. Participating organisations are designated by the individual nation and may be from the public or private sectors, such as government laboratories or company contractors. Because this type innovation must be entered into on a formal basis between governments, the goals must be clearly established and supportive of national interests. Multinational innovations are therefore generally problem or need oriented rather than driven by technology opportunity. In addition, because of the requirement to obtain prior approval between governments, such as through a Memorandum of Understanding (MOU), specific budgets are set up, and the innovation time typically has a well-defined conclusion. Detailed planning becomes of paramount importance in order to set specific objectives that contribute to the overall goals within the time-frame available.

The Airbus consortium is a multinational organisation which develops and produces commercial aircraft for the world market. A major factor in the formation of the consortium is the need to pool limited national resources for the large investments necessary to research, design, develop, test, manufacture and market modern commercial aircraft competitive with those from Boeing and McDonnell Douglas.

Where the costs may be prohibitive for any single European nation, a multinational arrangement such as the Airbus consortium allows participating nations to share know how as well as costs in a collaborative venture in pursuit of a common goal. Potential financial rewards, while important, are often overshadowed by the opportunity to share and acquire practical skills and technical know how that would continue to build a nation's science and technology base. Multinational innovations therefore provide avenues by which individual nations could build their science and technology base in an affordable way. This facet of multinational innovation is especially appropriate for CP nations as the process of nation building and integration into the industrial world unfolds.

Multinational innovations, however, have their disadvantages too. Language barriers are typical. Because the principal participants are from different nations with generally different goals, perspectives, and capabilities, subordination of national interests to the co-operative pursuit of a common goal is often difficult if not impossible to achieve. Prolonged discussions and negotiations are more the norm. For major decisions to be made, consensus is not enough. Unanimity must be reached. Work flows necessarily become more bureaucratic. As a result, the time needed to pursue a multinational innovation typically is much longer than one involving a single nation or organisation. Yet, the alternative may be even less attractive--an inability to pursue the innovation at all.

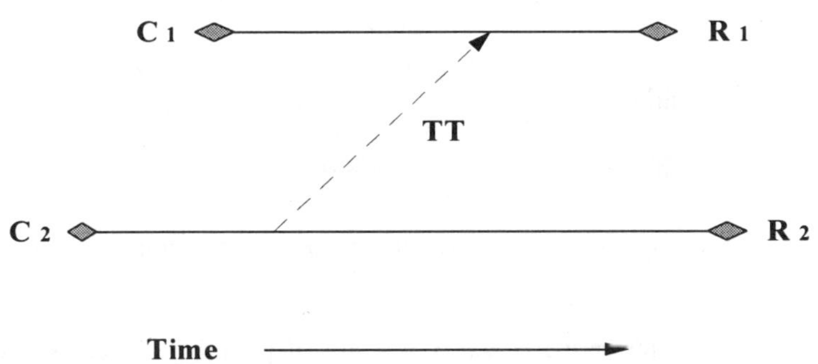

Figure 1. Technology transfer (TT) between two independent, primary innovations, from innovation 2 to innovation 1

It is well known that innovations can benefit from technology transfer originating from other innovations.[1] As shown in Figure 1, technology is transferred from innovation 2 to innovation 1. Both innovations are independent and primary. The innovation process is bounded by the point of conception C and the point of realisation R. That is, C1 and R1 bound innovation 1, and C2 and R2 bound innovation 2.

Perhaps less well known is that major innovations may themselves be dependent on successful realisation of secondary innovations. Secondary innovations are innovations that are initiated and realised as part of, and in support of, the primary innovation. This situation is depicted in Figure 2. Consider the parties involved in the primary innovation are nations a,b,...m, designated by (a....m) in the figure. Participants in a secondary innovation could consist of only a subset of the parties involved in the primary innovation, such as for example (a,c) shown in the figure. During the innovation process, problem solving can result in secondary innovations that are useful not only for the primary innovation but also for meeting external market needs also. In this case the secondary innovation becomes a spin-off innovation, depicted by rs in the figure. Often, secondary innovations produce or benefit from the transfer of technology, as represented by the broken arrows in Figure 2.

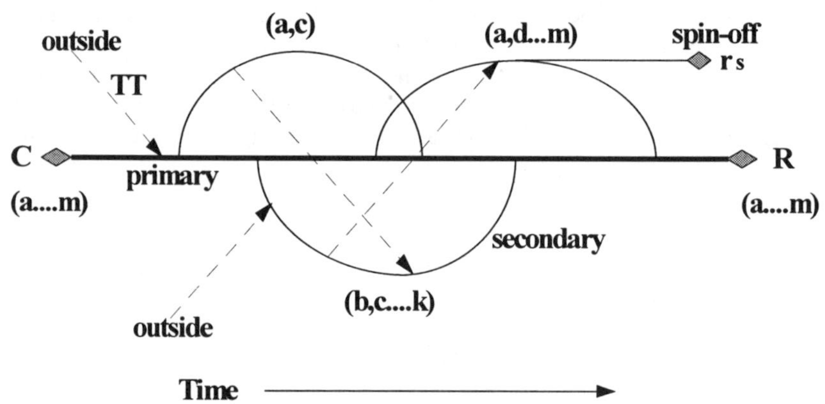

Figure 2. Technology transfer for primary, secondary and spin-off innovations.

These technology transfers are frequently a result of planning rather than serendipity because detailed planning is essential for technological innovations to be successful in a multinational framework. For this reason, technology transfers in multinational innovations arise more from need orientation rather than from technology opportunity. They could well represent cross-border transfers as different nations are involved in different secondary innovations. These transfers also generally result from recurrent interactions among the parties involved, and can be part of a larger process of recurrent or cyclic technology transfers.

3. Cyclic Cross-Border Technology Transfer

Cyclic cross-border technology transfer is the recurrent interaction among technologists from the participating nations and occurs as a result of effective multinational collaborative R&D. This form of technology transfer is especially suited to today's world of the Internet, where e-mail has made rapid communications available literally at our fingertips. The Internet's World Wide Web also makes available immense volumes of data that would otherwise be unavailable, or at best would be much more difficult to access without the Web. It is interesting to note that the Internet is itself an outstanding example of a defence technology that was transferred to, and has gained wide acceptance in, the civilian world.

Cyclic cross-border technology transfer stands out from the conventional notion of technology transfer in several ways. Conventional technology transfer typically involves the movement of know how, products or processes. It may or may not include training needed to use effectively the technology, such as initial training of personnel associated with transfer of turn-key plants to a recipient nation. The relationship between the source and user is similar to a teacher-student relationship. Technology transfer does not necessarily include the insightful knowledge or experience fundamental to an understanding of design rationale chosen, or the source of performance limitations of the technology transferred. On the other hand, cyclic technology transfer can occur early in the innovation process at the R&D stage where acquisition of technical knowledge is accompanied by collaborative hands-on learning experience and valuable peer-to-peer interactions. The peer interactions continue through all phases of the innovation process, and enables phased advances in design, prototypes and products. The interactions help build creditability and effective working relationships, and produces a cyclic increase in knowledge and understanding that contributes to successful development and realisation of the primary innovation. Cyclic technology transfer can also be an important factor for achieving secondary as well as spin-off innovations.

A good example of cyclic technology transfer is found in the interrelationship between the civilian and the defence sectors of the economy. Several distinct cases are mentioned here:

Case 1: Commercial-off-the shelf (COTS) products to be applied to military applications may require enhancement to meet military needs. Military needs may be fed back in terms of new or revised specifications for consideration by relevant commercial standards-setting organisations. These specifications in turn could result in new or revised commercial standards that are also relevant to the military. The newer standards are finally reflected in improved products and services satisfying military needs. (CSNI: commercial network routing protocols)

Case 2: In the same way, defence technology transferred to the civilian sector is most often not directly applicable to the civilian application. More typically, some adaptive engineering, where recurrent interactive activity between the defence developer and the potential civilian user, must take place to make the technology more suited to the commercial market. Considerations of affordability and competitiveness are major considerations in this regard. A good example is the Internet which originated in the defence sector and after many years of research eventually found wide acceptance in the public sector. Internet standards are even today still being assessed for military use.

Case 3: Dual-use technology intended for defence and civilian applications may require several iterations before suitability for both sectors are achieved.

An important factor that promotes cyclic technology transfer is the more flexible attitudes and willingness to learn that typifies early stages of the innovation process such as during the R&D stage. In contrast, potential technology transfers occurring late in the innovation process involve products where designs have been fixed, development costs have been incurred, and where production engineering has possibly started. Naturally, under these conditions there is less receptiveness to change by the developer. In addition, the strong company or national identification with the maturing product invokes national pride that is difficult to overcome when changes or adjustments are desired by the potential transferee or user. Therefore, the conditions for cyclic technology transfers are more favourable earlier in the innovation process. Yet, cyclic technology transfers can still occur at later stages, especially in cases where sufficiently strong good-will and understanding has developed as a result of mutually beneficial co-operation in earlier stages of the innovation process.

The interactive activity in cyclic technology transfer at early stages of innovation produces notable advantages beyond facilitating the learning process. The in-depth discussions and co-operative work arrangements tend to lower inhibitions and raise communications effectiveness. It helps establish a working rapport and degree of camaraderie that leads to a sense of shared responsibility and pride of development. It produces a relationship that enables the insertion of further cyclic improvements in the technology should the need arise. The mutual investment of time and money also imparts to the participants an inescapable recognition of having acquired a vested interest in the innovation. Most importantly, it provides a strong incentive on all parties to work together towards successful realisation of the innovation.

4. The NATO CSNI Experience

A recently concluded NATO project called Communications Systems Network Interoperability (CSNI) provides an excellent example of the importance of cyclic technology transfer in multinational innovations. The project participants were the governments of Canada, France, Germany, Netherlands, UK and the United States, along with the NATO C3 Agency (formerly SHAPE Technical Centre). For a long time NATO communications was possible only if the same radio equipment were used (often referred to as communications stove-pipes). NATO forces were not able to communicate if different nations used different equipment. This lack of interoperability understandably affected the operational effectiveness of NATO forces. The CSNI project therefore addressed the problem of providing NATO with interoperable communications despite the fact that different equipment were employed by NATO forces. Multinational collaboration extended from the R&D stage--through design, prototyping, and testing--to demonstration and adoption in military acquisition and operations. During the five year life of this large-scale $25 million project, many technical accomplishments were achieved, and are detailed in [3].

The primary innovation was the development and demonstration of a new capability for global, multiservice (voice, data, messages) communications across dissimilar national systems without the need to develop new communications systems and networks. Instead, internet work technologies were developed that enabled the interconnection of different, already deployed systems and networks. In essence, the primary innovation was to knit together the diverse national systems and networks into a global one that effectively supported a variety of communications traffic on demand (Figure 3). Routing of traffic is responsive to conditions of the network and the quality of service (QOS) requested by the user. For instance, traffic could be automatically routed along the path that best balances the communications load in the network while simultaneous meeting QOS constraints. Communications resources are shared by

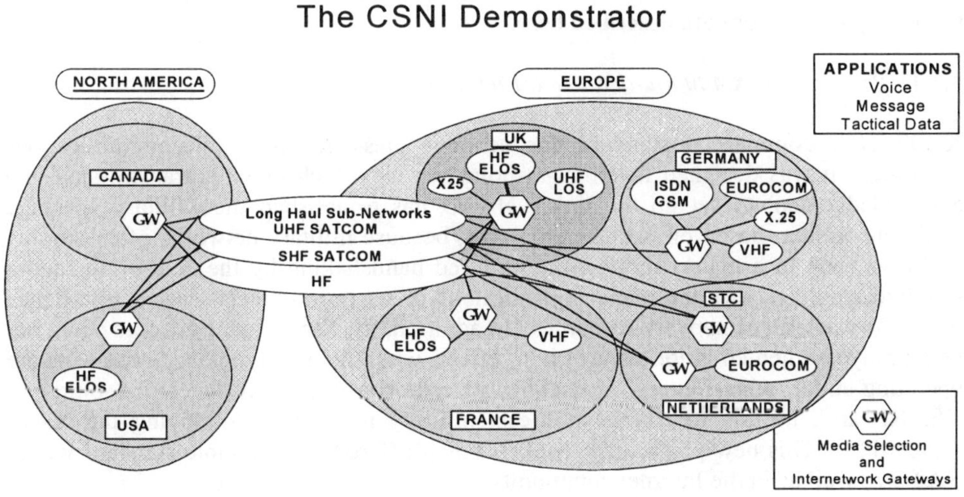

Figure 3. The communications sub networks for the CSNI demonstrator.

multiple users, giving rise to (1) more efficient use of limited capacity, (2) greater survivability through ability to re-route traffic around points of congestion or failure, and (3) lower costs through use of commercial products and standards to the maximum extent possible.

Civilian telecommunications networks differ from military networks primarily from the fact that the former typically have much higher capacity, while the latter typically has much lower capacity and is more bursty in nature. Yet because of the high level of commercial investments in telecommunications, the rapid advance in the state of technology, the desire to avoid redundant military developments, and the need to reduce developmental and life-cycle costs, COTS products and standards were employed wherever possible.[4] As a result commercial Open Systems Interconnect (OSI) standards were tested and demonstrated in a practical NATO military environment for the first time. In addition to OSI, other commercial standards and protocols were also employed as industry began adopting more Internet protocols and standards in its commercial products.

4.1. SECONDARY TECHNICAL INNOVATIONS

Secondary technical innovations were also produced that were deemed essential to realisation of the primary innovation. In many cases the secondary innovations involved extending and developing enhancements to civilian products and standards, because the COTS products and standards frequently were found to be inadequate for the military environment. These secondary innovations were incorporated in the primary innovation and were also made available to relevant commercial sources. Some of the major secondary innovations are described here.

4.1.1. *Protocol for X.400 Multicast and Emission Control*

X.400 is a commercial standard for sending messages over telecommunications networks. It was designed for point-to-point service which requires transmission and acknowledgement to set up the link. NATO has established the Military Message Handling System based on X.400. However, because military networks often employ multicast (one to a many) service, the required handshaking by the X.400 messaging standard would require too many transmissions to set up multicast service, and hence would be wasteful of network resources. In addition, the handshaking process could not be employed when emission control is in effect, when potential message recipients are operating under radio silence. Consequently, a new protocol was developed jointly by GE, NL and the UK to enable multicast military messaging employing the X.400 standard.[5] This new protocol, P_Mul, is being offered as a possible standard within NATO as well as in the Internet community.

4.1.2. *Network Layer Security*

The incorporation of network security is particularly important in military networks in order to provide services such as authentication, access control, confidentiality and integrity. Upon the recommendation of the NATO Ad-hoc Working Group on Security, network security was implemented by placing the (draft) international standard for Network Layer Security Protocol (NLSP), set forth by the International Standards Organisation (ISO), at the top of the network layer. Three separate implementations to provide end-to-end security were achieved by GE, NL and the UK, and all three implementations interoperated. Insights gained during the implementation and testing phase, and recommended improvements in the standard (e.g. adding a multicasting capability) were transmitted back to both the AHWGS and ISO, prior to issuance as the ISO 11577 standard.

4.1.3. *Quality-of-Service (QOS) Routing*
Quality of service routing refers to the ability of the user to specify conditions (e.g., capacity, time delay, error) to be met when network routing decisions are made. This QOS capability was not provided in the COTS routers that employed standard commercial routing protocols. Consequently, a routing metric generator was developed by CA and the US that would support QOS routing under changing network conditions typical of military networks. The dynamics of military networks arise from (1) greater use of over-the-air radio transmissions by the military, (2) mobility of military platforms, and (3) the potential for enemy disruption of network operations. QOS routing therefore enables route selection under variable network conditions and also supports precedence assignments needed under conditions of limited network capacity.

4.1.4. *Congestion Control*
Congestion control is required to adjust traffic flow in order not to exceed available capacity in military networks. Congestion control is less important in civilian networks because of the much higher capacity provided by fibre-optic networks. For this reason available COTS routers did not include the congestion control capability. As a result a technique for slowing the data traffic from the source was developed by the NL that would prevent message queue overflow in the sub networks. Whenever message queues neared overflow, source quench messages are sent from the sub networks to the source to reduce the data flow rate.

4.1.5. *Sub network-Router Interfaces*
The CSNI network consisted of a widely different set of sub network technologies ranging from HF point-point radio to SHF satellite broadcast services. COTS routers needed to interface to all these diverse sub networks. However, the civilian network standards existed only at the network layer and above. No standard interfaces existed below the network layer to interconnect COTS routers with the diverse sub networks. As a result, standard sub network access protocols were developed by several nations and the NC3A to provide the required router-sub network interface. This interface therefore provided the standard connection between COTS routers and the different radio sub networks.

Additional secondary innovations that were developed included the ability to transmit both voice and data packets, introducing a compressed header for the ISO standard connectionless network protocol to conserve bandwidth, and a multicast extension to the ISO standard connectionless transport protocol. Suffice it to say that a number of secondary innovations occurred that supported the primary innovation. In addition, secondary innovations often spin-off into potential new applications or use. Examples are the CSNI system architecture, integrated voice and data technology, the sub network access protocol, and the P_Mul protocol for X.400 multicasting and EMCON operation.

The work load for the broad scope of these activities was distributed among the participating nations. In the brief descriptions of the secondary innovations above, nations cited played leading roles but are not necessarily the only ones involved. No work was attempted unless a minimum of two nations agreed to participate. Though a subset of participating nations may be involved in any specific task, progress and results were shared by all nations.

4.2. METHOD OF WORK

The method of work can generally be placed in three major categories.

4.2.1. *Parallel Development*

Parallel development refers to two or more independent developments addressing the same problem but employing different design approaches or techniques. Though the methodology may be different, information nevertheless was shared concerning design choices and test results. Completion of parallel development may result in selection of one design over others, or implementation of all designs but ensuring that essential interoperation was achieved. An example of the former is the voice application where both real-time, circuit-switched voice was considered compared to an interactive packet voice approach. An example of the latter is the three independent approaches employed to implement NLSP.

4.2.2. *Partitioned Development*

Partitioned development describes a particular development that is divided into smaller work areas with different nations responsible for their own work area. In this case suitable interfaces need to be incorporated in each nations work module in order to ensure the eventual integration into the final product. For this reason partitioned development requires effective technical exchange and co-ordination among the nations during the software development process. In fact, the topic of how to conduct multinational software development effectively has been a research area in itself for many years. Partitioned development was employed, for example, in developing X.400 multicast and the sub network access control.

4.2.3. *Product Modification*

Often a nation may possess special expertise or had performed prior development that is relevant and applicable to a particular desired development. In this case the nation might modify and test the development for the potential new application. After making the modified product available to other nations, a period of training and consultation may follow to help recipient nations become familiar with the product. An example of this is the tactical data generator provided by the US which was employed by the CSNI nations to introduce emulated tactical data in network testing.

Even though different methodologies were used to pursue the primary innovation, a common thread among them was the need to support the different methods with effective technical interchange. Although all nations were not involved in all work areas, primarily due to limited resources, adequate mechanisms were in place to enable all nations to share in the fruits of the technical work. These mechanisms provided the effective, interactive technical exchange essential for an environment where cyclic technology transfer can thrive. The very fact that not all nations were participants in all work areas constituted a major incentive for all nations to ensure that conditions for a technology-transfer friendly environment be set up and maintained during the life of the project.

4.3. CONDITIONS PROMOTING CYCLIC CROSS-BORDER TECHNOLGY TRANSFER

The primary conditions promoting cyclic cross-border technology transfers are described here.

4.3.1. *Project Organisation*

As mentioned earlier, tasks were not attempted without a minimum of two nations involved. A task typically involved only a subset of all the participating nations and was led by an international task co-ordinator with the primary responsibility for ensuring that there is effective co-ordination among the nations working on the particular task. In addition, there were national task leaders who served as the primary focal point for their nation on the task. Progress and problems were reported by the international task co-ordinators to a technical co-ordinating body composed of all participating nations. The international task co-ordinators from all tasks were also in fact the members of the technical co-ordinating body. Major technical decisions and technical exchanges occur within this body involving all nations. This technical co-ordinating body in turn reported to a steering group composed of officially designated representatives from all nations. Guidance and direction was provided by the steering group to the technical co-ordination body and to the international task co-ordinators for implementation. In this way all nations took part and shared in all technical aspects of the work.

4.3.2. *Commitment to a Firm Schedule*

Because the primary innovation involved multinational demonstrations of the interoperability between dissimilar national communications systems, an overriding consideration was to have all parties commit to a firm schedule which would allow all the secondary innovations and developments to converge at the proper time into system-wide on-air tests and proof-of-concept demonstrations. Not insignificant also was the knowledge that the Memorandum of Understanding, which was the official authorisation for the international collaboration, expired after five years. The project therefore benefited from a strong goal orientation which served as a major incentive for all parties to work together in order to progress the work rapidly in accordance with the planned schedule. The flow of technical information across national borders was fundamental to this process.

4.3.3. *Compatible Prototyping Environment*

Compatibility of hardware and software is a prerequisite for effective collaborative R&D if technical progress is to be made jointly among a number of partners. Test results need to be able to be replicated and interpreted in a consistent manner. For this reason compatible operating systems, software tools, and applications programs were agreed upon at an early stage to facilitate the planned work and to avoid unnecessary delays. The actual computing platforms became transparent to the work flow. The transfers of technical data in various formats--large data files, e-mail messages, and graphics--were generally done smoothly. As a result the recurrent cross-border interactions were not impeded by incompatibilities in hardware or software.

4.3.4. *Availability of the Internet*

The work was greatly facilitated by the use of the Internet for e-mail communications and large file transfers. The Internet was also used to conduct some preliminary testing prior to full-scale, on-air tests. Its use produced considerable savings in time and expense, and was a major factor in the successful accomplishment of project objectives within given time constraints. The Internet effectively compressed time and distance and made possible continual, close technical interactions among investigators on both sides of the Atlantic. It thus served as an important enabler for cyclic transfers of technical information and technology across national borders.

4.3.5. *Lack of Alternatives*

It was widely recognised that to progress NATO communications beyond the stove-pipe era, a multinational venture was essential to harness modern technology to provide the long-sought capability for communications interoperability. No one nation could expect to develop this capability on its own. Even if one nation had the will, resources and technical expertise to do so, one nation could not perform testing with other national systems very readily. It would also be much more difficult for one nation to convince others within the Atlantic Alliance to adopt the technology than if a multinational venture was established for the purpose. Given that a multinational venture was required, then the need for developing effective cross-border working relationships was essential to achieve project objectives. This goal orientation and commitment was therefore strong motivation to encourage recurrent cross-border technical exchanges and transfers of technology.

It is also in the self interest of all nations to ensure that cyclic cross-border technology transfers be sustained during the life of the project. Because of the broad scope and complexity of a project like CSNI, one nation can expect to be directly involved in only a portion of the technical development. Cyclic cross-border technology transfers is the only viable means by which one nation benefits from technology developed by other participants during the life of the project. Furthermore, to achieve the project goal, all the technical developments must converge into full-scale, on-air, system-wide tests and demonstrations. Hence, only through recurrent cross-border technology transfers can all the secondary innovations evolve in stages from baseline through upgrades to completed modules and finally reach integration into the complete system. Cyclic

cross-border technology transfers are clearly indispensable if large-scale multinational innovations are to be successfully realised.
To conclude the CSNI story, international demonstrations were conducted in July 1995 and December 1996. In addition, each nation has proceeded to incorporate the technical developments within its own national communications infrastructure. Some of the products and architecture had already contributed to NATO operations in Bosnia and in the Persian Gulf. The CSNI architecture and technologies are also being adopted in nations outside the NATO community. The future will undoubtedly see more spin-off innovations based on the CSNI experience.

5. Relevance to CP Nations

Multinational innovations patterned along the lines of the CSNI experience not only can benefit industrialised nations but could also be particularly suited to developing nations such as in the CP community. Multinational innovations offer the opportunity to pursue technological innovations that otherwise would not be possible or affordable. The particular innovation of interest often requires technical expertise that is beyond that found within a single nation. In addition, the financial resources required may far exceed that available within a nation's budget. By entering into a multinational collaborative arrangement, each nation can contribute their special areas of expertise which, taken together, would make possible the technical progress needed to support the primary innovation. In the process, cyclic cross-border technology transfers will necessarily take place as technical co-operation requires the sharing of technical information, results and products. Each nation would also be responsible for funding its own participation in the multinational venture. Research, development and marketing costs can be shared equitably in accordance with prior agreements. Each nation can therefore benefit from substantial leveraging of its own investment in the venture.

In this way multinational collaborations can effectively speed the process of building up a nation's science and technology base and can create more opportunities to produce technology spin-offs into new directions and markets. Most importantly, multinational ventures can accelerate a nation's drive to industrialise by more rapidly developing the technological means to be more responsive and competitive in changing world markets.

6. Conclusions

Multinational technological innovations can benefit developing as well as industrialised nations as long as they are planned to reflect the capabilities and common interests of the participating nations. The process necessarily provides natural incentives to share technical resources between nations. Hence, multinational innovations by their very nature produce strong stimuli for cyclic cross-border technology transfers. They are particularly suited to CP nations in efforts to build up national technical and manufacturing capabilities in pursuit of modernisation and industrial competitiveness. As a result participating nations have greater opportunities to realise national aspirations earlier than would otherwise be possible.

References

1. Gee, S. (1981) *Technology Transfer, Innovation and International Competitiveness*, John Wiley & Sons, New York, NY.
2. Gee, S. (ed) (1979) *Technology Transfer in Industrialized Countries*, Sijthoff & Noordhoff (Kluwer Academic Publishers), Dordrecht, NL.
3. Gee, S. (1995) Internetworking for Interoperability: The CSNI Accord, IEEE Conference MILCOM 95 proceedings, San Diego, CA.
4. Kettlewell, N. (1997) Commercial Technology Ensures Diverse Multinational Operations, *Signal*, 55- 57.
5. Riechmann, C. (1997) An X.400 Message Transfer Protocol for EMCON Network Situations, to be published, IEEE Conference MILCOM 97 proceedings, Monterey, CA.

CHAPTER II

THE BEST PRACTICES OF KNOWLEDGE TRANSFER (NATIONAL AND ENTERPRISE LEVEL)

EVOLUTION OF A SUCCESSFUL WESTERN EUROPEAN TECHNOLOGY TRANSFER ORGANISATION

E J Duff and Jan Hilton, Vuman Limited
Manchester, England

ABSTRACT

Vuman Limited is the technology development and exploitation company of the University of Manchester The University of Manchester is one of the older and largest universities in the UK. It is also in the forefront of adding value to its research with a vigorous and growing technology transfer programme. This contribution will describe the development of technology transfer in the University of Manchester over the past 50 years, give examples of successful exploitation of University technology, and outline the exciting new developments which will achieve realisation before the Millennium.

1. INTRODUCTION

The Higher Education Institutions (HEI's) of the Western world have been founded on the traditions of the ancient seats of learning, such institutions as, for example, Heidleberg, Germany and the Sorbonne, France, Trinity College Dublin, and, in the United Kingdom, the Universities of Oxford and Cambridge, Glasgow, Edinburgh and Aberdeen. These seats of learning were founded on the traditional and fundamental task of furthering knowledge by teaching and research. They produced "scholars", universally educated on the basis of available knowledge at the time. This was changed by the first Industrial Revolution of the late 18th/early 19th century when man's quest over the previous centuries into the unknown realms of what we today call "science and technology" gave rise to the automation of the traditional artisan industries and the creation of the basis of the modern transport systems`. Amongst the first industries affected was the textile industry with the emergence of machinery, usually operated by complex belt drives driven by water or later by steam. Steam traction, the foretaste of modern means of mass transport of people and goods, first appeared in 1804 and during the next 20 years steam power had a major impact on virtually all manufacturing processes. In Europe, one of the leaders of this was the United Kingdom where the fortuitous juxtaposition of coal, ironstone and limestone enabled metal smelting and engineering to grow up, firstly utilising the canal system which emerged in the 18th

A. Inzelt and J. Hilton (eds.), Technology Transfer: From Invention to Innovation, 103–110.

century for transport and, later, the rapidly emerging railway system which, in the space of a generation, enabled all sections of the populace, regardless of class, to travel the length and breadth of the Kingdom. In this way, by the 1860's the United Kingdom had become the workshop of the world, with manufactured goods, frequently based on imported raw materials, being transported, largely in steam driven ships, to all parts of the world.

To design, build and maintain the industrial plant which was at the heart of this massive enterprise, the Universities played an essential part in what was to become known in the latter half of the 20th Century as the Technology Transfer Industry. In most industrial towns, Mechanics Institutes grew up to train the engineers and technicians required.

The tradition became established of young employees attending "night school" to hone their practical skills with that of theory. Universities established Faculties of Science & Engineering from which trained scientists and engineers emerged to continue the impetus created by the pioneers. By the 1860's, not only were new Universities springing up around Europe, America and elsewhere, but polytechnics, Grandes Ecoles, Technischen Hochschuelen, Institutes of Technology and similar institutions were being established to train technologists to establish and develop industries to compete with many of those in the U.K. Inevitably, close and long lasting links between industry and the HEI's resulted. This is seen today in many professors having consultancies, directorships and equity status in companies associated with the HEI's. Some institutions, such as Massachusetts Institute of Technology, have a well-structured technology liaison/transfer programme, which benefits both the participating academic and the company. In the UK, interaction between academics and industry was slow to materialise, especially within the more traditional Universities. A few, chiefly engineers, had consultancies with major companies.

A crucial element in this was the decision in 1919 to divorce the newly founded Industrial Research Associations from HEI's - in direct contrast to events in the USA and elsewhere, where such research institutions were associated with and, in some cases, established as part of prestigious academic institutions. Thus scientists, engineers and technologists worked frequently side by side with cross fertilisation of ideas. By this means, innovation naturally followed invention. The lack of involvement with industry, and the feeling that science should not be "tainted" by commercialisation, led to inventions made in the UK being developed and moved into manufacture in e.g. the USA. (Penicillin antibiotics are an example). In this process, an apparent arrogance emerged within British HEI's that science was "pure" and anything "applied" was for industry. It was not until the 1939-45 war that the innovative potential of the British HEI's was recognised by Government and Industry alike. It was to be another generation before the barriers to HEI's and industry working together really started to crumble. What follows describes the progress of one major UK University – the Victoria University of Manchester (to be called henceforth "Manchester") - in its progress along the road to realising the potential to be obtained from bringing the results of its research to the market place.

2. THE VICTORIA UNIVERSITY OF MANCHESTER

Manchester was the first of the new Universities to be founded outside London in the mid 19th Century. Established in 1851, it became rapidly one of the major "general" Universities in the United Kingdom. It has a reputation for invention – the modern nuclear industry started in Manchester in 1908 when Rutherford split the atom. For Manchester, the source of its present technology transfer activities can be traced to 1938, when a young electronics engineer, F C Williams (later Sir Frederic Williams FRS) joined the staff of the University and started to research the electronic storage properties of phosphors. Seconded to the Government Radar Research Establishment during 1940-46, F C Williams and his research student, Tom Kilburn, successively invented improved information storage devices and, in conjunction with industry geared to the war effort, developed vastly improved radar systems. These in turn provided the basic tools which enabled them, on their return to Manchester, to develop the first stored memory computer to be operated nearly 50 years ago, on 21 June 1948. The work of Kilburn and Williams had other spin-offs, not recognised until much later; for example the magnetron, an essential component of modern microwave ovens. Yet another application is seen in the radio-telescope designs erected at Jodrell Bank where the world's first steerable radio-telescope was commissioned in time to track the progress of the first orbiting satellite, SPUTNIK I in 1957.

The modern computer industry rests on this pioneering work of Kilburn and Williams. There was however, one essential ingredient which enabled the University to benefit. The work leading to the operation of the computer was funded by the British Government, which in 1948 had established an organisation – the National Research Development Corporation - to develop and exploit public funded research in the UK. For upwards of 20 years, the research carried out in Manchester in the fields of computer engineering provided the basis of much of the world's computer industry. Some 50 inventions were patented widely and licensed to virtually all computer companies under a series of complex licence and cross licence deals. The royalty income shared with the University – and shared by the University with those responsible for the development - was, in today's money, of the order of US $50M. More importantly, the significant royalty income generated from computers and dental cements (to mention but two) resulted in key figures in the University identifying the potential of harnessing research output to industry through technology transfer at a time when public funding for British HEI's was beginning to decline despite the emergence of Government policy to increase student numbers. This pattern persists today.

3. EMERGENCE OF A TECHNOLOGY TRANSFER POLICY

Until 1972, Manchester had no focal point for interaction with industry. It was becoming aware that research contracts and consultancies were increasing, funding from government schemes to catalyse interactions was becoming available and inventions made by staff of the University were being patented, usually by industry. The University determined that it needed to pump-prime these activities by providing dedicated professional advice. The Research Consultancy Service was established to act as this focal point and rapidly became involved in the management and exploitation of intellectual property. In one respect, this development arose through the accidental interaction of one of the authors (EJD) with a Manchester lawyer who at the time was President of the Licensing Executives Society, UK and Ireland. This in turn catalysed not only increased activity in licensing of inventions from the University and the laying of the foundations of an intellectual property strategy, but it led to the joint venture between the University, the City of Manchester and four Manchester based companies to establish Manchester Science Park on land adjacent to the University Campus. Amongst the first tenants of Manchester Science Park when it opened in 1985, was the University's technology development and exploitation company Vuman Limited (VUMAN: from **V**ictoria **U**niversity of **MAN**chester), which had been established in 1981 to develop and bring to the market certain selected technologies arising from the University. Vuman's mission was simple: identify good technologies, invest in those technologies and the people behind them and, by careful management, develop products, processes or services which could be marketed directly or through others. Vuman was not and is not a source of research funds; it is a source of investment funding (arising from its working capital and re-invested profits) and skilled management. It operates through a series of "spin-off" companies, presently numbering nearly 20, the majority of which are wholly owned. Potential of the technologies in which investment is made is realised through profits on trading operations and growth in asset value through careful management which can be realised when all or part of the spin-off is sold on to a third party. Vuman is presently beginning to realise the assets of its more mature subsidiaries.

Invariably, just as there have been successes, there have been failures. It required, on occasions, acts of faith by the University to increase its investment in Vuman at times of turbulent trading, times which coincided with growing financial pressures on British Universities as a direct result of government policy.

The original share capital of the Company was £2 sterling i.e. Vuman subscribed £2 for the issued share capital of the company in 1994. The growth in value of the company by applying intellectual property from the University to products development etc has lead to an external valuation (10/97) of £4M sterling.

4. PATENTS & OTHER INTELLECTUAL PROPERTY

1985 saw Vuman taking responsibility for the protection and exploitation of intellectual property. Patenting is a major activity in which investment is approaching US $0.75M per annum. Over 600 patents are held for the University of Manchester and the number of inventions, currently 50-60 p.a, is increasing rapidly. Provided that the University and Vuman are able to afford this "patient money" the rewards can be considerable. An investment of US $ 300K in one patent portfolio over six years has recently yielded US $ 800K AND a 35% (dilutable) equity stake in a single spin-off *company (dilutable means that the percentage of equity held by the University will diminish as external investors fund the growth of the Company).* Patent portfolios provide also the basis of major research contracts to the University where at least some of the costs of establishing and maintaining the patent portfolio for such contracts are met by the research sponsor. Such patent property also establishes the background intellectual property to any research contract upon which Vuman would expect the research sponsor to pay royalties should any of the contract outputs be exploited. With an extensive patent holding as is held by Manchester, policing and prevention of infringement is important and from time to time it may be necessary to resort to legal action to protect Manchester's interests. That this is done has established a reputation for the prudent management and exploitation of Manchester's Intellectual Property, which has made it possible for Vuman to obtain substantial venture capital investment in new spin-out companies *ab-initio*. Approaches have been received from venture capitalists seeking to invest in recent start-ups – in one case providing a valuation of US $5M for a company established three years previously with capital of US $5 – a 10^6 growth over two years.

In summary Vuman Limited is a company limited by its share capital, currently US $2.5M, all of which is owned by Manchester. Manchester is a relatively rich and financially well managed institution, which invests large sums of money in property and in the stock markets of the UK, Europe, the USA and the Pacific Rim. Its investment in Vuman is part of its large equity holding. Vuman is managerially separate from the University, with its own Board of Directors, and its profits from ongoing activities enable it to be financially independent of the University. However, in the interests of tax efficiency, a complex series of financial transactions maximises the funds available to Vuman for re-investment. Recently, the smaller, growing companies have been transferred to a Vuman "clone", Manchester Technology Developments Limited, managed by Vuman, which will form the vehicle for future spin-out activities.

Annual profits are but one way in which the University benefits from Vuman's existence. Through collaborative research between Vuman group companies, often funded by the British Government or the European Commission, research is "sponsored" through Vuman companies which benefits the University academically as well as financially AND results in exploitable technology. The interactive management of patents enables the University to maximise the value of major research grants from industry, whilst the availability of commercially tuned project management ensures that major research programmes within the University benefit managerially from Vuman's expertise which is complementary to the scientific skills of the University.

The University benefits further when Vuman exits from its subsidiary companies. Over the past few years, two Management Buy Outs (MBO's), a Trade Sale, and two external financings have occurred and an offer for another major subsidiary has been received. Again, the favourable tax treatment given to UK Universities has enabled the University and Vuman to maximise the capital gains received.

5. INCUBATORS

Licensing of technology and spin-out companies usually requires technology with significant value. Frequently, good ideas, properly nurtured and carefully managed, can result in the development and growth of specialist "niche" or larger businesses. The incubation of such companies requires special management skills, good quality accommodation, and patient money. These ingredients have been mixed together within the redundant Electrical Engineering laboratories in Manchester ***(which are no longer required for teaching or research)*** under the auspices of Campus Ventures Limited (CVL). This was a subsidiary of Vuman but has now separated as a non-profit making venture between the University of Manchester and Manchester Airport Ventures plc. CVL catalyses the start-up of new ventures, frequently originating from unemployed scientists and engineers.

An incubator of a vastly different sort is beginning to emerge – Manchester Biosciences Incubator. Modern bioscience, especially that with pharmaceutical or medical applications, requires a prolonged gestation period. Attempts to license such technology have not always been successful and, where licensing has been possible, has led to comparatively low returns in licence fees and royalty income. Biotechnology developments usually require substantial development capital and time to take a biotech invention to the market. To facilitate this, the University of Manchester has recently invested substantially, alongside the European Community and a leading Medical Charity, in Manchester Biosciences Incubator (MBI). In a project valued at US $30M, MBI is designed to take inventions from Manchester and other HEI's and from corporate and other research laboratories. Utilising high quality dedicated laboratories and staff, together with patient investment from its own Venture Capital fund of US $40M, MBI will take forward bioscience developments until they are capable of being encapsulated into a spin-out company in receipt of external investment or are ready to be floated on an appropriate stock exchange. In addition, fully equipped space and access to animal facilities can be provided to pharmaceutical and healthcare and biotech companies on short to medium term leases.

6. MANCHESTER SCIENCE PARK

The policy adopted by the University in the late 1970's early 1980's in outreaching to industry culminated in the establishment in 1982 of what is now regarded as the highly successful Manchester Science Park. Adjacent to the University Campus was 5.5 hectares (15 acres) of vacant land owned by Manchester City Council in a unique linkage between the City Council, the University and the two other University Institutions in Manchester together with four locally based multinational companies, the decision was taken to develop the site to provide high quality office/laboratory accommodation for technology based companies. No restriction was placed on the type of technology employed provided that the company was capable of undertaking research collaboration with one or other of the three Universities. Designed to accommodate a maximum of eight high quality landscaped planting of trees, shrubs and lawns, and with ample car parking space for employees of the tenant companies. The first building was opened in September 1995 and the fifth building was completed at the end of 1997. Available space amounts to over 20,000 square metres and the first four buildings are fully occupied. Vuman Limited was one of the original Science Park tenants and its group companies are mainly located on Manchester Science Park. Tenants enjoy on-line data links to University IT services and the University library (one of the major libraries in the UK). The proximity of the University, yet its separateness, is a major positive element in the attraction of high tech companies to Manchester Science Park.

Difference between spin-off and spin-out

Spin-off - a company which evolves from another but which remains owned by and managed by its parent.

Spin-out - a company which evolves from a parent which does not own a majority equity holding (or any at all) and therefore does not exercise managerial control.

7. FINALE

With its multi-tracked approach: Science Park, spin-out companies, intellectual property management, CVL and MBI, technology transfer at Manchester is becoming a core component in Manchester's strategy as it approaches the next Millennium – a business which by then should have a turnover of some US $15M p.a. on top of a US $100M annual research budget. Only at such levels is technology transfer meaningful and sustainable from its own efforts over the longer term. That Vuman is approaching these targets is possible only through a cultural change within Manchester itself and amongst it staff.

With hindsight, the slow initial growth of technology transfer operations within the University of Manchester and Vuman can be traced to the peripheral nature and lack of understanding – and confidence – in technology transfer activities by the University management. The complex nature of technology transfer operations, with its disparate mix of skills and activities, the entrepreneurial nature of many of which is alien to academic institutions, is rarely recognised by the provision of a "critical mass" of people and skills. Success all too often is followed by a "swamping" of available resources as others seek to join the exploitation bandwagon. Probably, a number of 5-6 professional staff engaged in patenting, licensing, business development, legal support and management is the minimum necessary for success. Currently, four senior managers, five professionals both in intellectual property and legal services and in market research and business development, together with administrative and financial management support, enable Vuman's business with an annual turn over (1996/97) of some US $ 11M (including licence income) to be supported and new business initiatives sustained. New skills need to be acquired continually and succession planning introduced as those who have been involved for many years move on.

In the United States of America, working with industry together with the management of intellectual property in association with a major technology licensing/company formation policy has for many years been regarded as a core activity of institutions such as Massachusetts Institute of Technology, Stanford University and the University of California (to name but a few). Much has been written about the "trail blazing" concepts and policies, and the capabilities for income generation, of these Universities. Universities in the UK have generally adopted a much lower international profile (perhaps as befit the National Character. The University of Manchester is now one of a small, elite group of 5 or 6 UK Universities who have demonstrated that in Europe, as well as the United States, academic entrepreneurship can pay handsome dividends.

BEST PRACTICE IN TECHNOLOGY TRANSFER MANAGEMENT

Strategies for the Promotion of Innovation in Eastern Economies

Thomas Gering
Technology-Licensing-Bureau (TLB) of the Higher Education Institutions in Baden-Württemberg, 76131 Karlsruhe, Germany

1. Introduction

For more than two decades technology transfer has been the "buzz-word" around which the political discussion about the effects of public R & D spending has developed. As early as in the 1960s, the National Aeronautics and Space Administration (NASA) told American Society that there is significant technological spin-off to be gained from advanced space programmes and that the TEFLON cover applied to the frying pan could never have been created without space research.

Since the collapse of the Iron Curtain between East and West, Eastern European governments have almost consequently started to reflect on the potential contribution of their public research base in the framework of economic restructurization and their countries' economic development process.

This paper is going to review technology transfer management between the public and the private sectors, especially with regard to the situation in higher education institutions.

2. Technology push or market pull

The traditional perception of technology transfer between public research institutions (including universities) and Industry was that of an inventive concept being explored in research institutions and made available subsequently for transfer into industrial development, application and market introduction. Although this process known as Technology Transfer has during the decades been (and still is nowadays) responsible for breakthrough inventions that have revolutionized technology sectors or have even opened up formerly non-existing markets[15], recent innovation research findings[16]

[15] Rosenberg, N.; Nelson, R.R.; 1994: "Once one sorts through the interviews (Government-University-Industry-Research-Roundtable, Industrial Perspectives on Innovation and Interactions with Universities, National Academic Press, Washington D.C. 1991) biotechnology stands out almost uniquely as an area where corporate managers look to university research as a source of inventions ... We would add that the technology itself was born in a university setting, which actually is quite unusual."

A. Inzelt and J. Hilton (eds.), Technology Transfer: From Invention to Innovation, 111–120.

suggest that the time frame for such type of innovation[17]-creating technology transfer appears to be rigid and rather long-term. Also the probability of success, that is of successful implementation of the innovative product and/or process in the market place, generally appears to be lower than that of other technology transfer mechanisms which try to focus on market needs when defining areas of R & D activities. This market pull approach is essentially based on Schmookler's findings who, in his basic research[18], came to the conclusion (see Scherer[19]) that invention generally follows investment which is triggered by market needs rather than the inventive opportunities creating such needs.

Market pull principles consequently created a large number of different technology transfer mechanisms between public research organisations, including universities and industry, which can be summarized as follows (see also Becher et al.[20]):

- collaborative and/or sponsored research
- continuing education
- consultancy
- exchange of personnel
- thesis work performed in enterprises.

Within this catalogue, sponsored and collaborative research are probably the most important technology transfer activities between public research institutions and industry. In fact, even national as well as European public research sponsorship has widely been made dependent on the participation of industrial partners. This followed the perception that industrial partners would integrate market pull forces into the R&D process and, indeed, past decades have shown that the most obvious effects of technology transfer have in fact resulted from joint research.

However as a consequence of the Treaty of Rome, European competition law has led public research sponsors to the widely adopted policy of non-exclusive access and exploitation rights for all partners involved in publicly sponsored projects. This evolved out of the basic consideration that research performed with tax payers' money has to be

16 Mansfield, E.; 1991: "The average time lag between the conclusion of the relevant academic research and the first commercial introduction of the innovations based on this research was almost seven years (and tended to be longer for large firms than for smaller ones)."

17 Schumpeter, J.A.; 1934: "As long as they are not carried into practice, inventions are economically irrelevant. And to carry any improvement into effect is a task entirely different from the inventing of it, and a task, moreover, requiring entirely different kind of aptitudes."

18 Schmookler, J.; 1966

19 Scherer, F.M.; 1984: "Specifically, Schmookler demonstrated that changes in innovative activity, as measured through a count of capital goods invention patents, are highly correlated with changes in investments. He has shown that peaks and valleys in inventive activity typically lag changes in investment, rather than preceding then in time. Schmookler argues that invention rises in response to a demand increase ...".

20 Becher, Gering, Lang, Schmoch; 1996

restricted to a pre-competitive environment and that results created through such research should not be exploited for the benefit of a single competitor because this would lead to market distortion and hence to a slow down in innovative activity.

It has been argued by numerous authors (see e.g. Thomas [21], Hersey [22], Gering[23]) that there is evidence that the existing policies do in fact hamper innovative activity rather than promote it. This is because of the fact that pre-competitive results need significant further development and hence investment before a marketable product or process is obtained. Experts in innovation management are of the opinion that the research investment devoted to a particular idea -on average- is only 1/20 of what the overall development and marketing effort needs until the corresponding product is finally introduced into the market place[24].

3. Principal objectives of higher education institutions

Society finances higher education institutions primarily because of two major intentions:

- First they should provide scientific education to the young people who will be responsible for the well being of the economy and hence society in the future.
- In order for the institutions to be capable of providing such education it is, according to Humboldt's higher education principles, mandatory to have university teachers perform science and research at the leading edge of technological and scientific progress.

How does technology transfer as an additional task fit into this set of overall objectives?

First of all, institutions and society should bear in mind that a large number of qualified and capable graduates will indeed form the nucleus for the future development of the human capital needed for our companies to grow and compete nationally and internationally. Even more so, as was discussed above, leading edge research will nowadays mean direct interaction with industries rather than the ivory tower type of science which formed the backbone of research activities in higher education decades ago.

21 Thomas, D.; 1989

22 Hersey, K.; 1996

23 Gering, Th.; 1995

24 Bremer, H.; 1989: "Considerable investment – usually substantially more than was made in the original research must be put into product and market development before the invention becomes commercially available. No company is likely to make this investment unless it can be guaranteed some exclusivity in the market if the product is successful."

We could therefore consider technology transfer activities to be perfectly in-line with the basic objectives of higher education. Some governments have already taken this reasoning as a motivation to go as far as to formally integrate technology transfer into the legal basis of the university sector.

However, technology transfer does also create a need for policies on conflicts of interest[25] and an overall set of guidelines, policies and procedures adopted within the entire institution if involvement in all the different kinds of transfer mechanisms is to be developed in a manner which will be compatible with the intentions of all decision making levels within the institution.

4. Value of technology transfer activities

However not only will there be a need for guidelines and policies within the institution but the technology transfer programme will have to adjust itself to the environment consisting of government, industry and science base. technology transfer mechanisms will always be dependent on the overall political scenario between government and industry. In other words, if industry succeeds to convince government that public sponsorship of certain technological and industrial activities is desperately needed, this will almost logically mean that industry will not accept it when higher education Institutions attempt to charge market prices for the research and services which they provide in these technology sectors. Especially in countries having high tax levels, like e.g. Germany and the Scandinavian countries, there is a serious unwillingness in industry to accept that there is a bottom-line of technology transfer project budgeting. This bottom-line is that the full costs of a particular project have to be borne by someone and if industry does not accept to pay full costs, subsidizing by public sources has to intervene or the project will necessarily become unfeasible.

In recognizing this bottom-line, the British National Committee of Inquiry into Higher Education recently recommended in its report[26] that research sponsors, whether public or private, "should pay the full indirect costs – as much as 90 to 120 percent above the direct costs" [27].

In short, government has a choice between two extremes. It can either provide its higher education institutions with generous budgets and set the framework for industries to tap openly into the know-how available in these institutions[28], or it can create a market

25 Severson, J.A.; 1993

26 http://www.leeds.ac.uk/educol/ucite/

27 News and Intelligence; 1997

28 Samuelson, P. (1987); "Reliance on the incentivizing power of a patent or patent-like monopoly to persuade the private sector to innovate is not the only possible solution to the perceived problem of keeping research and development expenditures - and hence the pace of innovation - at a high level. One available alternative would be for Congress to subsidize research efforts in academic communities, the fruits of which could then be made available to all who had need of them on an unrestricted basis".

around this know-how in which leading edge technology developed in the public sector can only be accessed by industry at a fair price.

These scene-setting decisions will be key in the framework of the determination of the inherent value of technology transfer. Is access free and open and is technology transfer heavily subsidized then there is a constant threat of devaluation of the technology transfer services provided. And what is more, since there is no or only a negligible price to be paid for these services, the receiving company will not be forced to make the best out of it internally, that is to say to create a benefit for the company in joining forces with the transfer service providers and in carrying the ideas through the development chain into market introduction.

5. The client base and a targeted approach

As discussed earlier market pull strategies can regularly ascertain high probabilities of success in technology transfer.

Technology transfer programmes should therefore primarily evaluate their potential client base in order to streamline their list of intended activities to existing needs.

In this framework technology transfer programmes in Eastern Europe are confronted with a particularly tough task. University scientists often, and understandably so, now concentrate on performing leading edge science and research, especially because their technical opportunities are now much better than they have been before the end of the Cold War. On the other hand, much of the re-structurization effort in these new economies desperately needs down to earth and sometimes relatively low tech development support which universities traditionally hesitate to provide. Higher education institutions therefore find themselves in a situation that demands both the creation of favorable conditions for scientific progress which will also mean increased international collaboration and on the other hand the contribution to the "social contract"[29]; that is to say, their efforts will also be measured by the local economy and the tax-payer in terms of their effect on local and national welfare.

Higher education institutions are therefore forced to mediate between their efforts in order to achieve these presumably contradicting objectives, at least in part.

Regarding the selection of appropriate technology transfer mechanisms it will help to bear in mind that different mechanisms can imply vastly different timeframes and that the local client base in a young Eastern economy presumably will be more interested in short-term benefits rather than in strategic long-term joint research, which could result in new products that reach the marketplace ten years from now.

[29] Leahey, H.S.; 1994: "Due to the investment by tax-payers, recipients of federal research funding have a fiduciary responsibility to ensure that the public benefits from such support .... In return for public support, there is a self-imposed obligation to make the benefits of federally funded research available to the public through traditional commercial channels."

6. Selecting mechanisms appropriate to the local scenario

Eastern economics generally are in a state of transition between large government owned industrial complexes and a newly developing culture of small and medium sized privately owned enterprises.

Where the old industrial complexes have already been transformed into private companies, the majority was split up into a larger number of "profit centers" which are now small and medium sized firms trying to become profitable independent ventures.

This situation determines that a very high percentage of the domestic technology transfer market will need to serve the SME-sector.

SMEs are, on the one hand side, generally considered to be more innovative than bigger companies[30], on the other hand they often lack

- their own substantial development and research capacity and infrastructure
- a substantial medium to long-term business strategy
- an intellectual property management strategy including the necessary resources to enforce intellectual property.

Hence, they often tend to concentrate on short-term effects. Technology Transfer between research institutions, including Universities, and Industry can generate such short-term effects if mechanisms are selected appropriately.

The transfer of know-how in the form of continuing education, training activities, Consultancy and student placements (including thesis work in companies) can create such short-term effects because these mechanisms create low hurdles before real interaction between representatives from both sides commences.

Also, this kind of activity can work as an appetizer for more concrete collaborative activities, such as contract research.

Restructuring the economy is regularly accompanied by high unemployment levels. Public recognition of universities in the new Eastern economies could therefore benefit greatly from job creation efforts and a company start-up programme in particular. There is a wide variety of experiences available in Western Europe as well as Northern America regarding the creation of start-ups, the role of incubators, science and technology parks[31]. It appears that higher education institutions in Eastern economies could be well advised to consider these experiences.

[30] Acs. Z., . Audretsch D.; 1990

[31] Tornatzky, L., Waugaman, P., Casson, L., (1995)

7. Marketing

Whichever array of mechanisms is going to be selected, the university will have to undertake a serious marketing effort in order to position itself as a service provider in the technology transfer market.

This is not a scientific activity, we are talking primarily about down to earth marketing experience which needs to be applied. Moreover, we are not in the field of product marketing but in the very special arena of technology marketing[32]. This arena is a special one because, compared with the field of product marketing, the supplier and the potential customer start their interaction without having a joint appreciation of the properties of a new product or process idea which the University is trying to sell.

Moreover, university technology transfer programmes not only have to perform an external marketing effort, they are also constantly marketing the idea of technology transfer as well as their services relating to different technology transfer mechanisms to faculty members, for whom this will regularly be a third tier activity besides their responsibilities in teaching and research.

The technology marketing effort will have to be geared towards the relevant decision makers in the target companies and transfer professionals will be well advised not only to rely on printed information materials, online activities (INTERNET etc.) but to act proactively in order to get as much information as possible to the potential client by personal interaction[33].

8. The legal and regulatory framework

As discussed above, regarding the field of European competition law, the legal framework can play a decisive role in the promotion or the restriction of technology transfer initiatives. There is a general catalogue of criteria which decision makers

[32] MacWright, R., Ritter, J., Willey, T., (1995), "These materials, and the workshop for which they were prepared, take a radical view: that a rational structure can be fashioned that will (1) give both the novice and the experienced artisan a framework for marketing that will work in most if not all situations, and (2) serve as a platform for exploration and assessments of technology and circumstance-specific refinements in approach."

[33] MacWright et.al., "The approach that we have taken to technology marketing is to make extensive use of the telephone, each call being followed up in writing. This approach is well grounded on principles of marketing theory, and in our collective experience, has proven to be the best. Although direct mail communication with potential licensees is perhaps the least costly approach, response rates are extremely low, and there is no way to clarify questions the recipients of such mail might have. The same can be said for computer databases and bulletin boards."

should take into account when developing policies, procedures and regulations in the following sectors:

- faculty members as entrepreneurs
- consultancy activities of faculty members
- conflicts of interest
- ownership of inventions, copyrights and trademarks generated during the course of scientific work at the universities
- preconditions and general research sponsorship policies regarding the granting of user rights to research results
- costing of university services provided to external parties.

This list is non-exhaustive. Technology Transfer will prosper if all parties involved are given their fair share of profits and as long as motivation can be kept alive in the entire process chain.

9. Conclusions

The opportunities for technology transfer between higher education institutions and industry are manifold. Universities in Eastern economies should develop a carefully selected array of transfer mechanisms which will

- be built on the individual institutions' strengths
- take market demand into consideration
- create a mix of short-, medium- and long-term strategies
- use short-term activities for interaction with a large number of local and regional players
- develop medium- and long-term activities with a view to create assets for the institution which can be used locally but also on a national and possibly international basis
- be built on the general consent within the institution that the different mechanisms will involve different probabilities of success and to decide whether or not transfer services should mainly be provided on the basis of full cost budgeting or if these should be subsidized.

Only if the latter decision will be backed by the majority of decision makers within the institution and the government can a self-sustaining activity become a realistic opportunity[34].

[34] Hull, C., Sellar, F. 1990 "Strong leadership at the highest level in an institution is crucial for winning support, intra- and extra-mural, for its technology transfer activities. A clear statement of objectives and a subsequent business plan, conscientously implemented, are prime requisites for a self-sustaining activity."

References

1. Rosenberg, N.; Nelson, R.R.; American Universities and Technical Advance in Industry, Research Policy 23 (1994), Elsevier Science B.V. 323-348.
2. Mansfield, E., Academic Research and Industrial Innovation, Research Policy (1991) 1-12.
3. Schumpeter, J.A.; The Theory of Economic Development, Cambridge, Mass. 1934.
4. Schmookler, J., Invention and Economic Growth, Harvard University Press, Cambridge, Mass. 1966.
5. Scherer, F.M., Innovation and Growth, Harvard University Press, Cambridge, Mass. 1984.
6. Becher, Gering, Lang, Schmoch; Patentwesen an Hochschulen, Studie im Auftrag des Bundesministeriums für Bildung, Wissenschaft, Forschung und Technologie, Bonn, Dezember 1996.
7. Thomas, D., Intellectual Property Rights and Wrongs, An University View, Industry and Higher Education, In Print Publishing Ltd., Brighton, March 1989.
8. Hersey, K.,Strategies for Managing Intellectual Property in University-Industry Research Centers and Consortia, AUTM 1996 Annual Meeting in Charleston, South Carolina, Febuary 1996.
9. Gering, Th., Förderung des Technologietransfers durch universitäre Patent- und Lizenzpolitik – vergleichende internationale Bestandsaufnahme und Folgerungen für die Europäische Union, TII Luxemburg, April 1995.
10. Bremer, H., University Technology Transfer: Where have we been? Where are we going? Journal of the Association of University Technology Managers, Vol. 1, No. 1, Spring 1989.
11. Severson, J.A., A summary of conflicts of interest policies, AUTM Technology Transfer Practice Manual, Vol. 1, Part III, Chapter 4.1., Norwalk, CT, 1993.
12. http://www.leeds.ac.uk/educol/ucite/
13. News and Intelligence; Technology Access Report, University R & D Opportunities Inc. Novato, California, August 1997
14. Samuelson, P.; Innovation and Competition: Conflicts over Intellectual Property Rights in New Technologies; Science, Technology and Human Values, Volume 12, Issue 1, MIT Press, Winter 1987
15. Leahey, H.S.; Philosophy of Licensing for Non-Profit Research Institutions; AUTM Technology Transfer Practice Manual Part II, Chapter 1, Febuary 1994.
16. Acs, Z., Audretsch D., Innovation and small firms, MIT Press Cambridge Mass. 1990.
17. Tornatzky, L., Waugaman, P., Casson, L., Benchmarking best practices for university-industry technology transfer: Working with start-up companies, Report of the Southern Technology Council, Research Triangle Park, North Carolina, 1995
18. MacWright, R., Ritter, J., Willey, T., A systematic approach to technology marketing, Educational Program Workshop, AUTM Annual Meeting, Nashville, TN, February 1995
19. MacWright et.al., op.cit.

20. Hull, C., Sellar, F. 1990, Technology Transfer between Higher Education Institutions and Industry in Europe, European Association for the Transfer of Technology Innovation and Industrial Information, Luxemburg

RE-DEVELOPMENT OF A FORMER EAST-BERLIN MILITARY SITE INTO A SITE OF SCIENCE AND TECHNOLOGY

Helge Neumann,
WISTA-MANAGEMENT GMBH, Berlin-Adlershof

AIMS:

to define the decline of Adlershof during the change
Need to create a vision and model for a city of science and technology and for finding key-partners in order to actualize the project
how to get support from regional and national associations and from the European Community for the construction of the carrier- organisation
to check the core competences and define steps to actualize the project
to establish the necessary alliances
support to establish the innovative SME and to create synergy within the scientific insitutes

1. The situation in Berlin-Adlershof before the fall of the Berlin wall

Adlershof, a site with long tradition
The site Adlershof-Johannisthal, situated in the Berlin district of Treptow, which is 25 minutes away from the city centre and 15 minutes from the future major international airport at Berlin Schönefeld, has had a long tradition as a site of research and technical development. Created in 1753, by order of Friedrich II in the south-east area of what is today Berlin, this location became the birthplace of the German aircrafts industry at the beginning of the 19th century. It was here that the first aircraft company was formed, and in the following years other aircraft industries, such as Albatros, FOCKER, Harlan and Rumpler-Werke have marked the site (1).

In 1912 the German Experimentations Center for air-transports was founded on the Rudower Chaussee Straße and was the very first research institution. Before and during the Second World War, this location was used by the Wehrmacht (National Socialist Army) as a development location for their aircrafts. On this site the first full-metallic aircraft in the world was developed, the famous three engined Junkers Ju 52. In the following years other industries were developed in Adlershof-Johannisthal and the surrounding areas (2). These included the chemical and pharmaceutical industries (Schering), the metal industry (cables, engines and material for trains), as well as the construction's material industry (Eternit), which are still very important in this industrial environment.

A. Inzelt and J. Hilton (eds.), Technology Transfer: From Invention to Innovation, 121–134.

After 1945 and until the fall of the Berlin wall, this site (see figure 1) had served three main purposes :

- The **Television Center** of the former GDR was built in the south area of the Rudower Chaussee, between the railways and Agastraße;
- The biggest part of this aeronautical site was given to the NVA (GDR Army) after 1949, and later to the **STASI** (Security Police), and was used as a site for the Barracks of STASI troups, as a store for the technical military equipment (transports etc.), and also as a training location for military exercises (former air-ground)
- Other small parts of the site, including the former buildings of the Experimentations Centre for air transport were however retained by the **Academy of Science** of the GDR. Within the following years it became the greatest research centre for natural and technical sciences of the GDR.

By 1990, 16 Institutes and 4 Service centers were established on an area of 78ha, north and south of the Rudower Chaussee.

About 5500 scientitists, technical staff and other workers had been working on this campus simultaneously until 1989. The scientific field was able to take advantage of the concentration of research in the domains of chemistry, physics, aerospace science, informatics and construction of scientific equipment. The Center for construction of Scientific equipment was the largest institution, with more than 1000 employees.
The task of these scientific institutions was to carry out fundamental research in the afore-mentioned domains, being also responsible for technical developments and applications.

From 1969/70 onwards, the researchers' tasks were reinforced and focused on the future key-technologies, in order to help the GDR economy to become competitive, and to contribute in finding solutions to the economic problems of the country. Following this, the priorities in the field of natural sciences and technics were shifted from basic research to industrial research and development. This sector of application grew to be more than 50 % of total activities (3).

By this time, the centers of the academy had also started on the creation of scientific contacts with other research and development centers of industry, as well as with universities. Many of those centers had only ever participated in international research programmes of the former comecon-countries, and relations with the non-socialist countries were very rare, if existant at all.

2. Process of Berlin unification with priority given to vision and planning of an integrated site of science and technology

In 1990 the STASI-regiment was disbanded, and the site became property of the State. Immediately, different business, services and commercial enterprises settled on this part of the site in the former STASI-buildings. Moreover, the Television Center of the ex-

GDR was liquidated as well as The Academy of Science. The Academic institutions of the ex-GDR had to adapt to the new situation in the process of German unification and had to be restructured to fit into the existing scientific landscape of the FRG.

In July 1990, in accordance with the unification contract between GDR and FRG, the Scientific Council of the Federal Republic of Germany was charged to evaluate the research institutions of the GDR's Academy of Science: Every institution and every section were evaluated separately, and as a result of this, many recommendations were made , for the future of these institutions, and for the new priorities of the research in the east, including recommendations to close, reconstruct or to private institutions, sections or working groups of ex Academy (4, 5, 6).

The "Co-ordination and Liquidation Office" for the Institutes of the Academy of Science of the former GDR" (KAI-AdW) was founded in October 1990. Although the Academy was dissolved, the employment contracts were not broken until 31.12.91. Based on the recommendations of the Scientific Council, the Federal Administration, the Senate of Berlin, the Berlin Chamber of Commerce and Industry, the Berlin Economic Development Society, KAI-AdW and scientific "task-force groups" have worked on the conversion of the site. It was decided to create a separate branch-office of KAI-AdW in Adlershof and to organise it in conjunction with the Berlin Economic Development Society.

At the same time liquidation of the former institutes of the Academy was commenced, personnel were discharged and new scientific institutes were created in accordance with the recommendations of the Scientific Council. .This caused a dramatic loss of jobs during this period and also influenced the emigration of many of the young and creative researchers away from the site.

Despite large turbulence in this period, the reconstruction in Adlershof did not lead to effects of such a large scale as in other academic sites, for instance Leipzig or Dresden. So fortunately about 3.500 of the former 5.500 scientists and technicians could remain on the Berlin-Adlershof campus and 1,300 have received new contracts with new or reconstructed institutes. Roughly 300 scientists became involved in the „Scientists Integration Programme“, which aimed at transferring those groups into existing university structures. According to our estimations about 700 people, who worked mainly on industrial application projects, started to create their own enterprises, mostly characterised as innovative technology companies. A large proportion prepared spin-offs from former Academy institutions or tried to join already established, innovative companies.

One thousand two hundred people were gathered in various job creation schemes. In the special case of the former Academy , the personnel concerned were able to continue working on scientific research in their labs, whilst being available to start a new job in the newly created facilities of industry or research. They were paid by the Ministry of Labour about 65 percent of their former salary during this period. Additionally they were able to earn the difference in their former income by a Special Arrangement with the Ministry of Research and Technology. This Special Arrangement took place from March 1991 up to 31st of December 1991, both for the staff of the former Academy and for reseachers and technicians of former collectives combines (7).

Nine employment and qualification companies (BQG) were created to carry out the Job-creating measures, which were supported by Berlin's administrative authorities and assisted by well known service companies. They also began to organise re-qualification courses in science-specific domaines like I&C-technologies, environmental protection and in the field of R&D project management.BQG`s contribution was to retain the scientific skilled on the campus, in order to provide inverstors in new technologies with highly-qualified staff, as often these scientific skills are unable to be quickly learnt within a short time span. Even if it failed to transfer all the personnel into new enterprises, it can be said that these job creation measures stimulated the East-German R&D-landscape.

Regardless of the fact that large investment from external private companies did not occur immediately, many of BQG's personnel succeeded in finding jobs in newly created innovative SME's.

Such attempts at privatisation were at that time essentially supported, both by federal and local innovation, investment and technology programmes (for more details please see related Programmes of Ministry of Economics, Investment Programmes of the Berlin Senat, Special Programme for Technology Oriented Enterprises, issued by Ministry of Research and Technology).

Essential to the restructuring of Adlershof was the support given to spin-offs by KAI-AdW: Scientists or technicians who aim to establish their own company in the field of new technologies are usually characterised as having many good ideas, good technical experience, comprehensive creativity but lack of money and minimal commercial knowledge. KAI-AdW tried to prevent the same situation occuring in Adlershof by setting up support for these persons, e.g. special contracts with favourable conditions for lending laboratory space and technical equipment. In particular KAI-AdW offered a graduit letting of instruments and took over personnel costs for up to 6 month (8).

Numerous initiatives by the scientists in conjunction with the **Pprogramme „Future for Berlin-Adlershof - Development of the site into an integrated landscape of Technology“** were submitted by KAI-AdW, the Berlin Chamber of Commerce and Industry and the Berlin Economic Development Society in March 1991 and proposed a comparatively smart restructuring of Adlershof - in comparison to the other East-German sites (9).This programme was used in the following period by the Senate and Federal Authorities as a basis for the development of the site and contained the following proposals:

1. To continue with measures such as the job-creation schemes and the setting up of a branch office of KAI-AdW in Adlershof.
2. Formation of a project development society.

3. Working out a framework concept
4. Carrying out a framework programme for the area of Adlershof, actively supported by a well known consultancy agency and aimed at developing an integrated landscape of economy and science in Adlershof. It consisted of the following elements:
5. site for governmental and non-university research institutes
6. site for university research and teaching
7. site for large scale scientific facilities
8. site for innovation and foundation centres
9. space for settlement of enterprises and spin-offs in the field of innovative technologies.

In 1991/92 a significant recession was observed in the scientific sector on the former territory of GDR. In the industry and industrial sector research was practically at a complete halt. In Berlin-Treptow, for example, about 40.000 industrial working places were lost - among them nearly all of the industrial research of the former Collectives. (Kombinates)

Despite this tendency, the afore-mentioned proposals set the basis for avoiding such a break-down in Adlershof and moreover to create an example of how to tackle the urgently needed structural change in Berlin's industry.

With the "Future of the Adlershof" Programme and the recommendations of the participating institutions, a reliable consensus has been reached. Due to this consensus, which concerns all the significant regional partners, the future development of the Science and Technology Center Berlin- Adlershof has become a priority for the State of Berlin. One could say, that based on this document the foundations were set to create a top modern Science and Technology centre with a European dimension .

3. Foundation and Development of WISTA-MG and BAAG

After the document "Future of Adlershof" Programme was adopted by the public authorities (Senate of Berlin, Chamber of Commerce etc.) a development and management company was first founded by order of the Berlin senate, to co-ordinate the realisation of the process. This was the birth of todays WISTA-MANAGEMENT GMBH, which was created as the Development Society of Berlin-Adlershof at the end of September 1991.

It soon became clear that this project could not simply concentrate on restructuring the former campus of the academic research of ex-GDR but also should include the area of the former landing ground in Johannisthal, where the East-Berlin's STASI-regiment was based. Due to the fact that the STASI-regiment had used the old landing ground as drill ground, most of the area remained a construction-free greenspace. This led to the concept of creating an urban space of science and technology, consisting of

* university research and teaching (Transfer of Natural Science Departments of Berlin's Humboldt University was decided)

* non-university research facilities
* large scale research facilities
* private innovative technology companies
* setting up of an innovation and foundation centre
* space for the settlement of external innovative companies or R&D-departments of large companies
* space for housing and social communication
* green space for recreation

In a next step a plan for **infrastructure development** had to be thought out. That would including future land use, future transport connections, roads, communication systems, water, power and energy supply etc., needed for an up to date Science and Technology Park, which would be attractive for international science and high-tech Industries.

The plans which were actually used are attached as figure 2 and 3.

One of the biggest problems was the multiplicity of elements required, such as ;

* the restructuring of the former academy campus, including the provision of required services
* the creation and implementation of infrastructure on the former airfield
* the attraction of private industrial investors in the field of high technology

Likewise, with the handing over of the land and real estates from the State of Berlin to WISTA-MG it was necessary to deal with the questions and needs of potential investors to settle innovative enterprises and scientific institutes in an efficient way and to provide all the services requested.

The former STASI- location was well known for its lack of development and infrastructure (roads, energy, water, residual water).But ,on the contrary, the former AdW location had an overtaken infrastructure, although it was to be restructured and modernised. Out of 320 buildings, only about 70 could remain, for which the whole communication and approvisionment system had to revised, as was the case for the campus in general.

The Berlin authorities decided to create another development company besides WISTA-MANAGEMENT in order to cope with problems that arose as well as for the setting up of some sort of infrastructure on the undeveloped airfield. And so the Berlin-Adlershof Construction Society (BAAG) was born.

Over the next 15 years, the northern district of Adlershof will see on it´s 420 hectares the emergence of a brand new City of Science and Technology. Developed around a 70-hectare natural and landscaped park, it will comprise of a top-quality sports and recreational facility, residential quarters for 15.000 people, schools, kindergartens and shops. Trade and industry facilities are also foreseen on a large scale(10).The development of the Centre of Science and Industry project remained the responsibility of WISTA-MG.

It should be made clear that at this point not everything was running the way that was originally foreseen. For example, the long term negotiations concerning the handing over of land and real estate properties from the State of Berlin to WISTA-MG caused uncertainty for the potential investors, which hindered quick settlements and, in effect, slowed down the whole project. This also meant that prices for the renting of partial or whole units of space could not be offered clearly and immediately up until 1993. During this time it became clear that potential investors would demand the following points:

- clear and reasonable prices for either renting or purchasing partial or whole pieces of real estate
- excellent infrastructure especially transport systems and
- telecommunication infrastructure
- research structures, consisting not only of research institutes but also of university natural science departments.

4. WISTA, the core of the city of Science and Technology

WIStA-MG was set up to develop the WISTA-site and is an independent limited liability company of public benefit, in no way profit orientated and of which the State of Berlin owns the majority shares. This means that all earnings from renting or leasing have to be reinvested in the further development of infrastructure. As WISTA is located in the former Eastern Berlin it belongs to the regional development zone within the European Union and receives active financial support from State and Federal Programmes as well as from the European Union within the federal programmes "Recovering East" and the EFRE-Programme of the European Union. To participate in these development programmes 60% of the WISTA-site was dedicated to innovative business and the remaining 40 % to science. In this way WISTA could benefit from the above mentioned related programmes for the reconstruction of economic infrastructure. To meet industrial demands and in order to create efficient services for potential investors, all the land of the former Academy and several units of the former airfield became the property of WISTA-MG in 1993. As well as this the share structure of WISTA-MG was changed so that the State of Berlin became the majority shareholder with 51 %. The advisory board was upgraded and now chaired by the President of the German Association of Chambers of Commerce and Industry, and appointed by persons from science organisations and leading German industry, - like Daimler-Benz and Siemens.

These developments led to the first wave of the agglomeration of firms in the year of 1993, which resulted in the creation of many new jobs and in turn brought an increasing amount of private investment to the site. Following these changes the Adlershof project quickly gathered speed ans was now declared to be the most important Berlin project next to the transfer of the German government from Bonn to Berlin.

Reconstruction and improvement of infrastructure at WISTA started with :

- the setting up of a new communication system
- the modernisation of it´s energy and electricity supply systems
- the progressive reconstruction of the building scheme.

At the beginning of the WISTA-project , most of the new innovative firms were spin-offs of the former Academy of Science. This also takes into account those so called new "entrepreneurs" who are good scientists with good ideas, but without much knowledge of the free trade market and its mechanisms, and with little or no capital. That is the reason why it was necessary to start reconstructing not only the buildings and infrastructure but also to create and to offer marketing and other services to those newly established enterprises. A very important decision was taken, to assist innovative SME and to attract investors by creation of a Innovation-Business-Incubator-Center

5. Innovation-Business-Incubator-Center.

This center should help the new firms in their first steps of operation within the free trade market.

- advantageous charges for rental space, made possible by subsidies from development programmes.
- collective services such as a common reception, secretary, post and copying offices, conference rooms etc.
- advice for the drawing up of business plans etc.

The decision to build a business-incubator-Center was quickly made and SME could move into the new building of 12.300 m^2 in 1993.

6. Telecommunication infrastructure

It must be pointed out at this stage, that the improvement of the **telecommunication infrastructure** started with the implementation of an up to date campus net.

At its core a high-speed back bone was installed, based on an FDDI ring with transmission speed of 100 Mbit/s. The FDDI ring allows the connection of the WISTA building and this in turn creates a closed network-infrastructure. The setting up of an ATM-network and of a virtual LAN is in preparation at the moment. In the campus buildings, the users of the high-speed back bone were offered a plug in connection. This way the users can be linked to the FDDI and to the internet.

7. WISTA - an international Center of Science and Technology

After these vital prerequisites for a modern technology park were achieved, the next step was the planning of BESSY II and the transfer of the Humbolt Universtity department of communication science commencing in August 1998. The following units will move to WISTA with a total of 4,500 students and 800 scientists(HUB-investment;750 Mio DM):

*Physics
*Chemistry
*Mathematics
*Geography
*Biology

This will have an international impact on the future developments of WISTA. This also applies to BESSY II a third-generation high-intensity synchrotron beam source for the ultraviolet and soft X-ray vacuum ranges. It is the most modern large-scale installation of its kind and will bring scientists from all over the world to Adlershof (11).

The scientific landscape of WISTA today is completed by the 13 non-university research institutes, which were reconstructed on the advice of the German Scientific Council, given in 1991/92. Their research activities focus on key technologies and form the basis for application-related research and development work at WISTA.

Meanwhile the WISTA institutes have a staff of more than 1.500 scientists and technicians and more than 220 innovative companies with approximately 2.500 employees. So there is a total of about 4,000 employees working on the campus today.

8. Innovative industry

After the first innovative-business-Incubator was built in 1993 it was almost immediately saturated by companies. Today more than 220 companies are on the WISTA-site. New facilities had to be built due to the lack of adequate space to house all of the companies which were interested in settling on the campus.

The above mentioned development in science as well as the ongoing stabilisation within the innovative sector led to the question of the future development of WISTA and the decision on which kind of innovative industry should be concentrated on in the future. An analysis from a scientific as well as from an economic point of view was undertaken in order to get a clear picture in which fields of technology WISTA should concentrate on. and which technological sectors would be the most important for future economics. According to a study by the Fraunhofer Institute for System's Technology and Innovation Research (ISI), the next century key-technologies were compared with the crucial R&TD points in WISTA. An agreement emerged on the following sectors :

* Photonics
* Information and Communication Technologies
* Environmental research and protection
* System Technology/New Materials

An analysis of WISTA´s potentials proves that the concentration of scientific research and the econmic innovative sector is particularly high on the site. It appears that WISTA should concentrate development on the mentioned key sectors and should assist SME in the fields where they have developed various concepts.

9. Specific branch innovation-centers.

The main idea was not only to allow the innovative firms to be based near one another, but also to enable them to be within easy access of the research entities and in doing this to increase the synergy between science and innovative business. So plans were established to have specialised business incubation centers situated next to the scientific institutes. An advantage of this center is that their facilities are like the services in IGZ and also have scientific infrastructure. This concerns expensive scientific apparatus like electron microscopes, x-ray spectrometers and other equipment, which is too expensive to buy for a single SME but necessary for them to operate. The realisation of this concept was supported by the state, the Federal government and also by the EU.

The following branch innovation-centers in WISTA are under construction at the moment:

* Environmental technicologies
* Photonics
* Computer Sciences
* Production technology

Work commenced in 1995 on the Photonics center.

10. Photonics Center

The Photonics Center focuses on the developing and marketing of products, methods and services in the following fields:

Laser Technology (Solid-state, gas-, and dye-lasers, for material processing, analysis, sensors, and communication)
Medical Technology (laser systems for diagnostics, surgery and therapy)
Optoelectronics (light emitters and detectors on semiconductor technology, optoelectronic integrated circuits, optoelectronic components and systems)
Display Technology (electro-luminescent displays and video screens, liquid crystals for TV)
Photovoltaic Technology (silicon crystal and thin-film solar cells)
Optical Electron-Beam and X-Ray Analysis (Analytical techniques for semiconductor technology and materials`research)
Technical Optics
(optical systems for beam control, beam shaping ; micro-optic and integrated optical components)

The Photonics Center consists of five buildings. In the spring of 1996 the first companies moved into the now fully completed Photonics Center. In the spring of 1998 the new buildings will be available for occupation.

11. Environmental Technology Center(UTZ)

The Environmental Technology Center focuses on the development and marketing of innovative products, methods and services in the following fields:

Preventive Technologies
Redevelopment Technology Services

The construction of the Environmental Technology Center started in summer of 1996.The first construction phase, 25,000 m² gross-area will be completed and will be available for occupation at the beginning of next year.

12. Information Technology Center

This center focuses on the development and marketing of innovative products, methods and services in the fields of:

Multimedia
Perspectives of project support
Software-technology
Systems for product-planning, especially construction-lines and product-logistic
The construction started in 1996. The center with a usable area of 3.800 m^2 will be available for occupation in 1998 (see 11).

13. International Office for Technological Co-operation

WISTA aims to be an internationally attractive campus. In order to assist research institutes and especially innovative companies in their efforts to establish international relationships. WISTA-MG was created with active support from the Senate of Berlin and International office for technological co-operation. Starting with a German-French (1993) and German-Russian (1994) Office for Technological Co-operation and liaison, contact points were set up at WISTA for International Scientific and Technology-orientated alliances.

The vocation of the International Office is network based and is used for cross-cultural mediation in the field of high technology for small and medium sized enterprises, research centers or **institutions** from Germany, France, Russia and more and more from other countries (European Union, Central and Eastern Europe).

Supported international co-operation projects concentrate on the following areas

- Research & development and/or distribution
- Creation of companies and their development

- Training/Exchange
- Research and technological priorities in the scope of the office are the same as at WISTA in general.

The International Office offers the following services to the scientific institutes as well as companies:

- Providing partners for co-operation
- by setting-up of initial contacts e.g. organisation of / assistance during international technological meetings, fairs; visits
- Initiation and coaching of co-operation projects
- Organisation of research visits and training periods

- Partners are
- AFAST-DFGWT-Network for technology transfer and simular regional, national and international networks
- Innovation Agency of the Russian Federation
- Technology agencies, economic development corporations, chambers of commerce and of industry, associations, etc.

Due to increasing interest in WISTA both from Western and Eastern European countries and because of the fact that Berlin is located at the edge of the European community, it was not difficult to see that WISTA is an ideal location to set up an East-West Co-operation Center for innovative business. This plan came to life in 1997 and meanwhile the East-West Co-operation Centre (OWZ) invites companies from Central and Eastern European countries (CEEC), which are looking for market access to the EU or Western European companies going East. OWZ addresses companies which:

* are working in technology-orientated branches,
* are engaged in the import and export business in different fields of technology,
* work for the economy in the R&D-sector
* market new products and services.

The construction of OWZ was significantly supported, not only by local and federal administration , but also by the EU within the EFRE-programme.

This way OWZ will become a contact point for R&D- and economic co-operation and a place for the transfer of technology between Eastern and Western Europe. Companies from Russia, France, Ukraine and other countries have started their business in OWZ in September 1997

For its companies the OWZ offers potential rental area of approx. 6.000 m² in a modern new building, including rooms prepared for laboratory and development activities, for the setting up of companies, an appropriate infrastructure and service equipment in order to assist the companies with their start and development (12).

14. Service facilities

Besides research, science and business, WISTA offers a wide range of complementary service facilities like restaurants, sport and leisure facilities and guest houses. Information on these services can be found at the WISTA Business Centre.

The business center will house a lot of service facilities for SME like banks, consulting and tax offices, a travel agency etc. Furthermore WISTA-MANAGEMENT GMBH is developing the Adlershof Communication Center in collaboration with the Humboldt University.It has been planned to combine information processing and enhancement, a cutting-edge library, a computer center and technology transfer office, using latest communication technology. The Communication Center will be linked by high-speed networks to databases and information providers throughout the world.

To conclude, it should be pointed out, that WISTA took the challenge of changing a former centralised campus, including a former military site into an ambitious modern Technology Centre for Science and Industry. WISTA, far from being a finished project, is still developing. It can be said that it has found a place in a modern technological landscape and WISTA invites all interested parties to get in contact.

Bibliography:

1 G. Schmitt: Als die Oldtimer flogen, Oberhaching, 1995

2 U. Busch, W. Thimm: Tradition, Umbruch und Aufbruch; Entstehung einer integrieten Landschaft; in:Wirtschafts- und Wissenschaftspark Berlin Adlershof/Johannisthal, Heft 4 der Schriftenreihe Transformationsprozesse in der Wirtschaft, Hrsg. von Wissenschaftssoziologie und Statistik, Berlin, 1993

3 E. Brink: Anmerkungen zu aktuellen Problemen beim Aufbau des Forschungs- und Technologieparks Adlershof; unveröfentl. Material, Berlin, 1993

4 Empfehlungen des Wissenschaftsrates, Drs. 329/91, Düsseldorf, 5.7.1991

5 Bundesminister für Forschung und Technologie: zur Vereinigungspolitk im Bereich von Forschung und Technologie gemäß Entschließung vom 12.9.1990, Bonn, 1991

6 Kurzdarstellungen zum Stand der neuen Forschungs-Einrichtungen im Beitrittsgebiet, 16.8.1991, Drs. 22/1991, Hrsg. v. BMFT

7 Bundesminister für Forschung und Technologie: Wegweiser für ABM im FuE-Bereich, Bonn, 1991

8 6. Sitzung des Gemeinsamen Ausschusses AdW am 6.2.1991 in Berlin; Berlin, 1991

9 KAI-AdW: 10-Punkte-Programm:Zukunft für Berlin-Adlershof; Entwicklung des Standortes Adlershof zu einer integrierten Technologielandschaft, Berlin, 1991

10 BAAG: A new city for the future, Berlin, 1995

11 WISTA Science and Technology Centre Berlin Adlershof, Berlin 1996

12 OWZ Ost-West-Kooperationszentrum, Wissenschafts- und Wirtschaftsstandort Berlin-Adlershof (WISTA), Hrsgg. IZBM Berlin, 1996

FACILITATING THE GROWTH OF SMEs; THE STRATEGIC ROLE OF SCIENCE PARKS

PROF. GEORGE KIRIAKIDIS
Vice President,
Science and Technology Park of Crete,
PO Box 1447 Voutes, Heraklion, 711 10, Crete, GREECE

ABSTRACT

The major contribution of Small and Medium-Sized Enterprises (SME's) to the European economy is widely accepted. Their particular role on the enhancement of the local economic and social structures has been recognised. Science and Technology Parks (STEP's) have made a catalytic contribution on the facilitation for the growth of SME's in as far as providing the required business environment that boosts the credibility of SME's and their financial and market spectrum. On the other hand, Science and Technology Parks are offering an ideal environment for the creation of new spin-off companies associated with neighbouring Academic and Research Establishments.

Some of the most valuable tools any SME may use in order to sustain competitiveness in today's globalised economy are innovation and technology. Successful innovation is founded on a marriage of technical competence and market knowledge. However, the leading edge on which an SME can base its future growth is the wise exploitation of new technologies that are provided through the technology transfer mechanisms of STEP's. These have been proven of crucial importance offering the channels through which SMEs may identify and acquire know-how, in an effort to create sustainable and competitive infrastructures.

Tools that facilitate the growth of SME's will be reviewed in the contest of successful operation of a STEP. A competitive study between a "young" and an "established" STEP will be presented in as far as the framework of their geographical, political, economic, social and business environments they are operating in, are concerned.

Technological Innovation and Economic Development

Throughout history there have been many examples of contemporary commentators regarding themselves as living though times of exceptionally rapid change.

A. Inzelt and J. Hilton (eds.), Technology Transfer: From Invention to Innovation, 135–144.

The effect of technological progress over the last century is there for all to see, in terms of reduced hours of work for most people, generally improved working conditions, higher standards of health and greater life expectancy. It has to be recognised that these benefits are, in the minds of some, more than offset by the irrevocable damage to the ecological system. Inadequate policies on resources depletion and waste management, the risk of nuclear conflict and the problems of stress which this and other hazards impose upon those living in the late twentieth century are only a few examples. Whatever the end result, it is beyond any doubt that science and technology have made a decisive contribution to today's economic development and progress.

There is a long debate on whether technology develops because of 'science push' or 'market pull', or indeed a mixture of the two. 'Science push' or 'market pull' implies a monocausal, perhaps even linear process, a model of innovation which is found to be over-simplistic. In terms of the relationship between science and technology, it seems more likely that there is a two-way interaction than the unidirectional model implies. This interaction leads to technological innovation closely related to economic growth.

Innovation can be regarded as the total process of the inception of an idea, into the production of a product and finally to its ultimate sale or consumption. It therefore includes invention and the many stages of implementation such as research, development, production and marketing.

Schumpeter (1928) and latterly Freeman (1982,1986) agree with the key distinction between invention and innovation: '*An invention is an idea, a sketch or model for a new or improved device, product, process or system.*' The majority of inventions do not lead to technological innovations. '*An innovation in the economic sense is accomplished only with the first commercial transaction involving the new product, process system or device, although the word is used to describe the whole process.*

But how does technological innovation come about? What are the processes involved and are there generalised patterns recognisable in different cases and across sectors? Is there a model of the innovation process that can inform policy and direct future developments?

A 'linear model' of innovation assumes a progression something like:

BASIC SCIENCE	⇒	APPLIED SCIENCE	⇒	TECHNOLOGICAL DEVELOPMENT	⇒	MARKET PLACE

This model, as well as assuming that the progression is linear and proceeds in discrete stages, also suggests that science pushes technological innovation.

As Schmookler (1966) and Freeman (1982) have emphasised, technological innovation is a process requiring, on the one hand technological knowledge (possibly but not exclusively containing scientific knowledge), and on the other, the awareness of a need in the economy or the community. The word 'need' is value-dependent, and economists generally prefer to talk of market demand; but for radical innovations (for which there is no pre-existing market), the term "market demand" may be substituted. The process of innovation is crucially one that unites both technological opportunity with market demand (or potential demand).

It is the catalytic role of the commercial world to exploit the results and the products of innovation and to transform them into a wave of economic growth. Large corporations but even more Small and Medium sized Enterprises (SMEs), are particularly suited to accommodate innovation, as well as, to transfer new ideas into processes and products that contribute to sustainable Economic Development. Thus methods to facilitate the growth of SMEs must be at the centre of all government polices.

The last 20 years have seen two key developments occurring simultaneously. The first is the growth of new technologies, many of which are applicable in a commercial form and on a much smaller scale than previously. Second is the importance of 'leading-edge' research, much of which is initiated in universities and other institutions of higher education.

It was in North America that these two developments merged soon after the war. The foresight of individuals, such as Professor Terman at Stanford, lead to the realisation that industry needed access to high quality research in order to be competitive in international markets. He, and others, also appreciated that academics might wish to commercialise the results of their research, and so the concept of a ***Science Park*** (SP) was born.

"Science Parks are environments within which the process of innovation, unites technological opportunity with market demand".

It was not until the late 1960s that these ideas crossed the Atlantic and Cambridge University pioneered this development in Britain. At the same time other European SPs such as Sofia Antipolis, France (1969) and later TECHNOPOLIS NOVUS ORTUS in Bari, Italy emerged. In UK, Cambridge was quickly followed by Heriot-Watt, while no other British university developed a Science Park until 1981. However, over the next few years there was a fundamental change so that by 1991 only six universities in the UK had not established a SP and appeared to have no plan for doing so.

In contrast to Britain and America, science park development in the rest of Europe has been characterized by a much greater degree of overt government involvement. The evolution of a more integrated Europe has been prompted largely by economic factors, principally the need to respond to the competitive challenge posed by the United States and Japan. The traditionally fragmented European market widely believed to be at a disadvantage in an increasingly global economy faces the possible threat of annihilation by its major competitors, especially in the advanced technologies. Thus the European

Commission has developed a wide range of policies to broaden the base of the market and improve the competitive capability of industry.

The main focus of EC science park policy was the SPRINT programme of DGXVI set up in 1989 to promote ***innovation*** and ***technology transfer***, principally through the consolidation of existing trans-national networks between companies and the formation of new ones. It includes measures to facilitate the diffusion of new technologies to companies, and strengthen the '*European innovation and technology support service infrastructure'*. This infrastructure includes science parks and the SPRINT Science Park Consultancy Scheme designed to further the exchange of information and expertise at the development stage.

Science Parks also feature in the Community's structural funds programme, specifically the European Regional Development Fund's 'Article 10' support for inter-authority co-operation. Under this scheme groups of local authorities can obtain funding for innovative projects designed to promote the spread of know-how and experience across national boundaries and assist the implementation of Community regional and other policy objectives. One of the Article 10 projects funded in 1990 is focusing on ***the value of science and technology parks as a vehicle for local economic growth.***

Science Parks are key elements of economic and commercial developments for the regions they operate in. It is widely believed that Science Parks are serving an important function and that most universities should have such a Park focusing of linkages with the commercial world. Even so, there is considerable scope for improvement, with greater efforts needing to be devoted from the management of the parks. There is evidence that, where the park is able to provide business and financial support services, this provides significant value to its tenants.

Science Parks demonstrate the willingness and ability of universities to work successfully with the commercial sector. Policies must not destroy the strengths of universities they wish to exploit by substituting short-term economic criteria for longer term and wider objectives. Governments should recognise their importance and establish a policy that will direct additional resources towards institutions able to demonstrate a willingness and ability to work successfully with the commercial sector. A thriving Science Park is a key element in that demonstration.

It is indeed now widely accepted that '*it is the catalytic environment provided by SPs that is expected to contribute decisively on the growth of Technological Innovations leading to Economic Development'*.

Technological Innovation within Science Parks

A crucial aim for Science Parks is to attract 'high-technology' enterprises which operate at the 'leading edge' of technology. Not surprisingly, a high proportion of Science Park establishments consider that they have a 'leading edge technology' product, or consider that their products or processes involve the application of advanced technology or their research being radically new.

There is no doubt that Science Park establishments are relatively sophisticated technologically. But rather than being 'leading-edge' in the absolute sense, they appear to be more involved in new applications of relatively novel technologies, to be small innovators rather than involved in major innovative break-throughs, and indeed often operate as diffusers rather than as innovators.

Given the strong popular association between high technology and information technology (IT), it is no surprise that the IT sectors predominate. Sectoral classification of what comes out of establishments gives a relatively superficial idea of the activities going on inside them. For example, the computer hardware sector includes a wide variety of processes, from the manufacture of sub-assemblies, products and systems to the warehousing and distribution of hardware products. Science Parks are much less involved in manufacture than off-park establishments but more involved in design and development and software production. Park establishments seem to be more focused on sales activities, similarly involved in servicing and repair and, surprisingly, have more (low-tech) warehousing than off-parks units.

In examining the relative technological level of individual establishments, one approach is to look at inputs to research and development (R & D). A range of measures are typically used, particularly the proportion of highly qualified scientists and technologists and the level of expenditure on R & D. There is no doubt that Science Parks contain many firms engaged in work employing highly educated staff. The percentage of qualified scientists and engineers employed by park tenants is generally very high.

Another measure is the level of technology financing - usually measured as the percentage of turnover devoted to R & D and a reasonable intensity of R & D effort. It is possible to extend this and look at measures of R & D output, particularly ***patents registered*** and ***new products launched***. Looking at the patenting performance of Science Park units, it is not surprising that overall, Science Park units have a greater tendency to take out patents than does a group of off-park units.

Although patent statistics are often used as proxy measures of technological performance, there are problems with comparing establishments - the tendency of firms to take out patents varies between sectors, between firms and between countries.

The data are less clear-cut when it comes to product launches. There is considerable variation between Science Park sites in their tendency for establishment and launch of new products. In some sites over two-thirds of units have launched new products, in others (not necessarily the newest parks) under half have done so. Overall, there is no difference between Park and off-Park establishments. However, it does seem that those establishments lodging patents are more likely also to be involved in launching new products. Higher levels of R & D expenditure, on the other hand, do not correspond to higher levels of new product launch. One reason may be that some establishments concentrate on research and the early parts of development, rather than the production of new products for the market. Another justification may be that many Science Park establishments are relatively new and that products are not ready in significant numbers yet Overall, the quantitative measures of R & D inputs suggest that Science Park establishments have relatively high R & D expenditure and a large proportion of qualified scientists and technologists.

On outputs, the results are less clear. *Significant patenting take place but product launch data are less impressive.* However, the commercial launch of a product is rather far downstream in the innovation process and thus not so immediate an indicator. There is also a sizeable proportion of Science Park establishments not engaged in manufacturing products. One gets a picture of a varied group of establishments - some focus quite strongly on the early parts of the innovation process (R & D and design), others, with a software orientation, get involved in later aspects of innovation, including product launches, confining themselves to more upstream activities.

THE CHARACTERISTICS OF SUCCESSFUL SPS

In the preceding review, the role of Science Parks in promoting economic development for their region was highlighted. It must also be noted that in recent years a greater role has been played by smaller firms (SMEs), in promoting technological change, and that this, together with the catalytic role of the neighbouring Higher Education Institutes (HEIs) is suspected to be the major reason which underpinned the growth of Science Parks in the UK and other European countries.

It is highly desirable to attempt an insight into the objectives of SP investors (Universities, Local Authorities, Governments etc.) and undertake an appraisal of the effectiveness of SPs as instruments of public policy. Thus, estimating the added value that SPs might provide for their tenant companies, mainly SMEs, and the degree they facilitate their growth.

To assess the level of success which is directly related to the accomplishment of raised expectations from the tenant companies, results of a study on the comparable characteristics between a young (3 years old) Science Park located in an under development and geographically isolated part of Europe (STEP-C, Crete, Greece) and an established one (14 years old) operating in a well developed centralised area of the UK (Guildford, Surrey) will be presented as a case study.

These two SPs have been established following the initiatives of the University of Surrey and the Research Centre of Crete who are the major share holders of their respective parks. Both wanted to promote the link with industry in an effort to enhance technology transfer mechanisms and create new job opportunities for university graduates. Close proximity with industrial firms was anticipated to encourage the development of new spin-off companies and stimulate joint ventures for the exploitation of the results of Research & Development (R&D) from research groups in the university.

Both Science Parks are located inside a university campus and are managed by a private management company that is responsible for the development of the Park. Neither of the two management companies receive any subsidy or other form of financial assistance from their universities to which, however, they are fully accountable for their financial performance. The role of these management companies in both SPs is to manage the University (State) property (land, buildings, etc.) in a way that it provides added value for the tenant companies and generate revenue for the University. Furthermore, they have to promote the diffusion of technology, to facilitate the establishment and growth of new high-tech companies, attract new businesses in the area and create new job opportunities, particularly for young university graduates.

Following an extended survey, (Kiriakidis.G, 1996) results of which are listed in the following Table, the operational ownership and motivation characteristics of these two Science Parks were extracted and compared with previous similar studies, mainly within UK. The key results of the study are summarised as follows:

	Crete	Surrey
a) Motivation for establishment.	HEI's interaction	Business location
b) Entrepreneurial spirit	Very high/academic	High/business
c) Product range	Risky high tech.	Conventional/leading edge
d) Interaction with HEI's	Highly desirable	Secondary importance
e) Suport services	Lower than expected	Expectation level
f) Added value	High	Average
g) Job creation	Insignificant	Insignificant

The founder of a Science Park company is more likely to be led by a high entrepreneurial motivation, with high academic qualifications. Particularly in the young Science Park, the element of entrepreneurship is more profound, the founder is younger and his business experience is limited. It was found that more than a quarter of the

Park firms were characterised as true spin-offs of the university environment, attracting predominantly "local" firms. However, the age and the geographical location of Surrey make it of a more "International" character compared with Crete. Compared with Surrey, firms in Crete are more "high-tech" orientated in as far as utilising IT communications for their business activities, although their product range was more "conventional" in contrast with a relatively high number of "leading edge" Surrey firms.

In Crete, the level of R & D activities and the interaction with the neighbouring University and Research Centre was reported to be high. However at neither of the Science Parks was it high enough to be regarded as their predominant occupation, contrary to what is widely believed. Nevertheless, the levels of academic links were not much different from those reported in previous studies (Segal Quince, 1988).

Finally, through this study it was expected to compare the level of "satisfaction" as reported by the tenant firms from both Science Parks. This is closely related to the debate on whether Parks meet the challenge for the promotion of innovation, facilitation the growth of new start-ups, job creation and technology transfer. It was found that in general terms, firms from within an established Science Park were more self sufficient and independent compared with their counterparts at the young Park which were expected a greater level of technical and financial support from the Park authorities relaying mainly on their initiatives for technology transfer opportunities. At neither of the Parks job creation was found to be enhanced at any significant level for the neighbouring community. This was in line with earlier foundlings (Storey and Johnson,1987).

On technological, financial support, recruitment and effect on sales, on average, firms have reported to enjoy the amount of support they were expecting from their Parks. This implies that there is a successful level of communication between the Park authorities and the companies as to the amount of support that can be provided in contrast to their expectations. Thus, these Science Parks may be characterised as potentially suitable locations, particularly for new business opportunities.

Firms within these Parks recognised the "added value" provided for their operations independently of their level of technological development. Despite the clear attraction for predominately the "local" companies and the corresponding effect on job creation, no significant contribution to the economic development of the area could be identified. It is believed that the justification for the investment associated with the establishment of either Science park is more related with the element of added value that Parks provide through their "location" and infrastructure rather than their proximity to a pool of know-how. Although R & D was not found to be of a predominant character, it is hoped that governments and local authorities should continue to resource these establishments providing the opportunity for the creation of more high-tech spin-offs as well as the promotion of innovation and technology transfer.

Conclusion

The role of SMEs on the enhancement of the local economies and social structures has been elaborated and the catalytic contribution of Science and Technology Parks on the facilitation of SMEs growth, demonstrated. The two major parameters that can assist SMEs to sustain competitiveness in today's globalised economies, *Innovation* and *Technology* were shown as the key elements that are provided through Science Parks.

Tools that facilitate the growth of SMEs have been discussed and a brief account of a case study was presented through a comparative analysis of the operational and functional characteristics between the young Science Park of STEP-C in Crete, Greece, and the established Surrey Science Park at Guildford , UK.

The environments within which Science Parks are operating in were analysed and the role of their neighbouring Higher Education Institutes (HEIs) was analysed and elaborated. Results on issues such as technology transfer, support for new start-ups, employment opportunities and technology innovation promotion were discussed in the context of the characteristics of these Science Parks and contrasted with past analyses of US and UK Science Parks.

It was found that younger Science Parks are characterised strongly by their links with their HEIs and are more technically oriented. Both Science Parks were successful in attracting "local" firms, but the older had a distinct international character. The choice of the right location was more important for firms in the older Science Park, while recruitment of new personnel from the local HEIs was considered of vital importance for the young Science Park.

The expectations of the firms from both Science Parks were contrasted with the eventual outcome and the level of satisfaction was found to be at a high level.

References

1. Freeman, C. (1982) "The economics of industrial innovation", 2nd ED., Frances Pinter, London.

2. Freeman, C. (1986) "The case for technological determinism", paper presented at ESRC Cambridge Conference 11-12 April 1986.

3. Schmookler, J. (1966)" Innovation and economic growth", Harvard University Press, Cambridge, Mass.

4. Storey,D.J. and Johnson, S.J. (1987) "Job Creation and Labour Market Change",MacMillan, London.

5. Schumpeter, J.(1939) "Business cycles", McGraw High, New York.

6. Kiriakidis,G. (1996) " A comparable study between a young and an established Science Park" MBA Thesis SEMS, Univ. of Surrey, UK.

TRANSFORMING INVENTIONS INTO INNOVATIONS AS A MAJOR CONCERN OF THE PHILIPS RESEARCH LABORATORIES MANAGEMENT: A HISTORICAL PERSPECTIVE

MARC J. DE VRIES
Eindhoven University of Technology
Faculty of Technology Management
P.O. Box 513, 5600 MB Eindhoven, The Netherlands

1. Technology transfer and corporate industrial research

The term 'technology transfer' is used for a wide variety of activities aimed at transforming inventions into innovations[xii]. These meanings range from transfer of technologies between countries (often between developed and developing countries) to the diffusion of technologies from industries to society, but also transfer of technologies between industries. In the context of industrial research the term 'technology transfer' can mean 'transfer from the laboratory to product development and manufacture within the corporation'[xiii]. It is with this limited meaning that the term will be used in this paper. By describing aspects of the history of the Philips Research Laboratories, famous for several important inventions, like optical recording (leading to the compact disc), an insight can be gained into the opportunities and difficulties for transforming inventions into innovations in the context of an industrial research programme. The issue of transferring research output to development and production has been a major concern and a point of many debates in the management of the Philips Research Laboratories. Even within the small number of industrial research laboratories that were similar in size and scope (e.g. Bell Labs, GE Labs, Siemens Labs, Du Pont Labs, IBM)[xiv] we can see differences in the formal position within the company that is given to the research laboratories. In the case of this paper this industrial research programme is of a corporate character (that is: in this corporate organisation the programme supports all product divisions within the company, and reports directly to the Board of Management[xv]. Until 1989 the total budget for the research programme was allocated at corporate level; thereafter the corporate funding became one third, the rest being funded directly by the Product Divisions): in 1914 Philips[xvi] decided to create a research programme that was formally independent from the industry groups within this company. In practice this meant that the laboratory could set its own priorities for doing research and determine its preferences for different types of research ('phenomenon-oriented' or 'fundamental' versus 'application-oriented', 'long-term' versus 'short-term'). In terms of technology transfer this is essentially different from a situation where research is done by a variety of laboratories that are found within the industry groups. As we will see, whilst it creates unique opportunities for inventing it also causes difficulties for transferring those inventions into innovations. By the end of the period

A. Inzelt and J. Hilton (eds.), Technology Transfer: From Invention to Innovation, 145–159.

we describe here, in 1990, the situation was as follows. The corporate research programme was carried out by a laboratory in Eindhoven, the Netherlands (the oldest one and still the largest) and other laboratories in Belgium, France, Germany, the UK and the USA. Outside the corporate research programme, development laboratories within the various industry groups[xvii] are focused on the 'D' of 'R&D', although the separation between 'R' and 'D' in the company cannot be defined clearly . The Lighting group is the only product division that has a large research facility, which, as we will see later on, caused a different technology transfer relationship with corporate research than in the cases of the other industry groups. The total amount that was spent on R&D, both in the corporate research programme and in the development laboratories within the industry groups, through the years varied from about 7 to 8 percent of the company's total turnover (about 1 percent of the total turnover is allocated to corporate research). The research topics range from phenomena oriented materials research to research into mechanics, various electronic devices, IC-design and -technology, systems, and software.

In this paper we will address the following questions from a historical point of view:

- what barriers for transferring research output to industry groups within Philips were perceived by the Philips Research Laboratories (section 3),
- what ways of overcoming them were sought in the course of the history of the laboratories and what effect did that have (section 4),
- were there differences with respect to a and/or b between the three main parts of the research programme (materials, devices and systems research) (section 5)?

As the research into the history of the Philips Research Laboratories so far has been focused on the period 1945-1990, this paper will have these same limitations. Also it should be pointed out that this research is still ongoing; therefore it is not possible yet to present firm conclusions. The character of this paper is rather descriptive and results in some hypotheses about changes in the way the issue of technology transfer has been addressed by Philips Research in the period 1945-1990. The aim of the paper is not to suggest optimal ways of transferring research outputs to product divisions within a multinational company like Philips. The research does not yet offer sufficient material for underpinning such suggestions. The value of the paper should be seen in the insights it offers into the complexity of the relationship between a (corporate) research programme and the product divisions in a multinational company like Philips. Before the questions stated above will be addressed it is necessary to give some background information about the historical research from which the paper draws.

2. Research into the history of the Philips Research Laboratories

The history of the Philips Research Laboratories will be described in three periods:

- 1914-1946: in this period the first director, Dr. Gilles Holst, was in charge of the research programme, that started with research into phenomena in light bulbs, and gradually diversified into many other areas;

- 1946-1972: in this period the famous physicist Dr. Hendrik Casimir was in charge of the programme; thanks to the favourable economical circumstances after W.W.II he was able to strengthen the research in more fundamental areas;
- 1972-1990: in this period first Dr. Eduard Pannenborg and after him Dr. George van Houten as a response to the changing economical climate took great care in establishing a more formal and systematic relationship between Research and the industry groups within Philips.

For the time being we use this periodisation as a working hypothesis; further research will have to show the validity of it. Although the description in terms of personnel seems to suggest that the periodisation is based on a the assumption that individuals could have a major impact on the research programme, in fact the proposed periodisation is largely determined on the basis of the changing economical circumstances. As we can see in the history of similar laboratories, the Fifties and Sixties were a period in which the economical circumstances were quite good for developing new markets for consumer goods and people had very high expectations of research, and in particular 'free' and 'fundamental' research. This allowed Casimir to do what at the same time fitted his personality as a physicist, specialised in quantum physics, very well, namely to stimulate the integration of new research fields that were challenging in particular from a scientific point of view. The Seventies and Eighties, on the other hand were characterised by a change in the economical climate, a growing impact of the Asian companies on the world market, a growing public awareness of the limitations and possible negative effects of technological developments, and as a result the need to increase the effectiveness and efficiency of research. Coupling research more strictly to the industry group's activities was seen as one of the ways to accomplish this. Pannenborg and Van Houten have done this and even though their way of managing perhaps can partly be explained because of their more entrepreneurial characters, they certainly also were driven by external (economical) motivations. Just as the year 1914 is a natural point for starting the history (this was when the central research laboratory in Eindhoven was initiated), the year 1989 is a natural ending point, because then contract research with the industry groups was formally introduced as a totally new way of financing the corporate research programme. From then on the Philips Research Laboratories were forced to acquire research assignments from the product divisions in order to finance this research programme and only a part of the research budget (about 35 percent) was allocated at the level of corporate management. This can be seen as a true turning point in the history of the corporate research programme. Of course this change also to a large extent affected the technology transfer from the laboratories to the industry groups (the intention certainly was to enhance commitment from both sides). In this paper we will not deal with those changes.

Finally, reference should be made to the meaning of the words 'in charge of', because this changes in the course of time. Holst was appointed as the director of the research laboratory. When he retired three directors were appointed for three different areas:

physics, chemistry and engineering (electrical and mechanical). This division of the research programme would later be rephrased into 'materials', 'devices' and 'systems' (also see our third question at the end of section 1), but there is suggested to be a relationship between this division and the original one in that 'materials' to a large extent consisted of chemists, 'devices' of physicists and 'systems' of engineers[xviii]. Casimir was one of the new directors, namely for the physics area. But in 1956 he was appointed as a member of the Board of Management of Philips. This meant that Research from then on would have its own representative on in this Board, while before that Dr. Frits Philips in a more indirect way took care of the Research interests in Board meetings. Pannenborg then took over Casimir's position as a director. In 1972 Pannenborg also succeeded Casimir in the Board of Management. In 1984 Van Houten succeeded Pannenborg. Apart from the people mentioned so far, there were others who were more directly in charge of (parts of) the research programme who also had important influences (here we can think of general managing directors, for example Rathenau, De Haan, Kramer and Bulthuis).

3. Barriers for technology transfer as perceived by the corporate research laboratories

It is striking that the minutes of the so-called concern research conferences, that have been held bi-annually since 1948 and in which all directors, co-directors and scientific advisors discussed the research programme, the issue of the transfer of research output to the industry groups from the beginning has been discussed by the directors of the Laboratories. Also in the minutes of the research director's conferences, the highest research management meetings in which only the directors participated and that were held more frequently (from 1965 to 1982 and from 1987 to 1990 three or four times per year and from 1983 to 1986 up to seven or eight times per year) the relationship with the industry groups is a frequently reoccurring issue. In the minutes of both concern research conferences (crc's) and research directors conferences (rdc's) we find various barriers for this transfer mentioned and debated. The ones that we find most frequently are the following[xix] (each is illustrated by one or more examples):

- the industry groups did not have an adequate (long-term) product policy that enabled them to set priorities for research support and to make sophisticated selections out of the technological options that were offered to them as research output (which was what Research saw as one of its main tasks). A typical time horizon for an industry group is a two-year development plan and a four year marketing plan; Research wanted guidelines for a time horizon of at least four years, maybe even up to ten years.

An example of a product that was not taken over by the industry group 'Elcoma' ('Electrical Components and Materials') was the index tube, a display tube for television. According to Research this tube was even better than the shadow mask tube that had already been transferred, but its use was limited to small sizes. Elcoma at that time did not see any potential for small size television pictures, which was regretted very much by some Research people[xx]. The same lack of long-term product policy was seen by Research as a barrier for transferring research output in the area of integrated circuits to the factory in Nijmegen in the early Seventies[xxi].

- The industry group does not feel responsible for research projects and therefore postpones the decision whether or not to take up the development work as a follow-up of it to a very late stage of the project.

An example of this is the work on SiC whiskers. By accident researchers had found that under certain circumstances semiconducting SiC crystals grow in the form of whiskers. The researchers immediately realised the potential of this to be used for application in composite materials[xxii]. In the 'Light' and 'Elcoma' Main Industry Groups the result was first received positively[xxiii], although neither of them were really involved in the market of fibre enforced materials. It took a long time before the final decision was taken and Research was informed that the topic would not be taken over for further development[xxiv].

- the industry groups did not spend enough effort to use the research output effectively. Either there was a lack of knowledge that hampers the industry group taking over the research output or there was not enough time devoted to serious elaboration of the ideas that were taken over from Research, because the industry groups first wanted to have some evidence of the feasibility of the proposed product or process before committing themselves to it. As a result the research laboratories had to spend a lot of time on supporting the industry groups after the research output had been transferred and sometimes pilot production even took place in the Research premises. Although a serious effort to support the further development of research output was seen as a responsibility of Research, the amount of time and money that had to be spent on this often led to complaints because of the danger that other activities received less attention that desirable.

An example of this is the Plumbicon, a very successful pickup television camera tube that was launched in the early Sixties[xxv]. Research had already transferred this tube to the Main Industry Group Elcoma, but later had to take it back and support quite extensively for pilot production by means of sending experienced personnel to the Main Industry Group, because the production yield remained very low[xxvi].

- The company structure caused a problem for technology transfer, in particular for those products for which more than one industry group had to become involved.

A typical examples of this was the role of the Main Industry Group Elcoma, that produced devices for other system producing Main Industry Groups. Several examples are known in which Research and a Main Industry Group on the systems level had already established the need to develop a certain device, but Elcoma refused to take up this development, because for them the development and production of the device only - not the whole system - was financially unattractive[xxvii].

We have to keep in mind that there is a certain bias in this list of complaints because the minutes of the concern research and research directors conferences only present the view from the Research people and were focused on the ongoing programme rather than on reviewing of already transferred research topics. It must be admitted that Research people often did not sufficiently realise the risks for an industry group to take over a research topic because of the investments being much bigger in development and manufacturing than in research. Yet when we study the minutes of meetings in which the industry groups were also represented, we find the same type of barriers (see the next section). It appears for example that industry groups do pose research questions, but mainly on very specific issues rather than on long-term issues. At the same time, it is striking that the structure for the research programme (the division into 'materials', 'devices' and 'systems' groups) was not questioned at the level of research management until the Nineties. Of course from the point of view of Research it makes a lot of sense to structure the research programme according to disciplines and the choice to use this structure was no doubt taken quite consciously. But this discipline-oriented structure did not yield an optimal match with the product oriented division of industry groups that characterised the rest of the company and thus possibly created a barrier for transfer of research output[xxviii]. At first sight this seems to be an important point, as it refers to the research programme as a whole. But to see if it is really a serious barrier for technology transfer we need to explore it further. In the next section we will discuss this division into the domains of 'materials', 'devices' and 'systems' in more detail.

4. Differences between the areas of materials, devices and systems

Philips as a company started as a device producing company (first light bulbs, but soon also other devices like various sorts of tubes for electronics, radio tubes, tubes for telecommunication and X-ray applications). When it became evident that in telecommunication it is important not only to consider separate devices, but even more to consider the communication system as a whole, Philips, like many other companies (e.g. Bell AT&T), started profiling themselves as a 'systems' company rather than a 'devices' company. In the period we report on in this paper, we can see this reflected in the main division of the research programme of Philips Research into three domains: 'materials', 'devices', 'systems'. It is an interesting question if Research was pro-active or re-active in this respect (this will probably become clear when the first period of the history of Philips Research, 1914-1946, is studied in more detail).

From the material that has been studied so far we get the impression that there are differences between these three main areas of research, 'materials', 'devices' and 'systems'. In the annual Concern Research Programmes we find the number of people working in these three areas[xxix]. By studying these numbers we can see that the amount of people working in the 'materials' area has been constant in the period 1956-1990 (about 25%[xxx]), the amount of people working in the 'devices' area grew from about

30% in 1956 to about 40% in 1968 and from then on decreased again to about 20% in 1990. In the 'systems' area about 30% of the people were working until the early Eighties and then this area grew to about 45% in 1990. A more detailed view is acquired when we look at the 'birth' and 'death' of groups within each of the domains. In the years 1956 to 1976 we see several new groups starting, most of them in the 'devices' and 'systems' domains (for 'devices' examples are: 'applications of solid-state devices' and 'optical semiconductor devices', for 'systems' examples are: 'systems engineering' and 'energy systems'). In the years 1976 to 1986 several groups in 'classical' areas disappear from the programme, most of them in the 'materials' and 'devices' domains (like 'biology', 'gas discharges', 'glass', pick-up tubes and cathodes'). In the years between 1986 and 1988 we see a number of new groups in the IC-design and -technology area. The cause of this is the large effort that was dedicated to the Mega-project, aimed at developing a 1-Megabit SRAM (Static Random Access Memory) chip. Most of these groups disappeared again after two years because the Mega-project was stopped[xxxi].

In the 'materials' research groups we find a lot of research that is aimed at providing analytical methods and instruments for measuring the properties of materials, mostly for internal use in the research laboratories[xxxii]. For this part of research, concerned with the development of the methods and the accompanying application knowledge, the laboratories had a 'natural' transfer partner in the industry groups, namely the Main Industry Group 'Produkten voor Industriële Toepassing' ('Products for Industrial Application'; in 1975 renamed: 'Scientific & Industrial Equipment' and in 1984 combined with 'Electro-Acoustics', 'ELA' into 'Industrial & Electronics Systems')[xxxiii]. In the early years of the laboratories, ferrites and other magnetic materials were an important research field and in particular the work on ferrites has yielded very positive outcomes, because they were applied successfully in all sorts of products as magnetic materials (e.g. loudspeakers, antennae, magnetic coils in televisions). Several Main Industry Groups have gained from the transfer of this research to them and for a long time there was continued research support for working on applications. Later on the attention shifted to semiconducting materials. In the beginning of the period we describe here, this was already seen as the main issue in the 'materials' area[xxxiv].

For the 'devices' area there was again a 'natural' partner in the industry groups, namely Elcoma. As we have already seen in section 4, the fact that devices were produced by a separate Main Industry Group sometimes caused specific problems for transferring research outputs in this area. On the other hand, in the development of ICs contacts on the worker's level often was good[xxxv]. It could be expected that the devices research would draw from the materials research. In fact the possibility of transferring knowledge from different disciplines is one of the most important reasons for having a central, corporate research programme. This transfer indeed did take place[xxxvi], although the two fields remained separate and the relationships were not always clear even to the researchers[xxxvii].

According to one of the former directors in the 'systems' area the systems research by its very nature had to have close contacts with industry groups, because systems are directly related to users[xxxviii] and therefore technology transfer could be expected to be easier in this area. There was also contact with the 'devices' and 'materials' areas. As in the contact between 'materials' and 'devices', here too we find examples of 'walls' between the areas. Thus it is told that in the devices group started working on the VLP even though the systems people (the 'natural' place because the VLP of course was a system) had already decided not to start working on this topic[xxxix]. When the ICs became very complex and in fact were systems in themselves, as well as parts of a larger system, the boundaries between the 'devices' and 'systems' areas became blurred.

As mentioned before, it can be questioned whether the (quasi-)disciplinary division into 'materials', 'devices' and 'systems' research creates a barrier for transfer to industry groups, that are not disciplinary, but product oriented. Other industrial research laboratories have made different choices here. The programme of the GEC Labs for example was changed in 1960 from a discipline-oriented (physics, telecommunication, chemistry/engineering, mark the similarity with the Philips Research programme structure) to a more product-oriented structure[xl]. At the same time we see that there are several topics (of which ICs is the most striking example) for which the practical meaning of the division into 'materials', 'devices' and 'systems' in terms of transfer is hard to see and hence the difficulties in connecting research to development, caused by the seemingly discipline oriented structure of the research programme, should not be over-estimated.

On the other hand a discipline oriented organisation maximises the effect of synergy in the corporate research programme.

5. Efforts by Research and industry groups to overcome the perceived barriers

In the course of time, several strategies have been used to improve the transfer of research output to the industry groups. At this point the periodisation of the history, as described in section 2 is important to take into account.

In the period of Casimir the contact between corporate research and the industry groups was based on individual contacts, mostly at the level of researchers and group leaders, and to a lesser extent at director level. In the late Fifties and early Sixties, meetings between representatives from the Board of Management, from Research, from industry groups and from the patent department took place in the so-called 'Quo Vadis' meetings. The aim of those meetings was to discuss future developments in specific areas that were or could become of interest for the various industry groups. Examples of such areas are: television, radar, semiconductors, magnetic materials, batteries, electronic medical equipment, and household appliances. The participation of Research in these meetings shows that the Board of Management thought it was useful to have input from Research in the development of long-term product policy by the industry groups. In the course of time, this would be mentioned several times during crc and rdc meetings as a way of overcoming the technology transfer barriers. It is not quite clear from the available materials whether or not this research input really helped the industry groups in establishing a product strategy.

In the late Sixties so-called 'director's contact meetings' were arranged with each main industry group separately to discuss present research output to industry groups and make an inventory of specific research needs from industry groups. The minutes of these meetings show that most of the research needs as expressed by industry groups were rather specific and seldom at a more strategic level. This confirms the concern that was expressed during crc and rdc meetings that the industry groups had a product strategy that was too short in time horizon to derive long-term research needs. There are, however, some examples of director contact meetings at which more strategic reports, made by an industry group, were discussed (some examples are: a report on 'Boundary conditions for the electronic part of RGT equipment' by the Main Industry Group 'Radio, Gramophone and Television' in 1971, the report 'Future of Resistors and Capacities' by the Main Industry Group 'Industrial Components and materials', ICOMA, in 1971, and the report 'Future Trends in the Materials and Components Fields' by the main Industry Group Elcoma in 1973). No evidence has yet been found of the impact of the research input in these discussions. It appears from the minutes of several meetings, that mutual irritations about decisions in which the other party had not been involved, were not prevented by the contact meetings. Casimir, for example, in 1971 wrote a letter of complaint to Otten (Main Industry Group PIT) about the decision of PIT to stop all cyclotron activities without consulting the Research people[xli] and in 1968 Klasens (Philips Research) wrote a letter of complaint to Van Gijn (ICOMA) about the decision of Research to move all piezo-electrical activities to one of the laboratories in Germany[xlii].

Later, under Pannenborg, so-called Research-Product Division (R-PD) Management Committees functioned as a continuation of the director's contact committees (in fact this meant no fundamental change in the relationships with the industry groups). Pannenborg saw these as a means for Research to get more market information as an input for research[xliii]. In some of the meetings proposals for common projects were discussed, but the minutes show that it was not always easy to find research topics that were relevant in the eyes of both the Research people and of the industry people. For example, an extensive contact with the Main Industry Group 'Elcoma' was needed in 1971 to select a shared project on complementary MOS technology, and again in 1972 it was not easy to reach agreement about a possible common project on magnetic bubbles for computer memories, because the Main Industry Groups had doubts about the feasibility while Research from the beginning considered this as an important topic[xliv]. This case is an example of a topic of which the decision of the industry group not to take it over was justified by history: magnetic bubbles for computer memories did not become a commercial success anywhere.

Another instrument that was used to reach a better transfer of research output to industry groups were the so-called Concern Research Exhibitions, of which the first one was held in 1959, and that from 1973 on have been held annually. These exhibitions offered the industry groups an opportunity to see demonstrations of research topics that Research was working on and that might be of interest to them. By comparing the content of these exhibitions with the content of the publications by the researchers (as collected in the Registers of Publications and the Registers of Reports), we can see that a rather product oriented selection was made (much less materials research was presented) but that this selection is a good representation of those product oriented researches that were current in the research laboratories.

In 1987 in the rdc meetings discussions began about so-called 'Transfer Projects', in which Research co-operates with one or more Product Divisions to bring new product ideas into a phase of transfer[xlv]. This can be seen as a final effort to show that a fruitful transfer from Research to industry groups is possible without contract research, in which the industry groups, and not corporate Research, have the authority to decide whether or not money will be devoted to certain research topics. Apparently these transfer projects did not meet that expectation, because in 1989 the transition to contract research is announced by Van Houten. Some reasons for this may have been that nearly all of the Transfer Projects lacked market information and a business plan and most of the more than 50 proposed projects were rather small[xlvi]. It can be questioned if this problem was solved by contract research, but at least this was mentioned as a serious shortcoming in the Transfer Projects. After contract research was introduced, the existing management committees in which research and the product divisions met, remained and now had to result in agreement on contracts.

In general it is difficult to establish the effect of formalising the contacts with the industry groups. In both the period in which those contacts were rather ad-hoc and informal (mainly under Casimir), and in the period in which contacts were more formal and systematic (mainly under Pannenborg, although De Haan by some is said to have played a vital role in establishing the contacts[xlvii]), we find examples of successful and of unsuccessful transfers. In the first period, the LOCal Oxidation of Silicon (LOCOS) technology for ICs was developed with mainly informal contacts on the level of researchers rather than directors[xlviii], yet transferred quite successfully and has become one of the most successful Philips innovations in the IC field. In the same period, the Plumbicon was developed by Research without a great deal of contact with any industry group, transfer was quite problematic, yet this product too was an enormous success. In the second period, the Video Long Play (VLP) was developed in close contact with the Main Industry Group 'Video', was transferred without problems, but was not successful in the market. It was the follow-up in industry in digital audiodiscs (the compact discs) that were a very successful innovation. From these examples we can see that a more formalised contact with industry groups in itself does not guarantee a successful transfer of research output to industry groups as long as they do not result in mutual obligations, and furthermore a successful transfer does not guarantee a market success.

6. Types of research output that are to be transferred

In this paper we have focused on the relationships between Philips Research and the industry groups within Philips. This is only one of the aspects that can be explored to learn about technology transfer in the context of Philips Research. The limitation to this aspect has to do with the fact that the historical research into Philips Research is still ongoing and other aspects are yet to be dealt with. In fact a more comprehensive view will have to comprise the following aspects:

a. the role of Research in creating ideas for new products,
b. the role of Research in supporting the improvement of existing products and production processes,
c. the role of Research in creating and sustaining a strong patent position that provides the company a certain freedom of action in certain fields.

In addition one can think of the role of Philips Research in making the company stay in touch with technological developments outside the company (e.g. by means of participating in scientific and technological circles).

With respect to the first aspect we have seen that a certain formalisation of the relationships between Research and the industry group did not seem to make a great difference in terms of technology transfer. Our suggested periodisation is valid in that the contacts between Research and the industry groups have changed from ad-hoc and informal to systematic and formal. But the real difference in the transfer of research output to the industry groups seems to depend more on the economical situation. In the period 1950-1970 there were many opportunities for creating new products for new markets because of the economically favourable situation. In this period we see a diversification of the research programme driven by the assumption that Research would yield ideas for new products. Indeed some ideas appeared to be successful, like the Plumbicon for television cameras and the LOCal Oxidation of Silicon (LOCOS) for making ICs. In the period 1970-1990 the economical situation for Philips becomes more problematic and there is a feeling in the general management that 'free' research has not been shown to yield enough output to justify the amount of money that was spent on it, so Research has had to look for a closer relationship with the industry groups in order to work on products with which the company can operate in markets under heavy competition. If we consider all technological developments world-wide it seems that Philips in this period has missed a considerable number of technical opportunities and it is a challenge for our historical research to find out what factors within Research may have contributed to this. It can be expected that transfer of research output became more problematic in this period because industry groups would, even more than in the previous period, hesitate to take up research ideas for further development and production because of the considerable financial risks, in particular for a company like Philips that operates in a rapidly changing mass consumer market.

With respect to the second aspect, the support for improving existing products and production processes, this requires a constant and direct coupling between Research and the industry groups and we have seen that various sorts of committees, and finally the transition to contract research, have been used in the course of time to establish that coupling.

Undoubtedly the third aspect is a crucial one. In terms of historical research it is also a problematic one. Mowery complains about the lack of information on the effects of research in the historiographies of industrial research laboratories[xlix]. The difficulties we have been confronted with so far to find the resources that are necessary for making a survey of the patent position of Philips in the course of time - and in particular the contribution of Philips Research to that position - have made us aware of the difficulties to provide the sort of information Mowery asks for. Further efforts are needed to get a better understanding of this aspect. The importance of using the patents for cross licensing, which probably is one of the most constant contributions of Research to the company, and maybe the most important, certainly justifies these further efforts.

References

1. Cetron, M.J. (1974), Technology Transfer: where do we stand today, in: Davidson, H.F., Cetron, M.J. and Goldhar, J.D. (eds.), *Technology Transfer*, NATO ASI Series E6, Noordhoff, Leiden.
2. Clayton, R. and Algar, J. (1989), *The GEC Research Labs, 1919-1984*. London: Peter Peregrinus, London.
3. European Industrial Research Management Association (1987), *Role and organisation of corporate R&D*, EIRMA, Paris.
4. Fagan, M.D. (1975-1985), *A History of Engineering and Science in the Bell System* (7 volumes). AT&T Bell Laboratories, Indianapolis.
5. Fusfeld, H.I. (1994), *Industry's Future. Changing Patterns of Industrial Research.* American Chemical Society, Washington DC.
6. Gradstein, S. and Casimir, H.B.G. (1966), *An Anthology of Philips Research.* Philips Gloeilampenfabriek, Eindhoven.
7. Heerding, A. and Blanken, I.J. (1980-1977), *Geschiedenis van de N.V. Philips Gloeilampenfabrieken* (4 volumes). Martinus Nijhoff, Leiden.
8. Hounshell, D.A. and Smith, J.K. (1988), *Science and Corporate Strategy: Du Pont R&D 1902-1980.* Cambridge University Press, Cambridge.
9. Mowery, D.C. (1981), *The emergence and growth of industrial research in American manufacturing, 1899-1945.* Unpublished dissertation, Stanford University, USA.
10. Reich, L.S. (1985), *Science and Business at GE and Bell, 1876-1926.* Cambridge University Press, Cambridge.
11. Sarlemijn, A. and Vries, M.J. de (1992), 'The piecemeal rationality of application oriented research. An analysis of the R&D history leading to the invention of the Philips Plumbicon in the Philips Research Laboratories', in: Kroes, P.A. and Bakker, M (eds.), *Technological development and science in the industrial age.* Kluwer Academic Publishers, Dordrecht.
12. Wise, G. 1985), *Willis R. Witney, General Electric and the Origin of US Industrial Research.* Columbia University Press, New York.

NOTES

1. The author wishes to thank Dr. Feye Meijer , former director of Philips' Corporate Research Bureau for his comments on earlier versions of the paper, and Dr. H. van Bruggen and Mr. B. van Gansewinkel of the Philips Concern Archives for making available the archive documents that have been used for writing this paper.
2. Cetron 1974
3. Fusfeld 1994
4. Most research laboratories, comparable to the Philips Research Laboratories already have their history written (for Bell Labs, see Fagan 1975-1985 and Reich 1985, for GE, see Wise 1985 and Reich 1985, for Du Pont, see Hounshell and Smith 1988, for GEC, see Clayton and Algar 1989). For Philips a history of the total company is published every few years and so far four volumes have come out, see Heerling en Blunken 1980-1977. In 1966 a collection of representative articles and some historical introductions were published when the Philips Research Laboratories celebrated its 50th anniversary (Gradstein and Casimir 1966).
5. These two elements are also the basis of the definition by EIRMA 1987
6. Philips started in 1891 in Eindhoven, the Netherlands, as a light bulb producing company. Since then its product range has widened substantially to many fields. From light bulbs it was not a big step to various sorts of tubes (e.g. amplifier tubes, X-ray tubes) and from tube the next logical step was to develop whole systems, of which tubes were a device (e.g. radio, television, X-ray equipment) and thus Philips became involved in the whole process of processing (transmitting and receiving telecommunication signals. Devices remained an important area for Philips also started developing various technologies for recording such signals (gramophones, magnetic tapes, optical recording). Philips is now a multi-national company with a turnover of about 70 billion Dutch guilders (appr. 35 billion Am dollars) and about 250,000 employees (Annual Report 1996).
7. The formal Philips terminology for a long time was Main Industry Groups (MIG's), and later these were renamed to Product Divisions (PD's).
8. Minute of the research directors conference of March 1986.
9. These can be found in the minutes of nearly all concern research conferences since 1952.
10. Interviews Dr. E.F. de Haan, September 9, 1997 and Dr. H.J.G. Meyer, September 16, 1997.
11. Minutes of the crc in 1974
12. Interview Dr. W.F. Knippenberg, April 1, 1997.
13. Overzicht Samenwerking Nat. Lab. – Lichtgroep (Survey Co-operation Research Lightgroup), august 1969 (PCA, box NL-18)
14. Dr. K. Teer (interview September 9, 1997) mentions the videophone as an example of this.
15. Sarkemijn and De Vines 1992
16. Interview Dr. E.F. de Haan, September 9, 1998

17. Interview Dr. E.F. de Haan, September 9, 1998, and interview Dr. K. Teer, September 9,1997.
18. We find this suggestion e.g. in the Minutes of a meeting of personnel with directors (Contact Committee – directors' meeting) on October 20, 1981.
19. From these Programmes we can see that the definitions of what belongs to materials, devices and systems can change in the coure of time. Only in cases where it was evident that a group was shifted from one domain to another without any evident change in the content of the research, done by that group, we have moved back that group when calculating the number of people working in the "materials", "devices" and "systems" domains.
20. The Mega-project is an example of a project, financed partially through the European Community research programme, that was carried out in co-operation by Research and the "Semiconductors" Product Division, and yet did not result in a commercially successful product. More successful was Siemens "4-Megabit Dynamic RAM, that was developed in the same European project.
21. Although this research in the Concern Research Programmes was reckoned with the "materials" domain, the transfer was rather in terms of measuring systems.
22. Interview Dr. W.F. Knippenberg, April 1, 1997
23. Minutes second concern research conference, 1952
24. Interview Dr. I.J. Tummers, April 2, 1997
25. Interview Dr. W.F. Knippenberg, April 1, 1997
26. Dr. F. Meyer, personal communication
27. Interview Dr. K. Teer, September 9, 1997
28. Interview Dr. H.J.G. Meyer, September 16, 1997
29. Clayton and Algar 1989
30. letter H.B.G. Casimir to J.D. Otten, d.d. February 2, 1971 (PCA, box NL-18)
31. letter H.A. Klasens to G. van Gijn, d.d. August 12, 1968 (PCA, box NL-18)
32. Minutes of the crc in 1978
33. Minutes Meeting Research – Elcoma Materials & Components, February 17, 1972, Report "Magnetic bubbles" by E. Olsen of December 22, 1971 (PCA, box NL-18), and Minutes of a meeting of the Co-ordination Committee on May 27, 1971 (PCA, box NL-20)
34. Minutes of the 89th rdc. June 1987
35. Minutes of the 91st rdc, January 1988
36. Interview H.J.G. Meyer
37. Interview Dr. I.J. Tummers, April 2, 1997
38. Mowery 1981

THE FUTURE OF RESEARCH WITHIN THE BIOTECHNOLOGY REVOLUTION

R. FEARS
SmithKline Beecham Pharmaceuticals,
New Frontiers Science Park, Third Avenue, Harlow, Essex CM19 5AW
UK

1. Introduction

During the last 20 years, there has been considerable progress in the contribution made by novel medicines to healthcare, but there are still major, unmet challenges to face in selecting and delivering optimum healthcare strategies. Biotechnology, and in particular genomics R&D, will aid our understanding of the disease processes underlying both single gene disorders and multifactorial disease. In consequence, we will be better placed to develop curative rather than symptom-relieving drugs. This will be revolutionary. In addition to the value for the patient, there are significant economic benefits to be gained from a new approach to drug discovery. The annual cost of uncured diseases in the US is estimated at.$100 billion or more each for Alzheimer's disease, diabetes, cancer and cardiovascular disease; $40-50 billion each for arthritis and depression and $20-30 billion each for osteoporosis and stroke.

The challenges for healthcare delivery and the opportunities arising from advances in genomics R&D are revolutionary in several respects and will be discussed in the context of.

- investment and output by pharmaceutical and biotechnology companies
- innovation and competitiveness in Europe• what industry needs from public policy
- public-private partnership in R&D
- public confidence and ethical concerns

My perspective, from a global, R&D-intensive healthcare company, addresses only selected, strategic issues relating to the translation of invention into innovation. The process of innovation is a complex one, with multiple feedback loops, and it is now widely recognised that the early linear models imperfectly described the relationships between invention and innovation. Many of the other contributions to this Workshop have analysed practical aspects relating to the mechanics of technology transfer and provided case histories with which to explore best practice. Further discussion on the role of some of the key players in technology transfer in biomedical research - for example bioincubators for early work-up of assets, research-funding councils in catalysing collaborative research, venture capital providers, intermediaries in strategic partnering -and a full taxonomy of industry-academia relationships can be found in other recent publications [1,21

A. Inzelt and J. Hilton (eds.), Technology Transfer: From Invention to Innovation, 161–170.

2. Investing to Shape the Evolution of Healtheare

It has been described [3] how the future of healthcare will be driven by four factors:

- economics
- genomics
- informatics
- ethics

Boundaries between scientific disciplines are disappearing and technology fusion is becoming ever more important. The medical application of genomics will depend intimately on advances in informatics. Furthermore, the use of genetic information is raising several important ethical and social issues - relating to equity, autonomy, responsibility, privacy and confidentiality [41

Pharmaceutical and biotechnology companies make considerable, long-term investments into R&D in order to deliver novel products and services. According to data collected by the Centre for Medicines Research [51, current global pharmaceutical plus biotechnology company healthcare-related R&D expenditure (excluding capital investment) exceeds $40 billion annually with biotechnology companies contributing about $8 billion to this total. In the UK, in consequence of this investment, the pharmaceutical industry now accounts for more than 70% of the total healthcare-related R&D expenditure of about £3 billion, (the Research Councils provide less than 20%, the medical charities about 10%).

3. European Innovation and Competitiveness

Public policy-makers in most of the OECD nations have acknowledged the important role of science and technology in pursuit of national goals of wealth and health creation. The UK is one of the strongest performers in European biosciences and recent growth of the biopharmaceutical SME sector has been significant [6]. But, Europe still lags the US in commercialising ideas by biotech SMEs. Recent data collected in a benchmarking study by EuropaBio [61 demonstrate: fewer companies in Europe (700) than the US (1300); fewer publicly-quoted companies (50 in Europe, 300 in US); lower R&D spend (1,500 mECU in Europe, 6,300 in US); and much lower turnover (1,700 mECU in Europe, 11,700 in US).

Considerable attention is now being devoted in European policy circles to the issues involved in growing European competitiveness (Table 1).
TABLE 1. Ranking of importance of external
factors for competitiveness in Europe
Analysis modified from [61

Competitiveness factor	Biotech SME	Pharmaceutical multinational
Intellectual property rights	+++	+++
Regulatory framework	+++	+++
Entrepreneurship	+++	+
Equity capital	+++	+
Consumer acceptance	++	+++
Science base/skilled staff	+++	++

Clearly, SMEs and large companies employing biotechnology as an enabling technology share many of the same competitiveness drivers. Attention to these factors will be crucial in establishing a favourable operating environment for R&D at a time of ever-increasing pace of change in both technologies and markets.

4. Industry User Needs from Public Policy

Although progress has been made, there is considerably more to be done to understand the critical success factors in innovation, technology transfer and competitiveness [2,3,61. And, policy makers must recognise the growing complexity of sustaining an attractive environment for world-class R&D. The scale of the opportunities in healthcare research is such that it is now imperative to seek a coherent public strategy to link science and medicine, to generate a suitably trained workforce and an environment conducive to innovation. The view of the pharmaceutical industry user is articulated in greater detail in a recent manifesto [71 from a consortium representing virtually all interests in the UK pharmaceutical sector (Table 2).

TABLE 2 What industry needs from public policy
Modified from Pharmaceutical Industry Manifesto [71

View of the user

- prior consultation and integrated long-term strategy across government departments
- ,sound economic base and regulatory environment
- commitment to promoting best use of medicines
- evidence-based treatment
- supporting IPR
- improving education and training

In Europe, there is significant recent progress in sharing commitment to innovation in consequence of hard work by the Commission, its advisers, Members of the European Parliament and bioindustry. The Green Paper and Action Plan on Innovation set out a strategy to build competitiveness that has been greatly supported by the important parliamentary vote in favour of the proposed Directive on biotechnology patenting. This harmonising Directive allows us to retain the ability to invent and innovate in biotechnology and helps to build the encouraging environment for long-term investment in research and the commercialisation of innovation.

Development of the objectives and scope for Framework Programme 5 funding of European R&D has also been important in addressing key issues for biosciences research. In operational terms, strategy development for Framework funding identifies the importance of promoting selectivity and focus in outlook, integrated with streamlined management and clear communication. In terms of scope, there is still concern that the large, overarching themes may be unwieldy. There is a case to be made for separating biomedical and environment themes and for giving greater emphasis to medicinal chemistry at the interface with molecular biology. And, in terms of the needs of the user community, is there undue emphasis on SMEs? There is European added value in encouraging further collaboration by innovative pharmaceutical companies.

5. Technology transfer

The mutual and complementary interests of large and small bioscience companies can lead to specific partnerships for technology transfer [1,8]. Of the current population of 1300 US biotech companies, only 3 market their pharmaceutical product under their own label and stay independent (Amgen, Genzyme, Biogen). One-third of the companies in the biotech sector have less than one year of cash remaining and onehalf of total funding comes from partnership with large pharmaceutical companies.

What new types of strategic initiative might be foreseen? The genomics era was inaugurated by the strategic alliance formed in 1993 between SmithKline Beecham and Human Genome Sciences. This partnership has now expanded to embrace Schering-Plough (US), Merck KGa-A (Germany), Takeda (Japan) and Synthelabo (France) to maximise the use of genomics information in furnishing discovery leads. Multilateral collaborations of this type may represent a new paradigm in technology transfer and entrepreneurship. Another new paradigm for the biosciences is exemplified by the creation of a start-up company by the spin-off of excess assets from a multinational parent. AdProTech is a protein technology spin-off company created recently by SmithKline Beecham to capitalise on novel approaches - using complement protein as the primary target - to developing treatments for thrombotic and inflammatory disorders. SmithKline Beecharn contributed patents and laboratory equipment receiving, in turn, 10% equity stake and status as a preferred development partner. The advantage in creating this new entity is that additional funding for R&D has been attracted from venture capital providers.

6. Foresight: **Integrating Technology Push and Market Pull**

In addition to specific, individual collaborations, partnership for technology transfer and the translation of invention to innovation also requires strategic liaison between industry, academia and government. One example of an UK government initiative to develop joint research strategies to address shared goals is provided by the Technology Foresight Programme, commenced in 1993 [9]. The objectives of this exercise were to work towards national goals to increase wealth creation and improved quality of life by developing new working partnerships between science and industry and by commitment to prioritise, and inform, decisions on public funding.

Details of the Programme and the involvement of industry are discussed elsewhere [101 and we have concluded that this was a highly important exercise for health and life sciences R&D. Successful Programme achievements can be counted in terms of the formation of new partnerships (linking academia, SMEs and multinationals) and in the highlighting of generic priorities. Among the key topics areas identified, where further work is vital, are genetic and biomolecular engineering and bioinformatics. It is arguable that the Foresight exercise has been less successful, to date, in influencing the allocation of public funds for research and in shaping the policy of individual government departments and their cross-departmental coherence. There are continuing opportunities here both for biotech SMEs and pharmaceutical multinationals to engage in policy dialogue and in driving wider participation to ensure that momentum in implementation is not further reduced.

7. Emerging Product Classes from Genomics R&D

The first recombinant medicine - human insulin - was marketed in 1982. Since then, it has been possible to identify the emergence of distinct product classes: biopharmaceuticals, including hormones, growth factors, enzyme replacement therapies and modified proteins vaccines; the recombinant era was initiated by hepatitis B vaccine and is now developing to address new target populations (for example, travel market, adolescents) and beyond the defence to bacteria and viruses (for example, malaria) gene therapy, still characterised by technical and clinical challenges [11] molecular diagnostics, a new era exemplified by the SmithKline BeechamIncyte joint venture diaDexus to develop products for improved detection, disease staging and patient stratification and pharmacogenomic tests to optimise treatment selection.

But, genomics R&D will also accelerate the development of New Chemical Entities as innovative medicines in consequence of the identification of specific human genes that cause disease. One recent example is the work, to identify inhibitors of osteoclast protease cathepsin K [121, that contributes a new approach to the potential treatment of osteoporosis.

The therapeutic promise of genomics R&D will not be confined to the study of human genes. The identification of bacterial genes has inaugurated a new era in understanding virulence, with the prospect for combating the problems arising from bacterial resistance to the present generation of antibiotics. The pace of discovery in sequencing and characterising microbial genomes is dramatic [131.

8. Creating Receptive Climate and Strengthening Markets

Strategic decisions for R&D must be contemplated not only amidst dramatic technical and market change but also at a time of dramatic social change. The potential to apply genomics R&D to improve human health will be influenced as much by the social environment in which scientific advances occur and in which they are to be applied as by research progress *per se* [141. In this context it is important to:

- enhance the public debate on societal concerns - promoting public awareness of R&D benefits and countering anti-technology activist pressures
- establish public confidence by defining the appropriate balance of professional standards and regulations
- inform dialogue on ethical, legal and social issues arising from biotechnology R&D

Genomic advances and claims evoke a variety of social and ethical concerns. Among the recent issues that have generated controversy and challenges for both the scientific and policy-making communities are: biotechnology patenting, privacy of genetic information, insurance and employment discrimination, eugenics, germline gene therapy, transgenics, cloning. Ethical principles need to be translated into a science-based system of regulation to ensure enablement rather than inhibition of development.

Surveys of public opinion (for example 1996 Eurobarometer [151) indicate a lack of confidence in the capacity of national political institutions to regulate biotechnology. It is important, therefore, that national initiatives to promote ethical considerations in genomics, R&D and the use of genetic information are supported by the pan-national initiatives currently being developed by UNESCO (International Bioethics Committee), the Council of Europe and the European Commission and Parliament We must generate informed international debate on these issues for, without a supportive public climate, we cannot be confident about the future of research within the biotechnology revolution.

References

1. IBC Conference (1997) *Technology Transfer in Biomedical Research,* Scrip Reports, London.
2. Anderson, L, Fears, R., and Taylor, B. eds. (1997) *Managing Technology for Competitive Advantage, FT* Healthcare, London.
3. Poste, G. (1997) Preface, in J. Anderson and R. Fears (eds.) *Sustaining the Strength of the UK in Healthcare and Ufe Sciences R&D: Competition Cooperation and Cultural Change,* ScienceBridge, Surrey, p.4.
4. Fears, R., Greig, R., and Poste, G. (1994) Human genomic research and innovative medicine, *Science in Parliament* 51, 4-5.
5. Lumley, C.E., and Dorabjee, S. (1996) *Trends in Pharmaceutical R&D,* Centre for Medicines Research, Surrey.
6. Business Decisions Limited and Science Policy Research Unit (1997) *Benchmarking the Competitiveness of Biotechnology in Europe,* EuropaBio, Brussels.
7. The Pharmaceutical Industry Council (1997) *Pharmaceutical Industry Manifesto,* London.
8. Stevens, T. (1997) The gene machine, *Industry Week* 16 August.Technology Foresight Steering Group (1995) *Progress Through Partnership,* Office of Science & Technology, London.
9. Fears, R., and Poste, G. (1995) Enabling the future: An industry perspective on the Foresight science policy exercise in the life sciences, *Science, Technology and Innovation* October, 26-30.
10. Verma, IM, and Somia, N. (1997) Gene therapy - promises, problems and prospects, *Nature 389,* 239-242.
    Zhao, B., Janson, C.A., Amegadzie, B.Y. D'Alessio, K., Griffin, C., Hanning, CK, Jones, C., Kurdyla, L, McQuency, M., Qiu, X., Smith, W.W., and Abdel-Meguid, S.S. (1997) Crystal structure of human osteoclast cathepsin K complex with E-64, *Nature Structural Biology 4,* 109-111.

Pennisi, E. (1997) Laboratory workhorse decoded, *Science 277,* 1432-1434.
Fears, R., Ferguson, M.W.J., Stewart, W., and Poste, G. (1997) Life-sciences R&D, national prosperity, and industrial competitiveness, *Science 276,* 759-760.
Biotechnology and the European Public Concerted Action Group (1997) Europe ambivalent on biotechnology, *Nature 387,* 845-847.

(i)Most economic theories concerning the production and exchange of products and value, which were build with reference to industrial activities (the same way the physiocrats had been building their theories with reference to agricultural activities), with the exclusion of service activities (seen as non productive by A. Smith), are not or at least to a large extent not relevant any more when it comes to informational service activities (J.De Bandt, 1995,1996, 1997).

(ii) This last point is more complicated than it seems. While the above assertion is correct in current prices, data in constant prices usually show the opposite. This is because of the way the products of tertiary activities are usually measured in national accounting systems : because of the absence, in the case of many services, of a unit product (on the basis of technical specifications) and thus of a unit price, product price evolutions are more often than not approximated on the basis of labor input prices, which means that the passage from current to constant prices is rather arbitrary. Even if this may seem debatable form the standpoint of evolution, at one and at different points of time, the relative importance of the different activities is of course given in current prices.

(iii) This is so mainly because of differences in productivity growth. As a matter of fact, because of what is said in the previous footnote, official statistics are usually underestimating products and thus productivity growth in services. But I cannot go into this here (J.De Bandt, 1989)

(iv) Relative productivity is defined as the share of a specific unit or sector in GDP divided by the share of that unit or sector in total employment . This is the usual productivity measure, in which both terms are divided by the corresponding total .

(v) One has to take account that the differences are minimized, due to the fact that industrial activities include that part of « informational services » which is internalized. These internal informational services are known to benefit of more rapid pay increases than the average.

(vi) For example Baumol (1989) splits the « information jobs » in two categories : data jobs and knowledge jobs.

(vii) I don't want here to go into lengthy (and somewhat tedious) discussions around several more or less artificial distinction : between knowledge as a stock and information as a flow, or between different levels of knowledge, or between theoretical and practical knowledge.... We are speaking here of knowledge as referring to the cognitive elements needed for solving a problem in the range of doing or producing something.

(viii) In various countries or regions, there is much talk about the importance of some competence blocs or poles, which are defined in terms of the quantitative importance or volume of skills and university degrees which are concentrated. This is of course not enough. What is important is the effective possibility and the ways and means by which such skill resources are effectively transformed into the capability to solve new production problems.

(ix) Of course, big firms can adopt decentralized organizations, within which decentralised or strategic units have large degrees of autonomy. But, while there are of course exceptions, the tendency remains strongly in favor of the integration of all activities in rather centralized organizational set-ups, with huge bureaucratic rigidities and costs. Big firms can also create « virtual » companies, outside their organizational set-up, but such virtual companies are more than often used precisely in order to assemble complementary resources and competencies from different firms.

(x) Cases are well known of big firms (this is not totally excluded for small ones) whose strongly focused competencies and deeply entrenched culture make it nearly impossible to enter new technological fields or evolutions.

(xi) Cases of strong rivalries and power conflicts between individuals, groups, departments.. within big firms are of course not rare situations, making often that co-operations are difficult to organize and appear to be easier outside than within the firm.

The hypothesis is usually made that within firms co-operations among various people and competencies are automatic and result from the very nature of the firm. Even while the firm can be seen as a place where cooperation is induced (because of contracts and prescriptions), or even to a certain extent imposed (because of possible sanctions), and facilitated (because of proximity of objectives and activities, common language and culture, lower transaction costs), there is no reason why within a firm cooperation would be spontaneous or easy. The firm is essentially organizing some division of labor between the various actors, within a well defined production process. When it comes to producing complex knowledge for answering new problems or questions, the question is how to bring the competencies to work together towards some common goal which is not known beforehand and which the actors are not even sure they can attain. While it is easy to control that the actors are doing correctly that part of the processes which has been attributed to each of them, there is of course no reference basis for evaluating why some expected but unknown new (not predefined) result has not been obtained (J. De Bandt, 1996).

[xii] Cetron 1974.
[xiii] Fusfeld 1994.
[xiv] Most research laboratories, comparable to the Philips Research Laboratories, already have their historiography written (for Bell Labs, see Fagan 1975-1985 and Reich 1985; for GE, see Wise 1985 and Reich 1985; for Du Pont, see Hounshell and Smith 1988; for GEC, see Clayton and Algar 1989) For Philips a historiography of the total company is being published every few years and so far four volumes have come out, see Heerding en Blanken 1980-1977. In 1966 a collection of representative articles and some historical introductions were published when the Philips Research Laboratories celebrated its 50th anniversary (Gradstein and Casimir 1966).
[xv] These two elements are also the basis of the definition by EIRMA 1987.
[xvi] Philips started in 1891 in Eindhoven, the Netherlands, as a light bulb producing company. Since then its products range has widened substantially to many fields. From light bulbs it was not a big step to various sorts of tubes (e.g. amplifier tubes, X-ray tubes) and from tubes the next logical step was to develop whole systems, of which tubes were a device (e.g. radio, television, X-ray equipment) and thus Philips became involved in the whole process of processing (transmitting and receiving telecommunication signals. Devices remained an important area for Philips, especially in terms of transistors and later ICs. Once involved in sound and sight, Philips also started developing various technologies for recording such signals (grammophones, magnetic tapes, optical recording). Philips is now a multi-national company with a turnover of about 70 billion Dutch guilders (appr. 35 billion Am. dollars) and about 250 thousand employees (Annual Report 1996).
[xvii] The formal Philips terminology for a long time was: Main Industry Groups (MIG's), and later these were renamed into Product Divisions (PD's).
[xviii] Minutes of the research directors conference of March 1986.
[xix] These can be found in the minutes of nearly all concern research conferences since 1952.
[xx] Interviews dr. E.F. de Haan, September 9, 1997, and dr. H.J.G. Meyer, September 16, 1997.
[xxi] Minutes of the crc in 1974.
[xxii] Interview dr. W.F. Knippenberg, April 1, 1997.
[xxiii] 'Overzicht Samenwerking Nat.Lab.-Lichtgroep' ('Survey Co-operation Research-Lightgroup'), August 1968 (Philips Company Archives, PCA, box NL-17) and letter H.A. Klasens to A. Dros d.d. October 14, 1968 (PCA, box NL-18).
[xxiv] Dr. K. Teer (interview September 9, 1997) mentions the videophone as an example of this.
[xxv] Sarlemijn and De Vries 1992.
[xxvi] Interview dr. E.F. de Haan, September 9, 1997.
[xxvii] Interview dr. E.F. de Haan, September 9, 1997, and interview dr. K. Teer, September 9, 1997.
[xxviii] We find this suggestion e.g. in the Minutes of a meeting of personnel with directors (Contact Committee - directors' meeting) on October 20, 1981.
[xxix] From these Programmes we can see that the definitions of what belongs to materials, devices and systems, can change in the course of time. Only in cases where it was evident that a group was shifted from one domain to another without any evident change in the content of the research, done by that group, we have moved back that group when calculating the number of people working in the 'materials', 'devices' and 'systems' domains.
[xxx] All percentages are rough estimates, because in some cases it was not quite clear to which area certain groups should be reckoned. Besides that the number in the Concern Research Programmes of the Fifties deviate rather strongly from the numbers in the early Sixties for unclear reasons.
[xxxi] The Mega-project is an example of a project, financed partially through the European Community research programme, that was carried out in good co-operation by Research and the 'Semiconductors' Product Division, and yet did not result in a commercially succesfull product. More succesfull was Siemens' 4-Megabit Dynamic RAM, that was developed in the same European project.
[xxxii] Although this research in the Concern Research Programmes was reckoned with the 'materials' domain, the transfer was rather in terms of measuring systems.
[xxxiii] Interview dr. W.F. Knippenberg, April 1, 1997.
[xxxiv] Minutes second concern research conference, 1952.
[xxxv] Interview dr. L.J. Tummers, April 2, 1997.
[xxxvi] Interview dr. W.F. Knippenberg, April 1, 1997.

[xxxvii] Dr. F. Meyer, personal communication.
[xxxviii] Interview dr. K. Teer, September 9, 1997.
[xxxix] Interview d. H.J.G. Meyer, September 16, 1997.
[xl] Clayton and Algar 1989.
[xli] letter H.B.G. Casimir to J.D. Otten, d.d. February 2, 1971 (PCA, box NL-18).
[xlii] letter H.A. Klasens to G. van Gijn, d.d. August 12, 1968 (PCA, box NL-18).
[xliii] Minutes of the crc in 1978.
[xliv] Minutes Meeting Research - Elcoma Materials & Components February 17, 1972, Report 'Magnetic bubbles' by E. Olsen of December 22, 1971 (PCA, box NL-18), and Minutes of a meeting of the Co-ordination Committee on May 27, 1971 (PCA, box NL-20).
[xlv] Minutes of the 89th rdc, June 1987.
[xlvi] Minutes of the 91st rdc, January 1988.
[xlvii] Interview H.J.G. Meyer.
[xlviii] Interview dr. L.J. Tummers, April 2, 1997.
[xlix] Mowery 1981.

CHAPTER III

THE ROLE OF CODIFIED KNOWLEDGE (PATENTS) AND OTHER INTELLECTUAL PROPERTY RIGHTS

EVOLVING SYSTEMS OF INTELLECTUAL PROPERTY RIGHTS
COLLABORATIVE R&D AS A GENERATOR OF NEW IP STRUCTURES

Hugh Cameron
PREST, University of Manchester
Oxford Road, Manchester M13 9PL United Kingdom

ABSTRACT

Existing frameworks of intellectual property rights (iprs) are the result of centuries of evolution. Compromises between the rights of originators and users of newly generated intellectual property (IP) have resulted in diverse national systems of protection, resulting from different emphases on rights and objectives. New technologies have created the need for new forms of IP, and regulations have increasingly struggled to cope with the rapidly changing technological environment. However, the growth in size and importance of international trade, and the growth of "knowledge-based" industries, has created the need for harmonisation of frameworks, thus creating a "selection environment" in which unified regulations will be accepted throughout the world. But IP frameworks consist of more than the formal corpus described in law textbooks. This paper considers the growth of collaborative research and development as a generator of new frameworks. These rely on the existing formal ipr systems, but are at present creating a new, additional set of practices and procedures, themselves evolving to meet the needs of participants in collaborative projects and programmes.

Key words: intellectual property rights, collaborative R&D, foreground, background, European Union.

1. INTRODUCTION

Intellectual property (IP) is increasingly recognised to be of fundamental importance in the modern "knowledge-based" economic system. In the commercial and industrial sectors, companies have long devoted resources to the protection of their IP, but it is evident that the newer and faster growing "high technology" industries (especially bio-

A. Inzelt and J. Hilton (eds.), Technology Transfer: From Invention to Innovation, 173–193.

technology and computer-related enterprises) consider issues of IP to be critical to their existence. Public research institutions, such as universities and public laboratories, have in the last decade made strenuous efforts to protect and exploit their outputs, while governments have come to see international structures of IP protection as essential elements of economic competitiveness, or as necessary conditions for access to the global economic system. As a result, international trade negotiations have identified IP protection and harmonisation as issues in the forefront of international economic relations.

In parallel with this trend is the related phenomenon of **collaboration**. The term has become associated with the field of research and development over recent decades, but is still ambiguous: it has been used with reference to work carried out jointly between independent commercial enterprises, between universities and commercial enterprises, between concerns in different countries, and between defence and non-defence organisations (and also to any combination of these). This collaboration takes many forms; some is entirely on a customer-contractor basis, much is sponsored or funded from public sources. The incentives and the information transfers are usually complex, including the whole range of IP from the formally identified and legally protected to the informal, tacit knowledge and skills which are so difficult for economists and lawyers to define or quantify.

This paper has the intention of providing a brief review of the interaction between these two modern phenomena, describing the main issues which have arisen in the field of collaborative research and development, and the frameworks and structures which have been invented or have evolved to cope with the difficulties encountered by participants in this type of endeavour.

The next section reviews the legal and regulatory structures, much of which will be well known to most participants in the field, and, more unusually, the subsequent section will describe the system of practice which is evolving (using the European Union as the example) to provide a basis for collaborative work. The conclusion of the paper is that international collaboration requires as a necessary condition the provision of a stable, predictable and economic system of rights and enforcement, but that IP issues are not a purely a legal concern to be left to lawyers and patent agents; they are a major element of business strategy and should be recognised as such by regulators and policy makers.

2. INTELLECTUAL PROPERTY RIGHTS

Article 2(viii) of the Convention establishing the World Intellectual Property Organisation (14th July 1967) defined Intellectual Property to include:

> "....the rights to literary, artistic and scientific works; performances of performing artists: phonograms, and broadcasts; inventions in all fields of human endeavour; scientific discoveries; industrial designs; trademarks, servicemarks, and commercial names and designations; protection against unfair competition; and all

> other rights resulting from intellectual activity in the scientific, literary or artistic fields."

The main accepted varieties of ipr are: patents, copyright, designs, trademarks confidential information, and more recently accepted, plant varieties. Patents protect inventions generated by individuals and businesses, while copyright protects the products of the film, music and publishing industries. A body of intellectual property law has been developed at a national level for centuries in the countries of Europe, and this has spread to many other parts of the world, though with important differences between national systems

Legal protection of intellectual property predates the development of the capitalist system; in England patents were first granted in 1449 to protect the manufacture of coloured glass (a newly imported technology) for Eton College windows, lasting for 20 years[xlix]. Trade marks were used for centuries before this, though they received legal protection in the UK only in 1875[xlix]. Though legal monopolies in printing copies of works were established in England in 1529, the world's first true "copyright" legislation was enacted in 1709 [xlix],for:

> ".....the Encouragement of Learning by vesting the Copies of Printed Books in the Authors or Purchasers of such Copies...."

The common feature of these and the subsequent evolution of all intellectual property rights throughout the world, has been the need to respond to new innovations in technology, and to exploit them to the benefit of their innovators and society as a whole.

The rules have been framed with the intention of achieving acceptable compromises between the interests of those generating and those using the IP, in particular:

- to protect the "moral rights" of the inventor or originator,
- to allow the originator to realise a return on the activities producing the IP and prevent "free riders" from exploiting unfairly, by permitting a "temporary monopoly" over the use of IP,
- to encourage the efforts of individuals and companies in pursuing innovative activities which will benefit society as a whole, and
- to encourage the diffusion of new knowledge, techniques, concepts and innovations throughout society, contributing to its general improvement.

As any compromise may be viewed as a negotiated trade-off between the interested parties, comparison of the different national systems of IP protection reveals considerable diversity in provisions, as different emphases are given to each of these objectives. The U.S. and Europe have generally given more emphasis to inventors' rights, compared with Japan, which has concentrated on industrial development by encouraging the dissemination of technologies.

Different patent definitions and processes have also emerged (which will be considered below), the most well-known variation being the priority given to time of invention under U.S. law ("first-to-invent"), while the rest of the world follows a "first-to-file" system.

In addition to the national systems, international agreements have been made, beginning with the Paris Industrial Property Convention of 1883 and the Berne Convention for the Protection of Literary and Artistic Works of 1886. Subsequent international treaties and agreements have been made at intervals since then, addressing the problems of incompatibility between systems, so that international trade may be continued on an equitable basis. In recent years, particular milestones have been the Patent Co-operation Treaty (in force from 1978) founding the World Intellectual Property Organisation, and the Agreement on Trade-Related Aspects of Intellectual Property Rights (TRIPS), which came into force in January 1995.

The process by which we have arrived at the existing international system is evolutionary, in the sense that changes have occurred over time as a result of the generation of diversity in forms of IP and its protection, and there has existed a "selection environment" in which national and international negotiations have attempted to harmonise the provisions of formal protection by means of the elimination of idiosyncratic or outdated rules. In addition, existing or proposed rules have been abandoned or not adopted as they have been superseded by systems preferred by users; in Europe, national patent systems are gradually being replaced by the European system based in Munich, but the proposed Community-wide system has not at present been ratified by the required number of member states, and has thus not come into use

The main generator of diversity has been, over the centuries, the development of new technologies. Printing resulted in copyright law, and this has been applied to computer software, a technology with radically different technical and economic characteristics. As with other areas of law and policy, regulation has struggled to keep up with the changing technological environment.

On the selection side, there has been pressure on national authorities to simplify procedures and reduce costs of patents in particular, but the main pressure has come from the international arena. The increasingly internationalised economy has made IP protection a first rank trade issue, and produced lengthy and often acrimonious negotiations at the global level. In brief, the issues addressed by the negotiations on patents have been[xlix xlix xlix]:

- the language of applications
- examination of patents
- opposition procedures
- term of patents
- grace period (disclosure before application)
- scope of patents
- costs of procedures
- enforcement procedures and costs

2.1 An Example of Difficulties in International Patenting.

It is appropriate in this paper to give an example of the fundamental differences in national patenting systems using the example of Japanese patents. The scope of all patents distinguishes the invention from "prior art". Compared with U.S. practice, Japanese patent practices are aimed at restricting patent claims' scope as much as possible, including limiting the scope of protection to the specific examples provided in an application. These examples must demonstrate the claims and results, while U.S. applications include broader claims without examples. In Japan, these examples are particularly important in the fields of chemical, bio-technological, and pharmaceutical fields, where actual physical data is required for all compounds covered by a claim. The examiners will not accept theoretical examples, unlike their U.S. counterparts. The U.S. view is that this makes it virtually impossible for pioneering inventions in these fields to be adequately protected. Inventions in the mechanical and electrical fields do not require such detailed examples, where drawings are sufficient.[xlix]

These observations fit well with the general feeling that Japanese patenting practice promotes large numbers of incremental inventions, while U.S. firms tend to apply for fewer, but more radical invention protection. Though the case above has been from a U.S. view, it seems that the frictions are caused by fundamentally different approaches to patenting in the two business and administrative cultures. An alternative view could be given from the other side: Japanese patent experts consider that U.S. applications do not conform to Japanese styles and procedures. More than 90% of U.S. applications in Japan are virtually identical to their U.S. applications. It seems to be easier, and less costly to do this, and to revise them at the request of the JPO, thus lengthening the period of waiting.

In addition to these, the views of ipr litigation are very different in the two countries. In Japan, litigation is considered to be an extreme and unwelcome action, entailing a loss in business reputation. Infringement suits are rarely considered; negotiated settlements are almost always concluded. In the U.S., a far more litigious country in all areas of the law, actions are far more common. In 1990 there were 1,236 patent infringement suites in the U.S., compared with 141 in Japan. Little business risk is involved in bringing an action in the U.S., and to outsiders it may even seem that infringement, and litigation, are seen as legitimate and acceptable business strategies.

The damages awarded in the two countries also differ considerably. In the U.S. triple damages are sometimes awarded, while in Japan the awards are little more than the license fees which would have been payable. Companies are often better advised to settle for negotiated license fees rather than pursue litigation.
A brief comparison of the three main patenting systems is given below in Table 1.

TABLE 1: Comparison of patenting systems

Features of System	European PO	U.S. PTO	Japan PO
Basis	first to file	first to invent	first to file
Patents for discoveries?	no	yes	no
Breadth of claims	narrow	broad	narrow
Grace period	none	12 months	6 months*
Speed of processing claim	slow	fast	slow
Filing permitted in any language?	any European language	yes	no
Are patent examinations published?	18 months after filing	no, secret until patent is granted	18 months after filing
Can patent examination be deferred?	yes, for 6 months	no	yes, for 7 years after 18 month publication
Is there an opposition system?	yes, after patent is granted	no, but other parties may request examination	yes, before patent is granted
Patent term	20 years from first filing	17 years from grant	15 years from date of publication for opposition purposes

Negotiations concerning these problems have been difficult in themselves, however they have come at a time when national systems themselves are under considerable strain due to the new technologies becoming commercially important.
Two examples will suffice to demonstrate these problems.

In the bio-technology field, the different approaches to IP and protection for the newly emerging industry have led to conflicts between the US and Europe over the "patenting of life". In the EU itself there have been prolonged discussions and disagreements over the approach to take. The European Parliament rejected the European Commission's draft directive for the patenting of genetically altered organisms and other bio-technological inventions, affecting the development of new pharmaceuticals. The issues at stake were the alteration of inheritable genetic characteristics, the patenting of elements of the human genome, and genetic manipulation of animals.

In the software field, in general the copyright laws are used to protect IP, but there has been increasing debate over the extension of patents to cover software. The US has permitted patenting of certain algorithms, which were previously thought to be outside the scope of patent law.

The different capabilities of countries in these sectors, and forecasts of their future economic significance, are leading to conflicting negotiating positions in international discussions. Though the tendency to international harmonisation is powerful, it must be remembered that there are significant pressures which produce new conflicts in national systems.

In summary, the following issues in international patent conflicts may be identified:

- Conflicts between national systems, including some fundamental conflicts, e.g. first-to-file vs. first-to-invent. Though the US is the main exponent of the latter procedure, changes due to the TRIPS agreement have allowed foreign dates of invention to be considered. However, the differences in balances between the rights of inventors and the objective of diffusion are still in evidence.
- The costs of obtaining patents has been referred to above. While the increasing harmonisation of patents can be expected to reduce these disparities in some areas, others (such as the propensity to undertake litigation) may remain.
- Conflicts regarding territorial jurisdiction. There are inconsistencies between the systems regarding location of the inventive activity and location of the applicant for a patent. Some require their nationally registered firms to file domestically, while others require first applications when the invention is made under their jurisdiction. In an age where R&D facilities are distributed globally by multinational firms, and where electronic information exchange makes international collaboration a necessary and routine activity, these restrictions must be seen to be increasingly inappropriate, and unworkable.

2.2 The TRIPS Agreement

The TRIPS Agreement has made a major step towards addressing the issues of international patent harmonisation. Some of the principal conclusions were:

1. For patents, TRIPS imposes obligations on the minimum term of protection (20 years), protection for products and process, allowable exclusions from protection, prohibitions on some forms of discrimination, protection of plant varieties, guidelines for compulsory licensing, and on evidence for infringement proceedings.
2. Members of the WTO which are not signatories of the Paris Convention are required to comply with Paris Convention provisions.
3. Developing countries have been allowed several years (until the year 2000) to enforce the provisions of the TRIPS Agreement. Developed countries undertook to apply the provisions by the beginning of 1996.
4. Signatories to TRIPS must accord "most-favoured-nation" treatment to all WTO members, with no possibility of delayed enactment.
5. TRIPS extends the Paris Convention's prohibitions on the use of confidential data in a manner contrary to honest business practices. For example test data submitted to governments during the course of gaining approval for pharmaceuticals, may not be revealed or used for other purposes.
6. The TRIPS Agreement provides for obligations to ensure that effective enforcement of iprs is available to owners, and that enforcement procedures do not create barriers to trade. Certain remedies must be available, including injunctions, damages, forfeiture, performance measures of enforcement authorities, and criminal penalties for wilful trademark and copyright piracy.

In the two years since TRIPS came into force, many countries have made strenuousJ efforts to implement its provisions. For example Korea has instituted many measures as a result, including:

- harmonisation of durations of patents,
- expansion of the coverage of patent regulations,
- expansion of the scope of patents,
- arbitration for non-exclusive licenses,
- restrictions on non-exclusive licenses in the semi-conductor field,
- increased penalties for infringement,
- introduction of a post-grant opposition system,
- etc.

In addition, Korea has instituted measures to protect trademarks (including those only known outside Korea, which were previously excluded), protection of three-dimensional trademarks, and the establishment of a Patent Trial Institute, and a Patent Trial High Court [xlix].

Many signatories have, however, proven slow in their responses to the Agreement. Fundamental difficulties remain, and there is little likelihood of their being removed in the near future, for example the costs of litigation, and penalties imposed, in the US courts are an integral part of that country's broader legal system and will not be altered by international harmonisation agreements in the ipr arena.

This section has attempted to show that there are many factors generating change in national and international ipr systems, in particular the development of new technologies, and the need for international agreement. The next section examines one particular influence on ipr treatment.

3. IPRS IN COLLABORATIVE RESEARCH AND DEVELOPMENT

The European Union's programmes of support for collaborative research and technology development (RTD), collectively known as the Framework Programme (FP), are the largest systematic activity of its kind. The First Framework Programme had a budget of 3.75 billion ecu over its lifetime (1984 - 1988)[xlix]. The latest, Fourth Framework Programme (1994 - 1998) has a budget of over 13 billion ecu. It covers areas as diverse as manufacturing technology, computing and communications, new materials, nuclear research, climate and the environment, economic and social research, and marine technology

Over the years of its existence, the FP has evolved a set of practices concerning intellectual property involved in collaborative RTD. The sources of these are:

- the legal ipr setting, including national, European (EPO) and Community regulations, as well as international agreements

- the minimum requirements of the European Commission (EC), with the objectives of ensuring that the publicly supported work gives value for money in its outcomes, including commercialisation
- the objectives of the participants in projects, including commercial concerns and public research institutions including universities.

The overall economic and political aims of EU RTD policies are to improve the competitiveness of industry, by means of increasing collaboration between member states. Collaboration may reduce wasteful duplication of work and increase the size of effective efforts. After the decision is made to promote research in a particular filed, technical strategies in each programme area are defined, and proposals invited from sets of participants to carry out projects within the scope of the programme. Each consortium must (with few exceptions) contain members from more than one member state, and usually they may contain many members from several states. Non-European Union enterprises are also eligible to participate if they carry out significant R&D activities within the EU.

The commercial enterprises' motivations may be to achieve critical mass in areas such as semi-conductor development, to gain access to complementary (including university) research and expertise and to promote user/supplier dialogue. They may also wish to reduce the effectiveness of competition in some markets, or to gain information on competitors. Universities and other research undertakings will gain funding to continue their research trajectories in times of constrained public R&D budgets.

A brief note should be made here about the conflicts which may be produced by publicly-funded RTD programmes aimed at producing commercial and (privately appropriable) research results. In private collaborations, the use of ip would be entirely at the discretion of the participants, which freely enter into the projects. However, when public funds are used to subsidise private research (50% of costs are normally paid by the European Commission), then it may be expected that the results are more widely disseminated than would be the case for entirely private projects. This poses a dilemma for the companies, which would find unrestricted use of resulting ip unacceptable, and the Commission, which must ensure wider dissemination to some degree. The compromise involves allowing enhanced access to different types of participant in EU RTD programmes, the details of which are given below (see Table 2). Naturally, some potential participants, and some commentators on use of public funds, find the compromise unacceptable, while the participants endure their reservations in return for subsidies for the work which they hope to exploit.
Another issue, characteristic of collaborative research, is the extent to which the increasing interest in ip has affected the process of scientific advance in some areas in which basic research produces results which are "close to market". In particular, the field of biotechnology has produced the well-known phenomenon of academic entrepreneurs and quick returns to research results. Academic researchers have traditionally had the objective of public announcements of their work, in journals and conferences, and science had advanced quickly by the open exchange of such information. Commercial research has had a contrary objective: to keep results

TABLE 2: Rights and conditions of IPR access in the EC Contract

USER of IP	FOREGROUND		BACKGROUND	
	for R&D	**for exploitation**	**for R&D**	**for exploitation**
any contractor within the project	*Royalty free*	*Royalty free* *- unless owner unable to exploit directly*	*Transfer* *- if necessary to use foreground* *- if free from other obligations*	*Favourable* *- unless major business interests exist* *- if necessary to use foreground* *- if free from other obligations* *- unless commercially available*
any EU contractor in ESPRIT	*Favourable*	*Favourable* *- unless major business interests exist* *- unless commercially available*	*Favourable* *- unless major business interests exist* *- if necessary to use foreground* *if free from other obligations*	*[Commercial* *- unless major business interests exist* *- if necessary to use foreground* *- if free from other obligations* *- unless already exploited within the EU* *- unless commercially available]*
any contractor in any EU programme	*Favourable*	*Favourable* *- unless major business interests exist* *- unless commercially available*	*Favourable* *- unless major business interests exist* *- if necessary to use foreground* *- if free from other obligations*	*[Commercial* *- unless major business interests exist* *- if necessary to use foreground* *- if free from other obligations* *- unless already exploited within the EU* *- unless commercially available]*
any non-EU contractor in ESPRIT	*Favourable*	*Favourable* *- unless major business interests exist* *- unless commercially available*	*Favourable* *- unless major business interests exist* *- if necessary to use foreground* *- if free from other obligations*	*[Commercial* *- unless major business interests exist* *- if necessary to use foreground* *- if free from other obligations* *- unless already exploited within the EU* *- unless commercially available]*
any EU undertakings	*Transfer* *- unless major business interests exist* *- unless already exploited within the EU*	*Commercial* *- unless major business interests exist* *- [unless already exploited within the EU]* *- unless commercially available*	*[no obligation]*	*[Commercial* *- unless major business interests exist* *- if necessary to use foreground* *- if free from other obligations* *- unless already exploited within the EU* *- unless commercially available]*
any EU legal entity	*[Commercial* *- unless major business interests exist,* *- unless already exploited]*	*Commercial* *- unless major business interests exist* *- unless already exploited* *- unless commercially available*	*[no obligation]*	*[Commercial* *- unless major business interests exist* *- if free from other obligations* *- unless already exploited within the EU* *- unless commercially available]*
any EU research centre	*Royalty free*	*[no obligation]*	*[Transfer* *- if necessary to use foreground* *- if free from other obligations]*	*[no obligation]*

[conditions not explicit in Contract, but implied] ***Table adapted from EC information***

confidential until patents can be applied for. Collaboration between these two cultures has had the controversial effect of limiting the dissemination of research results, and thus slowing progress: the opposite effect to that intended by all ip regulations, as outlined in the opening sections of this summary.

The original expectation of many policy-makers was that projects would have a "horizontal" structure, that is, several companies with similar activities would combine together to achieve the scale necessary for competition with their larger US and Japanese adversaries. The Japanese government-supported programmes of the 1970s and 1980s (such as the Fifth Generation Computing Programme) were thought to be of this type, in which several of the large electronics corporations combined their efforts to achieve more economical and faster results ("pre-competitive research"), and would then take away the results and use them in competition with each other.[xlix] Evidence suggests that though this type of project is common, many more should be classified as "vertical". In these, concerns with different capabilities, markets, and objectives, combine together to take advantage of the diversity of expertise, so that research institutions collaborate with manufacturing companies, and users add their specific requirements and testing facilities. Yet more structures exist for specific purposes, such as the establishment of de facto industrial standards[xlix]

This leads to a variety of project structures, which we may call "architectures of collaboration" in which a complex set of motivations, outputs and objectives combine together. These can be classified into three "dimensions of collaboration":

⇒ **Types of participant:**
- commercial enterprises,
- research organisations,
- public authorities,
- academic institutions, , etc

⇒ **Types of output:**
- intellectual property of various types:
  - patents,
  - copyrights,
  - designs,
  - confidential
  - information, etc.

- embodied in:
  - prototypes
  - software
  - methodologies or mathematical approaches to problems, etc.
- commercially valuable information, such as process data, derived from associated companies
- tacit, or uncodified knowledge and skills
- relationships with other companies and organisations, such as customers or suppliers
- agreed, or de facto standards for use in industry or consumer markets

⇒ **Uses of results:**

- for incorporation in new commercially available products or processes
- to improve or alter the functionality of existing products and processes
- for use in internal company work, such as product development
- for further research
- for publication in academic learned journals, etc

Consideration of even this non-exhaustive list reveals the variety of issues which may be faced by collaborative projects when negotiating ipr agreements. Investigations of the characteristics of collaborative projects suggests **that these issues determine the composition of collaborative projects.** Studies by the author of a large number of collaborations, and also unsuccessful attempted collaborations, have made it clear that negotiation of collaboration agreements is a necessary and valuable part of the process of collaboration. When the details of iprs are addressed, it quickly clarifies the motivations and expectations of the putative participants. If any fundamental conflicts in ipr expectations should arise during these legal negotiations, the project will probably be still-born. In most successful collaborations, these issues are addressed even earlier (perhaps implicitly) because only compatible participants may be invited to negotiate. In this sense, collaboration agreements merely record the detail of successful collaborative architectures in a formal document. Put another way, collaboration agreements do not solve the problems of badly conceived collaborative projects.

For example, a vertical collaborative project may involve a hardware manufacturer, a software house, an academic research group, and a manufacturing "user". The hardware and software companies may intend to incorporate results in marketed products and a potential new customer (the user company); the user company gains from the focus on his specific requirements, and will gain from the improved products (as well as, possibly, a favourable market price). The academic group receives research funding, training for researchers, academic publications, and possibly some royalties or licence fees. Even in this simplest of projects the ipr status must be specified in clear terms, as early as possible, in order to facilitate the successful collaboration. For instance the conditions under which the academic members may publish results must be agreed, and the ways in which iprs will be legally protected must be detailed. Most importantly, ownership, access and rights of use of the iprs must be set out in an agreement between the parties. Will members have rights to other members' ipr, and for what purposes? Can third parties be licensed? Can marketing agreements be concluded?

Another example: a group of significant companies in an industrial sector may collaborate in order to produce an industry standard; perhaps the specifications of a physical interface between component modules, which could be patentable. This may involve user and supplier companies. The user companies may desire the widest dissemination of the specification in order to generate maximum benefits for themselves, perhaps implying free licenses. The supplier collaborators, however, will prefer to retain control over the standard in order to reap the maximum benefit, perhaps by selling licenses.

How have these have been addressed in recent experience in European Union programmes?

4. THE IPR STRUCTURE OF COLLABORATIVE PROJECTS

Each partner in a collaborative research project will bring to the project a bundle of competencies and knowledge and know-how, etc. Much of this is protected by iprs of one form or another, often of considerable value, and which are protected in several ways by their owners (who may not be themselves involved in the project). This **background intellectual property** brought to the project must be available for the use of other participants where appropriate, so this availability is one condition of the collaborative agreement.

Background intellectual property can be defined as:

> **information, patents, inventions, designs, software, proprietary information, and all other IP (excluding the results of the project itself):**
>
> - which a participant owns or has the right to use
> - has contributed to or made available to the project
> - is necessary to the progress of the work or to exploit the results of the project

In general, the ownership of background will not be affected by use in a project. Products marketed before the project begins would clearly not be affected, though favourable conditions may be negotiated for their use during the project's activities.

The last point is especially important. Any exploitation of results by one participant may be dependent upon the use of background owned by other participants or even non-participants. Use agreed only for the purposes of the project **does not** imply use for the purpose of exploitation. Subsequent use for exploitation may require a licence, and thus payment.

Contractual conditions will normally include, therefore:

- ensuring that all such background as may be necessary is disclosed and defined at the beginning of a project or as soon as its importance becomes apparent
- ensuring that the project has rights to use the iprs for the purposes of the research project
- agreement to maintain the confidentiality of other programme participants' proprietary information

In addition, participants may have to agree to a "Chinese Walls" clause. This means that the company may only use the background donated by other partners for the purposes of the project in question, and will not pass it on for other purposes, even within the company itself.

The nature of the background may cause some practical difficulties. Easily defined and identified ip, such as that protected by patents, may be relatively easily dealt with, however tacit knowledge, contributory proprietary knowledge, etc, may be more difficult to disentangle from the complex network of knowledge being manipulated within projects, even if all participants are acting in the utmost good faith.

"Foreground intellectual property" means

> **the information, patents, inventions, designs, software, proprietary information, and all other IP including all kinds of results which are generated by any member, or sub-contractor, of a collaborative project, in the execution of the specific contract.**

Normally, the foreground will form the basis of the eventual exploitation of the results of a project.

Various regimes may be used to deal with foreground ipr issues. Ownership may be allocated on the basis of generation of results, or in proportion to the funding by participants. Or, ownership may be pooled for attributability reasons. All participants may have the right to exploit these results (i.e. those owned by other participants as well as their own), as this is the purpose of the project (except in cases such as involving a [sole] buyer of the "product", e.g. a monopoly PTT operator). This may require the cross-licensing of the foreground results of each partner, however in some circumstances experience has shown that this is not appropriate because:

- it can be a very cumbersome and complex task to decide exactly which partner produced a particular result, when in a good collaborative project all the members will have meshed their work together in such a way that attempts to trace the exact origins and contributors to a particular result are not meaningful
- in projects where it is agreed that each participant will follow a different route to the solution of a particular common problem, it is not appropriate for the one which is successful to gain the rights over the main project results, while the others must pay for licences. This also leads to disagreements over which participant is allocated the most promising route to the solution. The usual form of agreement is for all participants to agree to share rights, whichever participant happens to be successful.

These factors have led many programmes to insist upon the free use of all project foreground by each partner.

FIGURE 1: Foreground and Background Information in a Collaborative Project

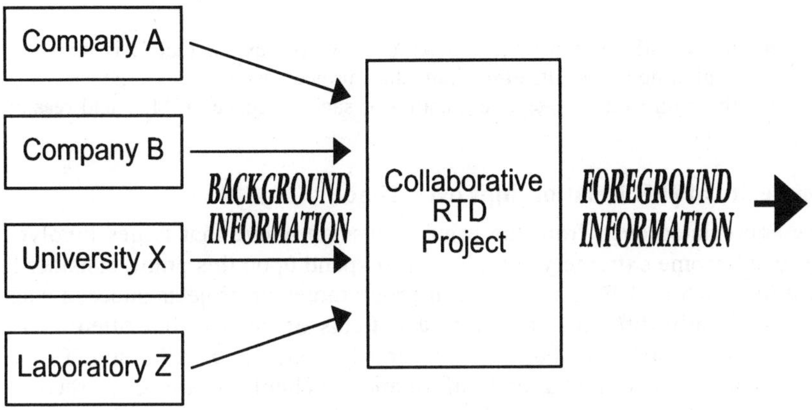

Several other complications arise from the exploitation of foreground. Not all participants have the intention, or capability, of exploiting results by means of marketed products, (e.g. small companies), or they may not wish to exploit, (e.g. academics), except insofar as they need to use the results for further research. "User" participants may wish only to have (preferential?) access to the marketed product. Such partners may be compensated in various ways, such as royalties from licensing, free use for research-only purposes, or reduced prices for products.

A frequent complaint of some companies involved in these programmes has been that in order to exploit results efficiently they must have **sole rights** to the foreground. For example, institutional investors may require proof of sole ownership of foreground before supplying the necessary funding for exploitation. The best solution to this is to predict the problems at the start of a project, and make the necessary arrangements, first in selecting participants, and next in formulating detailed clauses to agreements which can provide the security needed. However, in EU projects, due to the requirements for broader access to results than would be the case for privately funded work, it may not be possible to ensure sufficient protection for this purpose. Representatives of several companies, both large and small, have told the author that this is the main factor which prevents greater involvement in EU Framework RTD work.

In the case of **user/supplier collaboration**, other conditions may be required. Members not **directly** exploiting may wish to sub-licence or sub-contract production using the results. However such production may be in direct competition with other members of the project. This may be prevented by the exclusion of sub-licensing, or by a **"major business interests"** clause, which prevents members acting against the interests of other members.

As noted above, the background of other participants may be necessary to the subsequent exploitation of the foreground by one partner, and access to this must be ensured before the project begins, so agreements may contain rights to:

- use of all other partner's background to the extent that it is necessary in exploiting the results of the immediate project
- the right to negotiate the granting of such a licence on "fair and reasonable terms"

4.1 Complexity of IPR Relationships over Time

It can be appreciated even from the brief discussion above that issues involved in ipr contracts can become extremely complex. To expand upon this point, consider the long term case in which collaborative research programmes or projects succeed each other, each having slightly different objectives and fields of interest, but often overlapping. Several different sponsors may simultaneously exist (e.g. EC and EU national programmes in the computing and information technology area), or there may be privately funded projects. A set of projects within one of these categories may be succeeded by another set in a different programme, with the same members, plus some new ones, though they may have been "shuffled" to form new groupings within each project. Each programme may impose different ipr requirements, and individual projects may be allowed to negotiate additional terms internally. Differences in national ipr systems may also lead to complications and uncertainty.

Under such circumstances, **the foreground from one project may become the background for other new projects.** Previous non-participants may have to negotiate rights to this information. A single company may have several different levels of access to iprs is produced in various projects. The complexity of the ipr relationships involved here can readily be imagined.

5. THE CONTRACTUAL STRUCTURE OF EUROPEAN UNION PROJECTS

The contractual structure of EU projects which has evolved over 15 years is summarised in Figure 2. When a proposal has been accepted and a workplan agreed, contracts are signed between the European Commission and each of the participants on an individual basis. These contain details of the rights and duties of the participants, including a set of minimum IP conditions. The participants ("contractors") are encouraged to sign a separate collaboration agreement[xlix] with additional clauses specific to the project in question, some of which will usually refer to IP.

Under the standard Contract conditions, all foreground IP is owned by the contractor generating such information, including foreground patents.

Ownership of background information and background patents are not affected by the contract conditions, but granting of user rights for project purposes may be required as specified in the contract.

Where two or more contractors are involved in generating the foreground during the execution of the contract, they must agree between themselves on the arrangements for the ownership of foreground information and foreground patents, thus the need for the additional collaboration agreement.

FIGURE 2: Contractual Structure of European Collaborative Research Projects

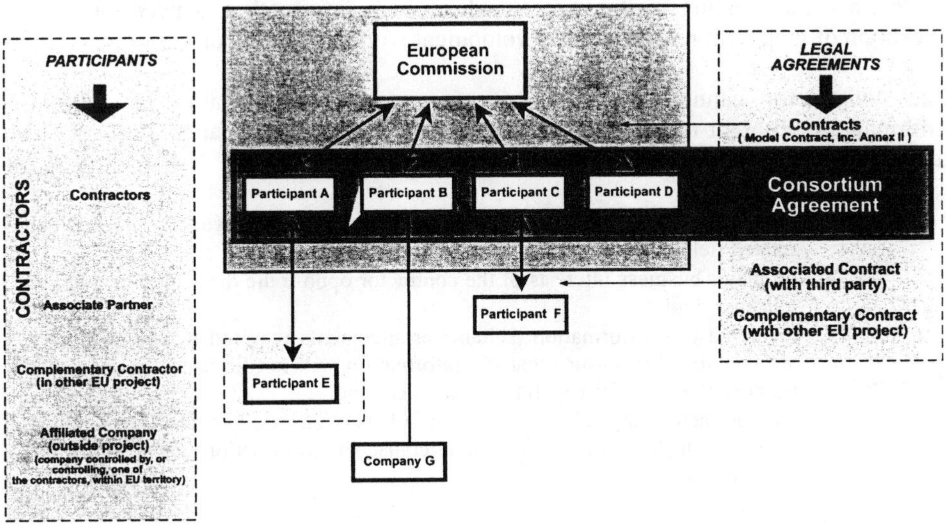

5.1 IPR in European Union Collaborative RTD Programmes

The ipr conditions are cast in three dimensions:

- background or foreground IP
- use in further research or use in commercial exploitation
- the status of the user (see Table 2).

We will outline the first two of these; Table 2 gives a comprehensive summary of Contract conditions for transfer of IP between participants.

5.1.1 Foreground Information and Patents: use for R&D purposes

Each of the contractors is required, on a royalty-free basis, make available its foreground information and grant licenses for its foreground patents to the other contractors participating in the project, where such information is necessary for the execution of their own research and development work under the contract.

Each contractor cannot unreasonable refuse to make available its foreground information and grant non-exclusive licenses for its foreground patents to Community undertakings, providing that;

- they are necessary for the execution of their own R&D work in the same or related fields in conformity with Community interests, and
- no major business interests of the contractor oppose the disclosure or grant of a license, and
- in the case of information, suitable arrangements required by the contractor are concluded to ensure that the information will not be used for any other purpose than that for which t was supplied, and
- the contractor may refuse if it, or any of its licensees, has taken or is taking adequate steps to exploit or commercialise the information or patents in the Community.

These rules are an example of the trade-off whereby the rights of ownership and exclusivity gained from patents is limited by the objective of encouraging diffusion of technologies in the wider community.

5.1.2 Background Information and Patents: use for R&D purposes

Each of the contractors shall make available and grant non-exclusive licenses for its background patents to the other contractors in the project for R&D provided that :

- the contractor concerned is free to disclose or license the use of such information, or grant licenses for such patents, and
- such information is, or such patents are, necessary for the execution of their own R&D work in the project

Each contractor shall make available its background information and grant non-exclusive licenses to its Patents, necessary for the use of foreground information made available to other contractors for the execution of a research and development contract concluded with the Commission

5.1.3 Foreground Information and Patents: use in exploitation and commercialisation

Each contractor is entitled to exploit the results of the contract (or Complementary Contract), or to have manufactured by third parties, products incorporating the results. Each contractor must be granted non-exclusive licences and user rights, on a royalty-free basis, for any foreground patents and foreground information generated. Any licenses and user rights shall not (unless expressly agreed by the owner) confer any right to sub-license, and may be subject to confidentiality undertakings.

Contractors may be unable or unwilling to exploit the results in commercial products, but may wish to use them for other purposes (e.g. non-commercial or small undertakings). In this case, in place of the royalty-free basis, these licenses and user rights may be granted on fair and reasonable financial conditions, agreed between users and owners, bearing in mind the contribution of the user to the results. The contractor in this case will have no rights to commercialise the results.

Contractors may not refuse to grant user rights and non-exclusive licenses to its foreground information and foreground patents to other participants in the specific, or in other, Community Research Programmes, providing that:

- the rights are necessary for the exploitation or commercialisation of the results of their contracts with the EC, and
- "no major business interests of the contractor oppose the grant of the rights or licenses, subject to these interests not abusively restricting the exploitation or commercialisation of the results", and
- the contractor may refuse if the rights relate to the products or manufacture which are, or are about to become, commercially available.

Similarly, rights to results should be granted to any Community undertaking finding these results necessary for exploitation or commercialisation of their own work in the same or related fields; the second and third of these provisos shall also apply.

If the contractor or its licensees do not taken steps to exploit or commercialise the results in a reasonable time, rights and non-exclusive licenses shall be granted to any Community undertaking with a legitimate interest in exploitation, again with second and third provisos above. This is of interest bearing in mind that "compulsory licensing" is one of the issues which has caused friction in international patenting.

5.1.4 Background Information and Patents: use in exploitation and commercialisation

Each contractor shall grant user rights and non-exclusive licenses for its background information and background patents to the other participating contractors, providing that:

- the rights or licenses are necessary for the exploitation or commercialisation of foreground information or foreground patents generated under the project, and
- the contractor is free to disclose and grant the rights or licenses, and
- the second and third provisos (above) shall apply.

Table 2 gives a summary of the conditions of access to project ipr, categorised by the three dimensions stated above.

6. SUMMARY

The description of IP conditions in EU collaborative projects given above is not, by any means, exhaustive. There is no coverage of the additional collaboration agreement, other related aspects of the legal documents, and no mention of possible conflicts with EU competition law, in particular Article 85 of the Treaty of Rome. Neither have the problems of coping with new technologies been addressed. But it should be clear that there has developed a considerable body of practice in European collaborative RTD programmes concerned with intellectual property.

Though this is set within the framework of formal IP rules and legislation, it is evolving and being disseminated throughout EU industry, and will heavily influence all collaborative activities in the future, including those not supported by public funds. This is one of the major contributions of the work since 1984: participants have had to acquire the expertise and competence to negotiate agreements in which it is accepted that the main outputs are forms of intellectual property. This expertise resides as much in the hands of managers and researchers as it is in the hands of IP lawyers and contracts specialists. Though the documents concerned have the status of legally enforceable contracts, their main utility has been in focusing the attention of managers at an early stage of projects on the potential outcomes, and planning for this. Reference to the agreements for litigious purposes, in Europe at least, is generally considered to be "unthinkable" [xlix].

Key words: Intellectual property rights, collaborative R&D, foreground, background, European Union, technology

Existing frameworks of intellectual property rights (iprs) are the result of centuries of evolution. Compromises between the rights of originators and users of newly generated intellectual property have resulted in diverse national systems of protection. New technologies create new forms of ip, and regulations have struggled to cope with the rapidly changing technological environment. However, the growth in size and importance of international trade, and the growth of "knowledge-based" industries, necessitates harmonisation of frameworks, thus creating a "selection environment" in which unified regulations will be accepted throughout the world. This paper considers the growth of collaborative R&D as a generator of new frameworks, relying on existing ipr systems, but at present creating a new set of practices and procedures, themselves evolving to meet the needs of participants in collaborative projects and programmes.

HOW USEFUL ARE PATENT DATABASES TO SMEs AS A SOURCE OF TECHNICAL INFORMATION?

Anthony ARUNDEL
MERIT
Maastricht, The Netherlands

1. Introduction

Patent databases contain an enormous amount of technical and other information. By early 1997, the European Patent Office (EPO) had published or granted over 850,000 patents, while the US Patent and Trademark Office (PTO) had published several million patents.

The technical description of each invention in a patent database can assist firms with their own innovative activities, while firms can use the information on the inventor to monitor the activities of their competitors. Patent databases also contain information of value for the management of intellectual property. A firm can search patent applications to check for infringement or to assemble information on the prior art as part of their own patent applications.

The potential value of patent databases has been recognised by policy makers. The recent Green Paper on Innovation by the European Commission [1] argues that firms, especially small and medium-sized enterprises (SMEs), should be encouraged to make use of the enormous variety of technical information available in patent databases. Yet the evidence from surveys of European firms shows that many firms seldom use patent data [2,3]. A survey by the European Patent Office reported that only 14% of R&D performing firms with less than 20 employees and only a slight majority of firms with between 500 and 1000 employees consult patent databases [3].

The purpose of this study is to improve our knowledge of the factors that are correlated with the use of patent databases and the reasons why SMEs use or do not use them. A better understanding of these issues should help policy makers in both developed and transitional economies to design programmes to encourage SMEs to use patent databases.

These issues are investigated through two surveys, both of which are limited to innovative firms because non-innovators are unlikely to find patent data of value. The first survey is the 1993 Community Innovation Survey (CIS), a multinational survey supported by EU member states, Eurostat, and DG XIII of the European Commission. This is the largest innovation survey to date in Europe and contains results on the value

A. Inzelt and J. Hilton (eds.), Technology Transfer: From Invention to Innovation, 195–206.

of patent disclosures for over 15,000 innovative firms, of which almost 13,000 are SMEs with less than 500 employees. The second survey is a combination fax and interview study of innovative Dutch SMEs in five high-technology sectors. The interviews are used to explore, in greater detail than possible in a survey, the factors that motivate SMEs to use patent databases.

2. Analyses Of The CIS Data

The CIS defines a firm as an innovator if it developed or introduced at least one product or process innovation over the three-year period between 1990 and 1992. The CIS questionnaire asked each firm to evaluate the importance to its 'innovation activities' of patent disclosures plus twelve other external information sources. 'Importance' is defined by a five-point ordinal scale that ranges from 'insignificant' to 'crucial'. The term 'information' is not defined and therefore could be interpreted by the respondents to refer to patent information of value for technical, competitive, or legal reasons. Useable results are available for manufacturing firms in 10 European countries: Germany, France, Italy, the Netherlands, Belgium, Luxembourg, Ireland, Denmark, Spain, and Norway. Further details on the methodology of the CIS is available in Archibugi et al [4].

Figure 1 gives the percentage of innovative manufacturing SMEs that give a score of 4 or 5 (very important or crucial) to the importance of patent disclosures and to a selection of five other information sources. The percentages are weighted to reflect the actual distribution of firms in each country by size and industry. The results for public research equal the highest score given to separate questions on universities, government laboratories, and technical institutes, while the results for suppliers equals the highest score given to separate questions on equipment and material suppliers.

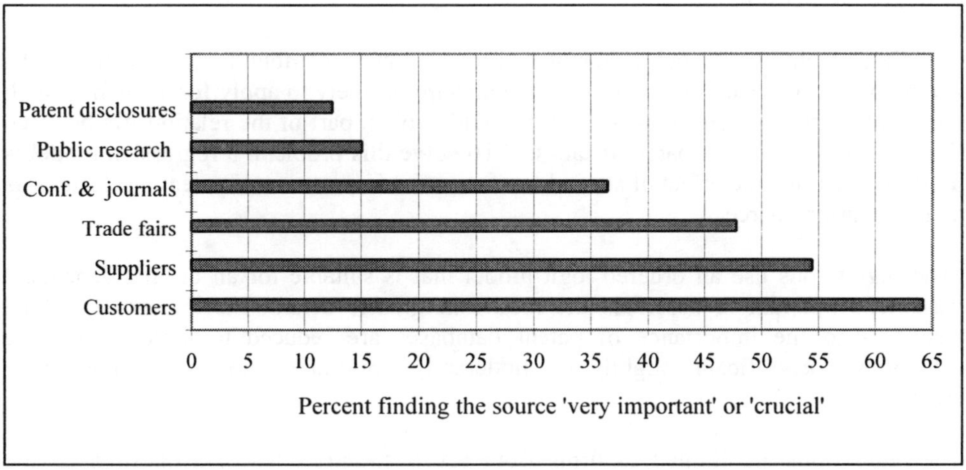

Figure 1: Importance of patent disclosures and other information sources to the innovative activities of 12,274 CIS manufacturing firms with less than 500 employees.

Only 12.5% of SMEs find patent disclosures to be a very important or crucial information source, which is slightly less than the percentage of firms that find public research of importance. In contrast, a much higher percentage find other information sources such as conferences and journals, suppliers, trade fairs, and customers of value. The percentage of firms that find patent disclosures of importance also increases monotonically with firm size, from 8.5% of firms with less than 20 employees to 49.2% of firms with more than 2000 employees.

The percentage of firms that find patent databases of value also varies considerably by sector, probably because of large sectoral differences in the value of patents as a method of profiting from innovation [5,6]. This will influence the propensity to patent in a sector and hence the amount of relevant information available in patent databases [7]. An analysis of the CIS shows that patent databases are most important in the chemical, instruments, and machinery sectors, where over 20% of the firms stated that patent disclosure was a 'very' or 'extremely important' information source. They are least important for firms in the non-automotive transportation equipment sector (shipbuilding, rail, and aerospace), food and beverages, and in the office equipment sectors. Less than 10% of the firms in each of these three sectors stated that patent disclosures were a valuable information source.

The CIS did not obtain data on whether or not each firm applied for a patent. However, the EPO survey reported that a higher percentage of firms that had applied for one or more patents regularly used patent databases than non-applicant firms.

3. Regression results

The size, patent status, and sector of a firm can act to confound some of the results given above. For example, large firms are also more likely to apply for a patent, which means that the applicant status of a firm could explain part of the relationship between firm size and the use of patent databases. To solve this problem, a regression model is used to determine the effect of several factors on the importance of patent disclosures as an information source.

JThe regressions use an ordered logit model that is suitable for an ordinal dependent variable. Appendix A describes the model in greater detail. The five-point ordinal responses to the importance of patent databases are reduced to three importance categories: insignificant, slightly or moderately important, and very important or crucial.

The regressions are limited to firms from seven countries where a question on the importance of patents as a method of profiting from innovation was asked. With the exception of interval variables for firm size and the proportion of sales from exports, all variables are dummy variables that are coded as '1' when the characteristic of interest is present and '0' otherwise.

The first three variables consist of the log of the number of employees (LEMPLOY), a dummy variable to identify firms based in Germany (GERMANY), and a dummy variable to identify firms that perform R&D (RDPERF). The variable GERMANY is included because an earlier analysis of the PACE data found that German firms are much more active patenters than firms from other European countries [8]. RDPERF is included because the EPO study found that R&D performing firms are more likely than non-R&D performing firms to use patent databases.

The proportion of sales from exports (EXPORTS) is included as a measure of the firm's exposure to global competition. The dummy variable PATIMP measures the importance of patents as a method of profiting from innovation. It is equal to '1' if the firm finds patents of value for this reason. Whether or not the firm is involved in cooperative research with other firms or institutes (COOPRD) is entered as a proxy measure of the importance of external sources of information.

Table 1 gives the results for three different regressions. A positive and statistically significant coefficient indicates that the variable increases the probability that the firm will give a higher rating to the value of patent databases as an information source. The second and third regressions include dummy variables for 16 sectors, although for simplicity these results are not shown here.

The regression results show that the probability that a firm finds patent disclosures of importance is greater for firms that perform R&D, for firms from Germany, and for firms that value patents as an appropriation method (PATIMP). The probability also increases with firm size (LEMPLOY), though part of the size effect is due to the importance of patents as an appropriation method (PATIMP), as shown by the higher coefficient for LEMPLOY in model 2 when PATIMP is not entered into the model. Both the firm's involvement in cooperative R&D and its export intensity are also positively linked to the probability of finding patent disclosures of importance.

TABLE 1. Ordered logit model coefficients for the importance of patent disclosures as an information source for innovative activities

	Model 1		Model 2		Model 3	
VARIABLE	Coeff	SE	Coeff	SE	Coeff	SE
Constant	**-2.430**	0.105	**-2.689**	0.129	**-2.847**	0.141
💻 (1)	**2.339**	0.047	**2.315**	0.046	**2.412**	0.049
GERMANY	**1.009**	0.064	**0.733**	0.059	**0.855**	0.068
LEMPLOY	**0.539**	0.048	**0.741**	0.045	**0.587**	0.049
RDPERF	**0.713**	0.076	**0.901**	0.071	**0.655**	0.078
EXPORTS	**0.658**	0.096			**0.494**	0.100
COOPRD	**0.301**	0.064			**0.273**	0.065
PATIMP	**0.889**	0.057			**0.832**	0.058
Industry dummies	No		Yes		Yes	

All models based on 5,148 firms. Results in bold if $p < 0.05$.

These analyses of the CIS show that patents are relatively unimportant compared to other information sources. The importance of patent databases are influenced by the sector of activity, firm size, whether or not the firm performs R&D, and the importance of patents as an appropriation method. Export intensive firms and firms that conduct cooperative R&D are also more likely than other firms to find patent databases of value.

4. Survey Of Dutch SMEs

A one-page faxed questionnaire survey on patenting issues was sent in 1996 to all innovative Dutch SMEs in five high technology sectors: information technology (IT), precision medical instruments, non-civil engineering, environmental technologies, and agricultural biotechnology (Agbio). These sectors were chosen for study because they encompass a range of new technologies that could be at the forefront of any changes in the importance of patenting. Two of these sectors, IT and instruments, are similar to the computing and instruments sectors in the CIS study. Responses were received from 191 of 273 firms that were sent the questionnaire, but the responses showed that 24 of these did not, in fact, innovate, leaving 167 innovating firms.

Detailed information on the use of patent databases was obtained through semi-structured interviews with responding firms from each of the five sectors and for both patent applicant and non-applicant firms. Interviews were successfully conducted with 22 firms that had applied for a patent and 9 non-applicant firms Basic information on the survey and interview respondents is given in Table 2.

TABLE 2. Dutch SME Survey and Interview Respondents

Sector	Questionnaire Responses	Average Firm Size[1]	% Patent Applicants	Completed Interviews
IT	28	111	39.3	4
Instruments	22	111	45.4	6
Engineering	53	86	28.3	6
Enviro Tech	47	94	85.1	7
Agbio	17	131	52.9	8
All sectors	167	100	50.9	31

[1]: Estimated from the midpoint for each of five size classes.

4.1 DUTCH SME FAX SURVEY RESULTS

The results of the faxed questionnaire survey show that 53.3% of the responding firms search patent databases at least occasionally, although there are large differences by sector, ranging from a low of 33.1% in the IT sector to a high of 76.5% in the agbio sector. Most of the difference by sector is explained by the percentage of firms that are patent applicants. As shown in Figure 2, sectoral differences in the percentage of applicant firms that search patent databases are minor, ranging from 70% to 100%. None differ significantly from the average of 76% for all patent applicant firms. In contrast, there are large, statistically significant differences in the use rate by sector for non-applicant firms.

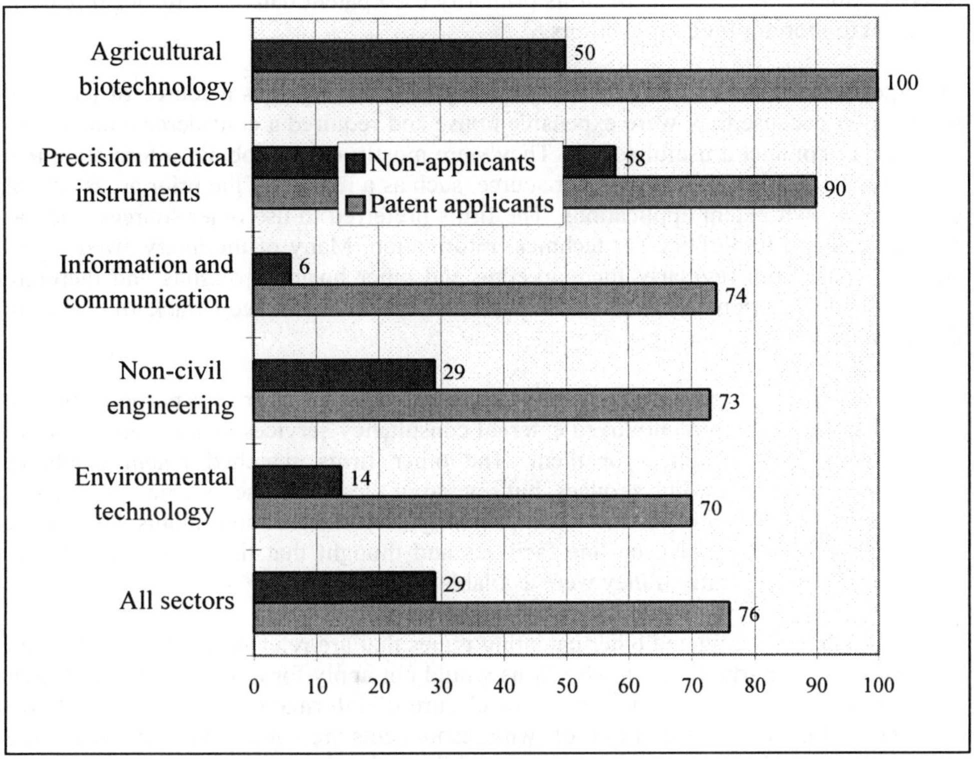

Figure 2. Percent of Dutch SMEs in five high technology sectors that search patent databases by patent applicant status.

These results suggest that many firms do not search patent databases to obtain technical information. If they did, we would expect little difference in the use rate between innovative non-applicant and innovative patent-applicant firms. The fact that a much higher percentage of the latter group search patent databases suggests that these SMEs could be using them for legal reasons connected to the management of the firm's patent rights, for instance to ensure that the firm is not infringing another firm's patent or to obtain the necessary information on the prior art to draw up its own patent applications. The reasons why SMEs use or do not use patent databases was explored in the interviews.

4.2 INTERVIEW RESULTS

The interviews showed that SMEs use patent databases for three basic reasons: for legal purposes, to monitor competitors, and to obtain technical information of value to their own innovative activities. The least important reason was to obtain technical information, which was a main reason for their use for only two of 24 firms that used patent databases. In contrast, 14 firms primarily used patent databases for legal reasons and eight to monitor their competitors.

The interviewees stated that they did not use patent databases as a source of technical information because they were expensive to use and required a considerable amount of expertise to conduct a useful search. They were mostly used to obtain information that could not be acquired from any other source, such as a review of the prior art as part of one of their own patent applications. The firms preferred to use other sources, such as publications and trade fairs, for technical information. Many of the interviewed SMEs attended trade fairs regularly for marketing and other business reasons and therefore only a small additional effort was required to use them to keep track of technical developments.

Half of the interviewed firms that did use patent databases for any purpose did not search them in-house but employed external consultancy services or innovation centres to conduct patent searches for them. The other firms searched patent databases themselves, three via on-line services, but several of these also use external consultants. At the same time, many of the firms that primarily used external consultants would have liked access to inexpensive on-line facilities and thought that they would use patent databases more frequently if they were available at low cost.

The SME study investigated other patenting issues that are relevant to the use of patent databases. An important reason why firms would not apply for a patent for one of their own innovations was concern over the disclosure of information once the patent was published. This raises the question of why respondents are concerned about disclosure when their own firm applies for a patent, while at the same time rarely using patent disclosures to obtain information about inventions developed by other firms. The most plausible explanation is that firms will not patent at all if the information disclosed in a patent will damage their interests, for example by showing competitors how to circumvent the innovation. Thus, the protection provided by a patent must be strong enough to outweigh these disadvantages. Otherwise, there are very strong disincentives for firms to reveal information by patenting. The effect of these strategies is to limit the technical value of the information available in patent databases relative to its cost of acquisition.

5. Conclusions

The analysis of the CIS survey confirms the results of other studies that have found that patent disclosures are a relatively unimportant source of technical information. The interview study showed that one reason why SMEs do not use patent databases is the high costs of searching patent databases compared to less expensive sources of technical information. SMEs primarily use patent databases to acquire information that is not available from any other source, such as during the formulation of their own patent applications. This partly explains why a higher percentage of patent applicant than non-applicant SMEs use patent databases.

Another reason is linked to the patenting strategy of firms. Firms are less likely to apply for a patent if the cost of disclosing the information exceeds the benefits to be gained from the patent. This strategy reduces the value of the technical information contained in patent databases.

What policy implications can be drawn from these results for transitional economies? First, and perhaps most importantly, it is clear that many SMEs benefit far more from attending trade fairs and from reading the technical and scientific literature than they do from searching patent databases. Trade fairs are particularly useful as a means of reducing the costs of obtaining information in sectors characterised by a large number of manufacturers and a larger number of buyers [9]. Policy makers in transitional economies should both strengthen existing local trade fairs and possibly subsidise local SMEs to attend trade fairs in other countries. Short-term subsidies to national trade and technical journals might also be worthwhile.

There are two general policy options to increase the use of patent databases. The first is to improve the value of the information held in patent databases and the second is to improve access.

The value of patent databases can be increased by improving the quantity and quality of the technical information. The quantity can be increased through policies to encourage firms to patent a higher percentage of their inventions. This is already under consideration in many countries as a means of improving the competitiveness of SMEs by strengthening their intellectual property rights. The disadvantage of increasing patent propensity rates is that it could limit the diffusion of innovations to SMEs by reducing the pool of technologies that can be adopted without paying license fees. Furthermore, the major barrier to patenting is the disclosure requirement, so it is not clear how more patenting could be encouraged. Reducing patent application fees, for example, is only likely to have a small effect on patent propensity rates because these costs largely discourage the patenting of marginal inventions or patenting by very small firms.

The quality of the technical information contained in patent databases could be improved through strengthening the disclosure requirement. The benefits, however, could be minor, since it would likely lead to lower patenting rates for inventions where the benefits of a patent are marginal and it would have no appreciable effect on inventions that can be easily reverse-engineered. These inventions will only be patented if sectoral and other factors ensure that the patent can block imitation.

Improvements in the ability of firms to access patent databases holds more promise. The most effective policy option is to reduce the relative cost of using patent databases compared to alternative information sources. The interview results show that many firms are interested in having less expensive access to patent data, with several firms commenting that their failure to search patent databases did increase the risk that they would invest in costly research to 'reinvent the wheel'. Unfortunately, current databases such as the Patent Register of the EPO are relatively simple to search to obtain information of value for legal purposes but are not of use as a source of technical information.

Primarily, SMEs need considerably simpler, faster, and more reliable methods of searching patent databases for technical information. Several respondents commented that less expensive and time-consuming methods of accessing patent databases, for example via the Internet, would probably increase their use of these databases. This will require the development of good search routines to reduce the time required to find something useful. There are two main problems. First, patent databases contain a substantial amount of similar and duplicative information. Software that could limit this repetition could be of great value. Second, an adequate search often requires firms to search several patent databases in the US, Japan, and Europe. Standardisation of search routines would be very useful.

The European Union has been actively trying to improve access to patent databases. An example is QUICK SCAN which searches patent databases world-wide for a fee. However, QUICK SCAN cannot be operated by the firms themselves and is therefore a second step. What is first required is an on-line system that is cheap and which is so simple to use that little technical expertise is required. It is possible that experience with very low cost on-line systems will lead firms to identify areas where they would like a thorough, professional search, as a second step, perhaps using a system such as QUICK SCAN.

This discussion suggests that policies in transitional economies to improve the use of patent databases should focus on three methods to improve access to patent databases that are currently available over the Internet. The first is to ensure good access to the Internet while the second is to prepare and disseminate information on how to use the Internet to access on-line patent databases. IBM, for example, has established a free Internet site that contains the front page, abstract and other information for two million PTO patents issued since 1971. Firms can use keywords to search the IBM site and order complete copies of interesting patents for a small fee. Innovation centres to provide search services at low cost could also be of benefit. Finally, policy makers in transitional economies should ensure that their own national patent databases are available on-line, accessible through easy-to-use software, and accessible at minimum cost.

The results of these two surveys also indicate that costs can be reduced by targeting programmes to those sectors and firms where patent databases are most worthwhile. The sectors which stand to benefit the most include chemicals, machinery, instruments, and biotechnology. Furthermore, these programmes are most likely to be of value to export-oriented and highly innovative firms.

6. Acknowledgements

The analyses of the CIS data was partially funded by DG XIII of the European Commission. The Dutch SME survey and interviews were funded by the Dutch Ministry of Economic Affairs under the programme Bedrijfsrelevant Onderzoek.

References

1. EC (European Commission) (1996). Green Paper on Innovation. European Commission, Brussels.

2. Horst, B. (1991) The use of patent information in innovation management: Results of an empirical study in the Federal Republic of Germany. In U. Täger and A. von Witzleben (eds) Patinnova '90, pp 121-125.

3. EPO (European Patent Office) (1994). Utilisation of Patent Protection in Europe, Volume 3. EPO, Munich.

4. Archibugi, D., Cohendet, P., Kristensen, A., and Schaffer, K.A. Evaluation of the Community Innovation Survey (CIS) Phase I. DG XIII of the European Commission, EIMS Publication 11, Luxembourg, 1994.

5. Arundel, A., van de Paal, G. and Soete, L. (June 1995) Innovation Strategies of Europe's Largest Firms. MERIT, Maastricht.

6. Levin, R.C., Klevorick, A.K., Nelson, R.R. and Winter, S.G. (1987) Appropriating the returns from industrial research and development. Brookings Papers on Economic Activity **3**:783-820.

7. Massini, S. (1996) Inter-sectoral flows of technology: an international comparison of technological specialisation profiles. Presentation to the Conference Econometrics of Innovation - Patents, Applied Econometrics Association, Luxembourg, November 28-29.

8. Arundel, A. and Kabla, I. (in press) Patenting strategies of European firms: An analysis of survey data. In S. Allegrezza and H. Serbat (eds) The Econometrics of Patents, Applied Econometrics Association Series, AEA, Paris.

9. Florio, M. (1994) Fair trade by trade fairs: Information providing institutions under monopolistic competition. Small Business Economics **6**: 267-281.

10. Greene, W.H. (1993) Econometric Analysis: Second Edition (MacMillan, New York.

11. Liao, T.F. (1994) Interpreting Probability Models: Logit, Probit, and Other Generalized Linear Models, Sage Publications, London..

Appendix A

In the ordered logit model, y is an unobserved latent value and μ equals the unknown threshold parameter separating the adjacent categories [10,11]. The observed value of y is equal to:

$$
(1)\qquad
\begin{aligned}
y &= 1 \quad \text{if } y^* \leq \mu_1\,(=0)\\
y &= 2 \quad \text{if } \mu_1 < y^* \leq \mu_2\\
y &= 3 \quad \text{if } \mu_2 < y^* \leq \mu_3
\end{aligned}
$$

The probability that y = category j is as follows:

$$
\text{Prob}\,(y=j) = F\left(\mu_j - \sum_{k=1}^{K} \beta_k \chi_k\right) - F\left(\mu_{j-1} - \sum_{k=1}^{K} \beta_k \chi_k\right) \tag{2}
$$

where K equals the total number of variables included in the regression and F equals the logit transformation:

$$
\log\left[\frac{p(y \leq j|x)}{1 - p(y \leq j|x)}\right] \tag{3}
$$

The probability that y equals a specific category such as 1, 2, or 3 is determined as follows:

$$
\begin{aligned}
\text{Prob}\,(y=1) &= F\left(-\sum_{k=1}^{K} \beta_k \chi_k\right), \\
\text{Prob}\,(y=2) &= F\left(\mu - \sum_{k=1}^{K} \beta_k \chi_k\right) - F\left(-\sum_{k=1}^{K} \beta_k \chi_k\right), \\
\text{Prob}\,(y=3) &= 1 - F\left(\mu - \sum_{k=1}^{K} \beta_k \chi_k\right).
\end{aligned} \tag{4}
$$

INTELLECTUAL PROPERTY RIGHTS
What they are and how to use them

KEITH HODKINSON CPA EPA MITMA Patent Attorney
Marks & Clerk
Sussex House 83-85 Mosley Street, Manchester M2 3LG UK

Innovation, technology transfer and intellectual property

At the heart of much technology transfer is the granting of interests in intellectual property rights. Indeed intellectual property rights are often the means by which the rights being transferred are defined. Yet unfortunately there is a significant degree of fundamental misunderstanding of even the most elementary points about intellectual property rights among SME's. This lack of awareness leads to inefficient use of the system and misconceptions about what it can achieve, what it costs and when it should be used.

As an introduction to the section of this workshop dealing with intellectual property rights and their role in innovation and technology transfer, this paper was requested to provide non-experts in the field of technology transfer with an overview of some of the basic points about the intellectual property system from a practical point of view.

Unlike many of the papers at this workshop, this presentation is not an academic study, nor a statistical analysis. Rather it is intended as a personal account borne out of an individual's experience of the practical impact of the intellectual property system on SME's which share, even in Western developed economies, many of the same problems as SME's in the transition countries, based on a lack of resources.

A warning: obviously this cannot, in the short time allotted, be a detailed account of what is an extremely complex area of business life and generalisations, of which there are many in this paper, are often misleading and dangerous. Detailed advice should always be taken on any specific problem.

The material in this paper represents the personal views of the author and may not be representative of the views of his organisation.

A. Inzelt and J. Hilton (eds.), Technology Transfer: From Invention to Innovation, 207–231.

What are intellectual property rights?

Intellectual property rights are legal rights granted by the state to protect inventions and certain other expressions of ingenuity or aesthetic effort from reproduction or copying (some are true monopolies like patents, others rights against plagiarism like copyright), to act as an incentive to such inventive and aesthetic activity by protecting the inventor from some competition and so to enhance the return on his effort, assuming that the invention has commercial appeal. no intellectual property right rewards invention per se, a point often forgotten by inventors.

Oddly, trade marks which protect goodwill and reputation and have nothing to do with inventiveness or aesthetic activity but do reward good quality and product reliability are also included in intellectual property rights.

A variety of intellectual property rights exist to protect different aspects of research and development and these will be briefly considered in turn. Trade marks, protect goodwill and commercial reputation are also briefly reviewed. The respective roles of these rights are often confused by SME's who may select the wrong type of right to protect their development.

Patents

A patent provides the legal right to exclude others from exploiting an invention which is a technical or scientific development having some practical applicability and which is described and defined by carefully worded "claims" in the patent. A broadly drafted patent may be able to claim the underlying technical concept behind the development, not just the embodiment developed by the inventor. As such it can be a powerful tool.

In most countries this right, once granted, will last for up to 20 years from the date the patent was applied for (although in the USA the period was for cases filed up to 8th June 1995 only until 17 years from the date of grant of the patent, which could be three or four years after the initial application). Patents are national rights, so that a patent granted in the United Kingdom does not give the patentee any rights in, for example, the USA. Therefore, for international protection separate patent applications must generally be filed in all the countries in which protection is sought.

Some of the consequences this and of a system with little international harmonisation are discussed elsewhere: fundamental differences in approach (e.g. between the USA and Europe); high administrative costs; complexity; difficulty of access to information; language issues and so on.

However, although these differences in substantive law are rightly pointed out it is in my opinion rarely the case that they make a significant impact compared with the administrative costs of filing in a variety of jurisdictions and persuade a variety of governmental organisations to examine the applications separately. It is in this sphere where the greatest progress can be made in reducing cost.

More local or regional changes such as a European Union single trade mark have occurred and a true European Union patent may soon be in effect if the proposals of the recent European Community Green Paper on the Community Patent Convention are taken up. Mutual recognition of searches, reduction of translation requirements may arise but all changes have their downsides: if a reduction in translation requirements is made for applicants this merely throws the cost of translation on those who need to read the patents to see if they infringe; and this can lead to the inefficiency of many separate and unofficial translations instead of one readily accessible and competent translation.

Furthermore use of international systems such as the European patent Convention (EPC) and Patent Co-operation Treaty (PCT) discussed below can defer for a considerable time the costs associated with international patent filings so as to give companies more time to consider whether there is a real likelihood of patents being required for the technology, reserving the possibility of protection in many countries whilst remaining uncommitted to the major national costs for up to 30 months from the first patent application filing date.

The number of countries which have joined these systems show one practical way in which patenting costs can at least be substantially deferred without major reform. These issues are dealt with in more detail in another paper and I do not want to spend too much time on them now, merely to place them in a context.

The other area where cost is a severe inhibition to use of the system is the cost of enforcing rights against infringers: this is a feature common to many areas of law and is by no means unique to the intellectual property system, although the large sums of money at stake do often exacerbate the costs of such proceedings as does the complexity of the legal and factual background to many cases.

There is perhaps no point in having a patent if you can never expect to enforce it, although in many areas there is a surprisingly high level of respect of patents even where they are held by SME's - other papers deal with this point. Practice differs from area to area. Often the poor plaintiff has an equally poor defendant equally fearful of costs. Out of such mutual fear settlements often arise.

Oddly enough there does not seem to be any inverse correlation between litigation costs and use of the system - the USA has amongst the highest take up of patents by SME's and some of the most expensive litigation.

This is a cultural matter in many respects and the reluctance of Japanese companies in Japan to litigate contrasts with their frequent filing of oppositions to the grant of each other's patents in the European Patent Office. The presence of specialist courts and a higher level of confidence in the decisions of such courts tends to determine where companies litigate if they have a choice of forum.

Streamlining of enforcement procedures and attempts to procure cross border recognition of injunctions may be more effective in the future but really depend on harmonisation of law and interpretation of patents. However these are again matters which require international agreement.

In the meantime, there are now a number of ways of cutting down on the expense and effort involved in such an international patent filing campaign, as a result of the European Patent Convention (EPC), which at the time of writing provides for a single European patent in seventeen member countries (with five possible extensions to certain Eastern European countries)), and through the Patent Co-operation Treaty (PCT), administered by the world Intellectual Property Organisation (WIPO) which enables co-ordination of patent applications in up to ninety countries through a single preliminary examining body (as of 5th September 1997), although eventually national patents are granted as a result.

Other measures which have been tried out to tempt SME's to make more use of the system include preferential lower rate official fees for small entities (SME's and universities) in the USA but this appears to have limited impact save at the margins. In many cases, indeed, minor tinkering with the system seems to impact only on minor incremental technology. It is not in my experience a determining factor in encouraging or discouraging patenting of major technical breakthroughs.

Table 1 gives a list of current members of the major patent conventions enabling patents to be filed claiming priority from a home filing or to defer patent costs (for convenience tables are presented at the end of this paper). The most significant country not included on this list is Taiwan.

Patentable inventions

There is much misunderstanding among SMEs of what can and cannot be patented, a confusion enhanced by undisciplined use of the term in many newspapers.

I will focus on the scene in Europe but for the most part there are not significant differences in what is regarded as patentable subject matter. Although there are areas of controversy, more ethical than technical in many cases, and although some areas of technology move very quickly, the overriding impression I have in practice is that the current system broadly caters extremely well for the majority of technical fields and innovations in those fields

In broad terms patents are granted in respect of inventions which are: novel, in the sense of not being previously known to the public; inventive, in the sense of not being obvious developments over existing technology; and capable of industrial application, i.e. not merely scientific discoveries, mathematical formulae or means of doing business or performing calculations.

Most novel and inventive processes, methods, devices and products will be patentable. However, most countries have certain public policy exclusions to deprive certain inventions of protection.

Computer programs are commonly excluded from patent protection, although if the program has a technical or industrial application, such as controlling the operation of a machine or robot, protection will often be available.

Another common exclusion is that of methods of treatment of the human or animal body, although a drug used in such a method will be patentable. Inevitably, national patent laws differ but the position in most countries in Europe is set out in Table 2.

It is highly dangerous for a person who lacks training in the field to make a judgement as to the potential patentabilty of an invention. This is a question on which professional advice of a patent attorney should be sought. The question is a technical one, depends often on subtle argument and is affected in many cases by complex case law. A patent attorney can conduct searches to determine whether the idea has been published before and whether there is any difference from existing knowledge that would give rise to a patentable claim and advise on these complex matters.

Furthermore, even if a patent is not available, some alternative right may be and this is often forgotten. In many cases a portfolio of different rights will be required to protect a product.

Applying for a patent

Obviously the decision to seek or to maintain a patent is a commercial one. You should look at what is going to come out of it and what it is going to cost and make a decision on that basis. You can always terminate the expenditure at any time and you should not be afraid to do so. But you should also remember that once you have withdrawn a published application you cannot opt in again - if the invention that was initially unsuccessful becomes a winner you cannot reapply - you have lost the monopoly for ever. When should you file a patent?

Patent applications are perceived as being relatively expensive without any guarantee of success in the market place or even success in obtaining a valid right: for the SME there is no doubt about that. Some practical questions which a company can ask itself before going down this road are set out in Table 3.

This Table represents a typical decision flow chart of the type a SME will go through in determining the usefulness of filing for patent protection. Again the point to be stressed is that there may be alternatives and also there may be situations where the need to patent is perceived differently at different stages of development: an initial filing may need to be reassessed later on.

One key factor often ignored is the practical advantage if any which the patent might give you. If your development is such that whilst you might be able to patent the specific embodiment you have developed, others would be able to product something functionally as good and within the technical concept but outside the scope of your claims, there might be little point in patenting it. What modifications could a competitor effectively make which would take him outside the scope, of your claims and which would render the patent redundant?

Another point is to ask how easy would it be to discover if competitors are infringing the patent. Can they perform processes behind closed doors which you would never get to know about?

There is of course a danger in keeping material secret if other people then invent the same thing and patent it. Whilst you may have in most countries an ongoing right to continue to do what you have done secretly in the past the right cannot in most cases be expanded, modified or transferred, thereby severely restricting your potential use of the invention. The position varies enormously from country to country and the solution to the legal problem cannot be assumed to be the same in all places where you wish to trade.

If you think you may wish to apply for a patent a number of crucial points must be made.

The first is that in almost every country any non-confidential disclosure of the invention to anyone before the date of filing an application to patent the invention will fatally prejudice the application. The golden rule is: file first and disclose after. The USA has some anomalous exceptions but these cannot be easily relied on by foreign inventors and only apply within that country. The USA requires much more in the way of inventor's notebooks, records of experiments and the like where applications come into conflict.

The second point is that in almost every country the first to file the application and not the first to invent will have the better right to a patent in most cases; if two inventors come up with the same idea independently of each other the first to file will have the better right to the invention irrespective of who actually was the first to make the invention (again the USA is an anomalous exception and there have for many years been difficult discussions regarding how the systems could be merged).

The third point is that if it is subsequently decided to keep the invention secret, a patent can be withdrawn at any time up to 18 months after filing (in most cases) without it having been published, so that secrecy can be maintained.

The mere filing of an application is not fatal if a decision is later taken to withhold the invention from public disclosure. In most countries after 18 months the invention will be published for all to read.

Again the USA is an anomalous exception to this rule and there are discussions about possible changes to this rule to avoid the risk of so called "submarine patents" which remain hidden from view for many years before being granted.

In fact published patents can be a fascinating source of information although the use made of patent information especially by SME's is not as efficient as it might be, for a variety of reasons.

Some of those reasons are explored by another paper in this workshop and I do not wish to trespass into that area: however, in my experience, much information is published in documentary form only in patents and many SME's seem to miss out on the information because of poor searching strategies which often miss relevant material and waste money, increasing the apparent cost of the access to this information.

In most countries a patent is obtained only after rigorous examination of the application for novelty and inventiveness, and compliance with legal formalities by the Patent Office. The application contains a summary of the existing state of the art, a description of the invention and how it is used, and claims to legal monopoly in those aspects of the invention that are new and inventive.

Although there are claims that patents are sometimes written to mislead (especially with respect to titles) this is more myth than reality because firstly, titles are reviewed by the Patent Office and classifications are applied by the Patent Office for searching purposes and, secondly, the dangers of not disclosing enough to enable a person to reproduce the inventions - consequential lack of validity - and the inability to add this new material later are severe inhibitions to concealment.

Indeed in some countries, such as the USA, the inventor must disclose the best method known to him or her of putting the invention into effect, whether alternative methods are available, and any prior inventions of which he or she has knowledge or ought to have had knowledge which affect the consideration of whether the invention is novel and inventive. If a patent is filed elsewhere with a view to filing later in the USA this best mode of performance must be disclosed in the original filing to enable priority to be claimed for the earlier patent and to meet US requirements.

It is true that filing can involve considerable professional fees which may intimidate SME's. However, the patent application is a complicated legal document as well as an exposition in scientific or technical terms of the invention.

Unless the applicant is well versed in patent practice it is extremely unwise to attempt to draft and file the application without professional assistance if the invention is likely to have any commercial significance.

A badly drafted application can lead to an otherwise patentable invention being rejected or, even worse, held invalid in subsequent court action; it can lead to loopholes in the scope of claims and thus narrower protection than the applicant would otherwise be entitled to.

An application which is too narrow cannot at any later stage - even before grant - be expanded to cover things not previously disclosed. Conversely it may disclose things which need not have been disclosed and that would be better kept secret.

Finally, in some countries, such as the USA, a failure to disclose all that should have been disclosed about the known prior art can lead to loss of the patent and even to liability for fraud on the US Patent Office.

If overseas protection is sought, the assistance of qualified attorneys in obtaining adequate translations into the local language and in dealing with patent office objections will be in practical terms essential and in many countries legally required. It is in the field of translation requirements where the greatest savings might be made in the future.

The scope of patents

The scope of a patent is determined by the claims which appear (usually at the end of the document) defining the monopoly protection claimed. These claims will be of varying width, ranging from broadly expressed claims to narrower claims covering specific features and embodiments, providing the patentee with a fall back position if the broadest claims are found to be invalid.

The claims are in many ways the critical part of the document but the problem of determining the scope of a patent is exacerbated by the fact that different countries follow different principles of interpretation ranging from close literal interpretation to broad "equivalence" doctrines. Local professional advice is in practical terms essential in any case of doubt.

Contrary to a common misconception among SME's, a patent is not an entitlement to use the invention.

Firstly there may be safety or other product regulations to acquire, for example in the case of pharmaceuticals.

Secondly there may be other rights which preclude use of the patent. Merely to have a patent does not mean that one is free from potential liability to others. A patent may have been granted to you because your invention is new and inventive but there may still be a broad patent prior to yours which precludes the exploitation of your invention without a licence from the prior patentee.

For example, I may invent and patent the wheel. You may invent and patent the tyre. I cannot sell wheels fitted with tyres without your consent but neither can you sell wheels with tyres - you can sell tyres to fit on my wheels.

I in turn may invent and patent a tyre with a tread pattern to improve grip by assisting in the dispersal of water from the tyre surface. I cannot sell tyres without your consent because of your patent but you cannot sell treaded tyres without my consent.

You may then invent a better tread which may be patented. I cannot make use of that improvement because of your patent, and so on.

In this lies the basis of cross licensing which is common in many industries. As products become more complex it will be readily understood that different aspects of a product may be protected by patents in the names of different owners.

Utility models

Some countries such as Germany and Australia offer "petty patents" or utility models which are meant to provide a cheap and simpler means of obtaining protection for functional, often incremental technical improvements. These may in some countries have the same nominal requirements as patents but have less (or no) examination, shorter terms or may have lower requirements of inventiveness.
There is very little harmonisation of these regimes which often incorporate peculiar local rules, such as local, not global novelty requirements, limited disclosure requirements, limitations on the number of claims made, etc.

These systems are often superficially attractive to SME's but ultimately have not become very popular because of doubts about their effectiveness as rights. Australia, for example is now considering reforms to its petty patent system and a new "innovation patent". Conversely Hong Kong has just adopted a new short term patent system with only formal examination as an alternative to full patents.

These "lower" forms of patent may co-exist with full patents in a number of countries. There has been considerable discussion of their merits or disadvantages which for reasons of time cannot be gone into here. They are mostly searchable in the same way as patents. The transition countries falling within the former USSR commonly had inventor's certificates which followed similar principles.

Table 4 sets out those major countries which still operate such systems

Design rights

For our purposes designs may be regarded as aspects of shape, configuration, pattern and ornament to industrially mass-produced goods. They are features of specific design appearance rather than underlying technical principle. As such, design registrations perform an entirely different function to patents.

In some countries both functional (non-patentable) and purely aesthetically appealing designs may be protected either by registration or even by reliance on copyright or allied unregistered rights in design drawings and prototypes. In others, only non-functional designs may be protected by design registrations or unregistered designs rights and this alone will be an option in the absence of a patentable functional feature.

There is very little international harmonisation of the law on designs, unlike that for patents, and professional advice is essential if international protection is desired. Searching of such designs is time consuming and is often very unreliable.

Systems to reduce the costs of international design registration have begun to appear, the most notable of which is the system of International Design Registrations administered by WIPO. A list of member countries appears in Table 5

Plant seeds and varieties

This is a highly specialist area of significance only to a small sector of the agro-industry. Some countries protect plant seed and varieties by a patent but others, such as the United Kingdom, maintain a separate regime to protect the reproductive material of the new variety from commercial reproduction.

Methods of cloning and other biotechnological methods will usually be protectable by patent. Recently in the United States and Europe genetic methods associated with genetic manipulation in animal breeding were patented and similar methods would be prima facie patentable in respect of plant life. There is no significant harmonisation of laws. These are detailed questions outside the scope of this presentation.

Trade secrets

These are not strictly intellectual property rights at all but remain an option, particularly for short-term protection, such as when deciding whether to patent or not.

Most countries give some degree of protection against abuse of trade secrets but do not protect the possessor from accidental disclosure or from the independent discoverer in the way that patent and designs do. These rights are normally based on personal (not property) rights founded in contract with the persons to whom information is disclosed, or in tort or delict if devious means are used to acquire information kept secret by another.

There may be a number of reasons for looking to trade secrets. The invention may not be patentable but may be useful commercially and easily concealed, such as in a black box or factory secure area, and the owner cofident that others cannot easily invent the same thing or reverse engineer it. Policing a patent may be thought commercially impracticable.

On the other hand, the vulnerability of the information to leakage may lead the owner to conclude that the longer-term security of a patent may be necessary. Equally, the licensing of trade secrets and know-how can be extremely difficult to accomplish satisfactorily. Not least among these issues is the fact that some countries where approval is required for agreements with overseas companies are not altogether enthusiastic about technology transfer agreements dependent only on trade secrets and may withhold state approval in the absence of evidence of real substance to these rights.

In other cases, whilst anti competitive terms may be tolerated for trade secrecy agreements in the same way as for patents, there are almost always requirements that the trade secrets be shown to be substantial and the duration of terms may be curtailed. The usefulness of trade secrets varies enormously from company to company. Whilst many use them on grounds of costs, most would, in my experience, given the choice (that is, independent of cost) use patents.

Copyright

Many SME's harbour an unjustified belief that they can "copyright" their ideas. Copyright exists to protect not ideas as such but only the manner of expression of ideas in certain forms, primarily literary, artistic, dramatic, musical works and modern means of recording and reproduction of such works. At the borderline copyright can in some countries protect some forms of industrial design.

Otherwise, for industry protection is limited to items such as brochures, operation manuals and similar publications. Protection extends to prevent copying of the text etc. and does not prevent use of the ideas expressed in the text.

Copyright remains in most countries the principal form of protection for material such as computer software, although increasingly computer firmware is being accorded special protection through autonomous regimes, such as the Semiconductor Chip Protection Act in the USA and similar regulations in the UK and elsewhere in the European Community.

In most countries copyright arises automatically and without formality as soon as the work in question is created but in the USA and most South American countries a deposit of at least extracts from the program are required for full protection in litigation. Expert advice should be taken in all cases. The period of protection differs markedly from country to country for the "new technology" rights and may not be the traditional period found for books and similar publications.

Most countries do not require any express claim to copyright, although the required claim in UCC (Universal Copyright Convention) countries of the copyright symbol © plus the name of the owner and year of publication is common practice throughout the world. It is certainly prudent to make such a claim on the program.

Some countries will give a degree of protection to industrial designs through the protection of copyright in the production drawings or an original concept drawing of such designs. Expert advice should always be taken on any particular case and the law is haphazard at an international level. Reliance on copyright is hazardous.

Trade marks

A trade mark can often be a useful commercial supplement to a patent or design. It will be associated with the product or process. Even when the product is not new, not patentable, not anything, it can still be a market leader because it is better built, better designed, better packaged or simply better marketed. The name then sells the product. The term "trade mark" is now generally extended to service marks, identifying the supplier of services, though not all countries protect service marks by registration.

Trade mark registrations, unlike most rights are capable of existing in perpetuity, provided that renewal fees are paid periodically (typically every ten years) and the marks remain in use so that a third party cannot obtain an order to cancel them for abandonment.

If marks could remain in force without use there would be an obvious problem with a crowded but useless register of rights of no real on going commercial relevance.

The trade mark symbolises the reputation for quality or reliability of the product or process and long after the other rights are out of force it may be a valuable right worth substantial sums. In the case of many western companies the trade marks are the most valuable asset on the balance sheet of the company's accounts.

Thus a well-known trade mark helps to preserve product differentiation even in a crowded competitive market and assists in the development of brand loyalty and in dealing with counterfeiters and other competitors seeking to cash in on a market developed and led by the innovator. The roles of trade marks should not be forgotten when planning commercial development of technology, even through licensing.

The scope of trade marks which are registrable has expanded significantly in recent times with many jurisdictions allowing registration for the first time of shapes, smells, sounds and the like. However, the vast preponderance of trade marks remain the most obvious - words and devices.

Trade marks should be registered whenever possible. Although in many countries there is no need to register a trade mark and rights can be acquired by use and reputation in the mark a registration is nearly always stronger and cheaper to enforce.

Compared with patents, registration of your trade marks is cheap and can operate as a very effective insurance policy against legal costs, because it enables you to cut the cost of litigation founded on your reputation and, in certain cases, to invoke the assistance of the public authorities, such as the Customs and Excise, in cutting off counterfeit goods at source.

Furthermore, the risk of accidental trade mark conflicts arising is reduced because registered trade marks are easier to search for than unregistered marks and a reputable company, especially a reputable foreign company selling into the country for the first time, will search for availability of the proposed mark before adopting it. Equally, of course, it is always wise to have a trade mark availability search conducted yourself before adopting a new trade mark.

Trade mark registration is a national affair and the period taken for a mark to be registered varies from as little as a few months to three or four years, depending on the country involved.

Once the mark is applied for there is in principle no reason to refrain from using it, unless you are concerned about possible prior rights (which might in any event be exercised against you even after registration) and in some countries early use before registration is extremely desirable.

Your patent attorney or trade mark attorney will be able to advise you of the requirements of particular countries. Some countries operate a first to use, and others a first to file, system, with only limited protection for well known marks of overseas countries in the absence of a registration in the home market.

Although international trade mark protection is again based on national rights there are initiatives such as the Madrid Agreement, the Madrid Protocol (both administered by WIPO and providing for cheaper extension of registrations of trade marks based on a home registration or application), the Community Trade Mark (providing for a trade mark applicable to the whole of the European Union) and the arrangements for a registration with the AOPI (which is a similar arrangement for certain African states).

Table 6 sets out a list of member states of the Madrid Agreement, Madrid Protocol and the Community Trade Mark.

Trade marks are of growing importance: alongside copyright protection for software they are receiving increasing attention from lawyers, accountants and investors. Accountancy practices throughout the world are having to adjust to the value of brands by accepting them increasingly as non depreciating assets on the balance sheet.

In the race to modernise, attention focuses on technology, but in the transition countries in particular one cannot avoid the impact of foreign originating trade marks on everyday life.

One phenomenon of the late twentieth century has been the explosion of trade mark licensing in general and the more fashionable terms franchising and character merchandising in particular. You have only to look at the number of franchised operations in transition countries which appear on signs between an airport and the city centre to observe this phenomenon: McDonalds, Burger King, Kentucky Fried Chicken; hotel chains such as the Marriott and Radisson groups; and many others where local production facilities operate under licence. This takes us to the next area for discussion.

Licensing patents and other rights

Even before the popularity of trade mark licences really took off, licensing-in has been a traditional way for developing, and now transition, countries to take rapid leaps. These deals range from taking older models and producing them more cheaply (in the case of the automobile industry) while building up expertise and production line experience, to buying in basic technology to fill gaps in expertise whilst producing directly competing products (many food production techniques); to advanced technology licensing making use of lower labour costs or existing equipment (laser technology for production of CDs and the like).

A short point should be made. The licensing of patents and other intellectual property rights is in many countries subject to specific provisions of national or regional competition law: the parties do not have full freedom of contract as to what they agree. Contravention of these competition laws (for example Article 85 of the Treaty of Rome) can lead to difficulties of enforcement and even to sanctions from the public authorities. Collaboration agreements give further difficulties when intellectual property rights have to be considered, for example in terms of the conditions imposed by government or research funding body grants.

Accordingly any licence agreement needs to be carefully vetted by professional advisers. Equally, legal doctrines limiting the effect of patents and other rights once a product has come onto the market with the consent of the patentee are in force in many places. In particular, in the European Union, the so-called "exhaustion of rights" doctrine, prevents the use of parallel patents in more than one country to extract payments twice from users. These doctrines have to be taken into account when licensing or exploiting the rights in question.

Dealing with competitors' intellectual property rights

It is essential to consider the intellectual property position of others before entering into the market place with a new product or indeed into a new territory. To enter into production with a product that claims to be subject to someone else's patents is sheet folly unless careful thought has been given to the matter.

The consequences can include injunctions to prevent further manufacture and sale, delivery up of existing stocks, damages for loss sales, costs and in extreme cases criminal sanctions.

If a conflicting right does appear there may be ways of dealing with it. The existence of a prior right may not be fatal to your development. If a right does exist there may be the possibility of a licence or even a cross licence if you have some complementary development. Equally, all may not be as it seems.

Many people are daunted by the apparent threat of a claim to be infringing a patent and do not take the trouble to investigate the matter fully, when investigation may disclose that there is no risk. The patent may not in fact be in force in the markets where you wish to sell, or may of course be invalid. Equally, the fact that you are free to sell in your own country does not mean that you will be able to sell by exporting overseas.

If there is one piece of advice to give to SME's it would be: please do not be afraid of patents and the other intellectual property rights. They should be regarded as an opportunity, not a threat. They have an aura of mystery that is quite unjustified. They are, when all is said and done, technical documents with a legal element. The bulk of a patent is a concise and clear account of an invention, with illustrative drawings complying with the usual engineering drawing standards.

Both patent and other intellectual property right documents can contain a great deal of useful and new information about technology and about competitors. You might need this information for a number of reasons. You might for example want to know what your competitors are up to. You might want to start to catch up with the competition and think of an improvement. You might want to look for ways of avoiding the patent. You might of course want to take the risk of copying and infringe another's rights: they might not sue you. You might want to look for ways of attacking the patent if you are sued. The patent will often be the only reliable source of information about what the patentee thinks his rights really are.

Summary

Even if patents are not thought to be appropriate for your institution or company, you ignore them at your peril. If you are not interested in a monopoly and just want to prevent other people from being able to stop you from doing something or to threaten you with such action the answer is cheap and simple. Just disclose the invention to the public. The same applies for registered designs. Unregistered design rights need not be disclosed to safeguard your position but of course much that is protectable by such rights will also be potentially patentable so that non-disclosure may be a risk.

If you want to be sure that your new product does not infringe third party rights then it is possible to do some very rapid and cheap searching of the patent literature to determine whether you infringe existing patent rights. It is also possible to undertake clearance searches for trade marks, but it is more difficult to check-up on registered designs and almost impossible to run any check on unregistered design rights.

If you do face a problem you can always get advice on avoidance engineering - for instance, you might have a prototype that is a bit close to some else's machine. You can obtain professional advice on changes to the design and on modifications to avoid other people's design rights. Again the important thing is to make sure you know what the true position is. Advice is relatively cheap and quick. Ignorance may, in the short term, be bliss but in the mid to long term it is often much more expensive.

Table 1a

Countries where patents may be filed claiming priority from an original filing up to twelve months after that filing by virtue of Paris Convention for Protection of Industrial Property, through World Trade Organisation (WTO) or to a limited degree by bilateral agreements with certain countries

Albania
Algeria
Antigua[a]
Argentina
Armenia
Australia
Austria
Azerbaijan
Bahamas
Bangladesh
Barbados
Barbuda[a]
Bahrain[a] (from 29/10/97)
Belarus
Belgium
Belize[a]
Benin
Bolivia[a]
Bosnia and Herzegovina
Botswana[a]
Brazil
Brunei[a]
Bulgaria
Burkina Faso
Burundi
Cameroon
Canada
Central African Republic
Chad
Chile
China
Columbia[a]
Congo
Costa Rica
Cote d'Ivoire
Croatia
Cuba
Cyprus
Czech Republic
Darussalam[a]
Democratic Peoples Republic of Korea (North Korea)
Denmark (including the Faroe Islands)
Djibouti[a]
Dominica[a]
Dominican Republic
Egypt
El Salvador
Equatorial Guinea
Estonia
Finland
France (including all overseas departments and territories)
Gabon
Gambia
Georgia
Germany
Ghana
Greece
Guatemala[a]
Guinea
Guinea-Bissau
Guyana
Haiti
Holy See (Vatican)
Honduras
Hong Kong[b]
Hungary
Iceland
India[b]
Indonesia
Iran
Iraq
Ireland
Israel
Italy
Jamaica[a]
Japan
Kazakstan
Kenya
Kuwait[a]
Kyrgyzstan
Latvia
Lebanon
Lesotho
Liberia
Libya
Liechtenstein
Lithuania
Luxembourg
Macau[a]
Macedonia
Madagascar
Malawi
Malaysia
Maldives[a]
Mali
Malta
Mauritania
Mauritius
Mexico
Monaco
Mongolia
Morocco
Mozambique[a]
Myanmar[a]
Namibia[a]
Netherlands (including Netherlands Antilles and Aruba)
New Zealand
Nicaragua[a]
Niger
Nigeria
Norway
Pakistan[b]
Panama
Paraguay
Peru
Philippines
Poland
Portugal
Republic of Korea (South Korea)
Republic of Moldova
Romania
Russian Federation
Rwanda
Saint Kitts and Nevis
Saint Lucia
Saint Vincent and the Grenadines
San Marino
Senegal
Sierra Leone[a]
Singapore
Slovakia
Slovenia
South Africa
Spain
Sri Lanka
Sudan
Suriname
Swaziland
Sweden
Switzerland
Syria
Tajikistan
Thailand[a]
Togo
Trinidad and Tobago
Tunisia
Turkey
Turkmenistan
Uganda
Ukraine
United Arab Emirates
United Republic of Tanzania
United States of America (incl. all territories, including Puerto Rico)
Uruguay
Uzbekistan
Venezuela
Vietnam
Yugoslavia
Zaire
Zambia
Zimbabwe

[a] World Trade Organisation (including the agreement on Trade Related Aspects of Intellectual Property Rights) member only.
[b] By virtue of bi-lateral agreements only.

Table 1b

Countries which are party to the Patent Cooperation Treaty whereby central preliminary examination by a International Preliminary Examining Authority may be requested and deferral of national filing costs in the member States may be achieved.

Albania
Armenia
Australia
Austria
Azerbaijan
Barbados
Belarus (Filed when asssociated to EPO only)
Belgium (Filed designating EPO only)
Benin (Filed designating OAPI only)
Bosnia and Herzegovina
Brazil
Bulgaria
Burkina Faso (Filed designating OAPI only)
Cameroon (Filed designating OAPI only)
Canada
Central African Republic (Filed designating OAPI only)
Chad (Filed designating OAPI only)
China
Congo (Filed designating OAPI only)
Cote d'Ivoire (Filed designating OAPI only)
Cuba
Czech Republic
Democratic Peoples Republic of Korea(North Korea)
Denmark (including the Faroe Islands)
Estonia
Finland
France (Filed designating EPO only)
Gabon (Filed designating OAPI only)
Georgia
Germany
Ghana(Filed designating ARIPO only)
Greece (Filed designating EPO only)

Guinea(Filed designating OAPI only)
Guinea-Bissau (as of 12/12/97)
Hungary
Iceland
Indonesia
Ireland (Filed designating EPO only)
Israel
Italy (Filed designating EPO only)
Japan
Kazakstan
Kenya
Kyrgyzstan
Latvia
Lesotho
Liberia
Liechtenstein
Lithuania
Luxembourg
Macedonia
Madagascar
Malawi (Filed designating ARIPO only)
Mali (Filed designating OAPI only)
Mauritania (Filed designating OAPI only)
Mexico
Monaco (Filed designating EPO only)
Mongolia
Netherlands (Filed designating EPO only)
New Zealand
Niger (Filed designating OAPI only)
Norway
Poland
Portugal
Republic of Korea (South Korea)
Republic of Moldova
Romania
Russian Federation
Saint Lucia
Senegal (Filed designating OAPI only)
Sierra Leone
Singapore
Slovakia
Slovenia
Spain
Sri Lanka
Sudan (Filed designating ARIPO only)

Swaziland (Filed designating ARIPO only)
Sweden
Switzerland
Tajikistan
Togo (Filed designating OAPI only)
Trinidad and Tobago
Turkey
Turkmenistan
Uganda
Ukraine
United Kingdom
United States of America
Uzbekistan
Viet Nam
Yugoslavia
Zimbabwe

Table 1c

Countries which may be designated in a European Patent Convention Application either as contracting states or as "extension countries" under agreements reached with those countries

- **EPC contracting States**

Austria	Greece	Portugal
Belgium	Ireland	Spain
Denmark	Italy	Sweden
Finland	Luxembourg	Switzerland (and Liechtenstein)
France	Monaco	United Kingdom
Germany	Netherlands	

- **EPC "extension" Countries**
-

Albania	Lithuania	Romania
Latvia	Slovenia	

Table 2

European patent requirements and exclusions

- **Fundamental requirements for patentability of invention in Europe**

  Novelty - invention must not be publicly known anywhere in the world

  Inventiveness - invention must be non-obvious to person skilled in the field

  Capability of industrial (including agricultural) application

- **Express exclusions from patentability in Europe**

  A mere discovery, scientific theory or mathematical method

  A literary, dramatic, musical or artistic work or aesthetic creation

  A scheme, rule or method for: performing a mental act; playing a game; doing business

  Computer programs (contrast an industrial machine driven by one)

  The mere presentation of information

  Offensive, immoral or anti-social inventions

  Varieties of animal or plant

  Any essentially biological process for the production of animals or plants (not being a microbiological process or product of such a process)

  Methods of treatment of the human or animal body by surgery, therapy or diagnosis (not a product for use in such activity)

Table 3

Potential decision chart for filing patents

Is the development new and unpublished?
Yes
⇓

Is the development clever?
Yes
⇓

Is the development industrially applicable?
Yes
⇓

Is the development saleable?
Yes
⇓

Is the product life long enough?
Yes
⇓

Would secrecy work just as well?
No
⇓

Where would a patent be useful?
Yes
⇓

Can we afford it?
Yes
⇓

Can it realistically be enforced?
Yes
⇓

File a patent!

Table 4

Main utility model system countries

AIPO (African Intellectual Property Organisation)
Australia (petty patent)
Brazil
Bulgaria
China
Chnia (Taiwan
Colombia
Costa Rica
Dominican Republic
Germany
Greece
Guatemala
Haiti
Hong Kong (short term patent)
Italy
Japan
Korea (South)
Leasotho
Malaysia
Morocco
Oman
Philippines
Poland
Portugal
Qatar
Saudi Arabia
Somalia
Spain
UAE
Uruguay
Yemen

Table 5

International design registration systems (Hague Deposit)

Belgium
Benin
Bulgaria
Cote d'Ivoire
DPR Korea
Egypt
France
Germany
Holy See
Hungary
Indonesia
Italy
Liechtenstein
Luxembourg
Macedonia
Moldova
Monaco
Morocco
Netherlands
Romania
Senegal
Slovenia
Spain
Surinam
Switzherland
Tunisia
Yugoslavia

Table 6a

International Trade Mark Registration Members of the Madrid Agreement

Albania
Algeria
Armenia
Austria
Azerbaijan
Belarus
Belgium
Bosnia-Herzogevina
Bulgaria
China
Croatia
Cuba
Czech Republic
Democratic Peoples Republic of Korea (North Korea)
Egypt
France
Germany
Hungary
Italy
Kazakstan
Kyrgyztan
Latvia
Liberia
Liechtenstein
Luxembourg
Macedonia
Moldova
Monaco
Mongolia
Morocco
Netherlands
Poland
Portugal
Romania
Russia
San Marino
Sierra Leone
Slovakia
Slovenia
Spain
Sudan
Switzerland
Tajikistan
Ukraine
Uzbekistan
Vietnam
Yugoslavia

Table 6b

The International Trade Mark Registration Members of the Madrid Protocol

China
Cuba
Czech Republic
Democratic Peoples Republic of Korea (North Korea)
Denmark
Finland
France
Germany
Hungary
Iceland
Lithuania
Moldova
Monaco
Norway
Poland
Portugal
Russian Federation
Slovak Republic
Spain
Sweden
Switzerland
United Kingdom

Table 6c

- **The European Community Trade Mark**

Austria	France	Italy
Benelux	Germany	Portugal
Denmark	Greece	Spain
Finland	Ireland	Sweden
		United Kingdom

R&D CAPABILITIES OF EASTERN ASIAN COUNTRIES - CENTERING ON ASIAN NIES -

TAKABUMI HAYASHI
Professor of International Business, Faculty of Economics
Rikkyo University, 3-34 Nishi-Ikebukuro, Toshima-Ku,
171 Tokyo, JAPAN

1. INTRODUCTION

For years, from observing the production capability of eastern Asian countries in their various aspects, in particular, those of Asian NIES, i.e., Newly Industrializing Economies, the author has located his basic viewpoint as follows: if we stop adhering to the concepts and ideas of OECD and analyse NIES in the flow of economic history, it may be interesting to look at the position in the world economy of Germany and France in the latter half of the 18th century when England was in the midst of the Industrial Revolution, or that of Japan ant the end of the 19th century when the United States was in the final stages of its own industrial revolution. In the same sense, as the 20th century nears its end, I would like to define those several eastern Asian countries that have been rapidly establishing their industrial infrastructure, as NIES(Chen & Hayashi[1995]).

In doing so, an extremely important issue is how we can evaluate the technological capability of these countries, which represents a key factor among various aspects of their production capability. When estimating technological capability, we may start by dividing such strength into two aspects: process technology and product technology.

In the area of process technology, the evaluation of technological capability to produce is quite important. Technological capability to produce here means how well, in the process of establishing a mass production system, these countries can imitate the technology of those products belonging to mature industries that have been developed and introduced to markets by other countries, to the extent that their products can be offered to the international market stable in quality and lower in price(Hirakawa[1993]). This is exactly where we can find what makes NIES what they are. This paper will not linger in this area, however. Rather, it will attempt further to verify that the NIES

A. Inzelt and J. Hilton (eds.), Technology Transfer: From Invention to Innovation, 233–254.

countries in eastern Asia have broken out from the level of imitative production technology and are beginning to acquire R&D capability in the domain of product technology. For the purpose of this paper, "product technology" is defined as "technological capability to develop a series of related technology elements that comprise a product." With this definition of product technology, what indices can we use to evaluate this technological capability represented by such product development capability? This paper will verify it by using two indices. The first index is the position of the eastern Asian countries in major technological sectors as estimated by the number of patents granted to these countries by the United States. The second index is the position of these countries as estimated by the number of scientific papers published in the United States.

The first approach to use the number of U.S. patents granted to verify R&D capability is viable for the following reasons: when a company has developed a new technology, the greater the strategic significance the new technology seems to have in the international market, the more likely the company is to apply for patents in the U.S., which has a larger market and more competitors than any other country, to ensure the exclusive right to use of that technology. Since the details of all patented technologies are disclosed, those newly developed technologies that relate to production processes where strongly-suspected infringement by a third party is difficult to prove are likely to be treated as trade secrets. On the other hand, those strategic technology elements comprising finished products where patent-infringements can easily be proved tend to be the subject of patents.

Therefore, a study of trends in technologies that have been subjected to technical examination and granted patents in the U.S. will provide measurements that are viable in identifying to a certain extent trends in R&D capability based on applying organizations or their countries. This means that patent analysis is an effective way of grasping to some extent the R&D capability in terms of product technology of a specific industrial area. The data used in this paper were sourced from PATOLIS issued by the Japan Patent Information Organization (JAPIO) and materials issued by the Patent and Trademark Office of the U.S. Department of Commerce.

The second index, searching through scientific papers to determine numbers by country and organization, is used to clarify where these eastern Asian countries stand in the ranking of the numbers of scientific papers published in the U.S. in major technological sectors. Since the authors of scientific papers are mostly researchers belonging to companies, universities, research centers or other similar organizations, standards that can be verified from the numbers of papers reflect not industrial technology itself but the level of science and technology on which it is built. The reason this paper restricts its target to papers published in the U.S. is that most papers have passed screening examinations and been qualified to be equal or better in quality and significance than a required level. Searching through these technological papers, therefore, is a viable method of verifying the research standards, in other words, science and technology standards, of these eastern Asian countries or their companies, universities and research centers, even after discounting possible bias resulting from the facts that the U.S. is the publishing country and that the language is English. This search through scientific papers was made online on the JOIS database of the Japan Information Center of Science and Technology (JICST).

2. U.S. PATENTS AND THE POSITION OF EASTERN ASIAN COUNTRIES

Studies on competitive R&D capability through analysis of U.S. patents have been done by some researchers(Cantwell, Etemad, Hayashi, OECD[2], Townsend,etc.). These studies, however, limited their targets to major developed countries; there seems to be none that consider the R&D capability of eastern Asian countries, in particular NIES, from the viewpoint of the acquisition of U.S. patents, or more appropriately, there has been no necessity for it. This paper will first confirm the position of the eastern Asian countries by looking at a ranking by country concerning U.S. patents granted. It will then consider the position of these countries by further breaking down this ranking into different major industrial areas.

2.1. TRANSITION IN RANKING BY COUNTRY

Table 1 shows a transition in the ranking by country of the number of U.S. patents granted. The cumulative total of patents granted by the U.S. from 1963 to 1974 for different countries is the largest for the U.S., followed by Germany, England, Japan, France, Switzerland, Canada, Sweden, Italy and the Netherlands, forming the top ten countries. According to a comparison of the results for individual years since 1975, the first major change is found in the decline in the share of patents granted to inventions made in the U.S. to slightly above 50% and subsequent recovery to 55% after entering the 1990s. The second major change is that Japan moved up to second place to replace Germany in the 1980s and has maintained this place ever since. The third major change is a sharp increase in the numbers of patents granted to Taiwan and Korea in the 1990s. In 1980, the numbers of U.S. patents granted to inventions made in these countries were only 65 and 8, respectively. In 1990, these numbers rose dramatically to 732 and 225, 9th and 12th places, respectively. In 1995, they further jumped to 1443 and 943 and have been increasing steadily since. In 1995, Korea becomes No.8 ranking and Taiwan moved up to 7th place, even surpassing Italy, Switzerland, the Netherlands and Sweden in the ranking of the numbers of patents. This means that, in limited terms of the number of patents on a year-to-year basis, or from the limited viewpoint of the flow of technology, Taiwan and Korea have reached, quantitatively, a stage in the progress of R&D capability that is comparable with that of major developed countries. In the next section, the makeup of U.S. patents granted to the two countries according to industrial areas will be studied.

Table 1

	Ranking	1963-74	1980	1985	1990		1995
US Invented	1	583080	37356	39555	47393	US Invented	55739
Foreign Invented		199272	24463	32106	42971	Foreign Invented	45680
JP	4	30562	7124	12746	19524	JP	21764
DE	2	51079	5782	6718	7610	DE	6600
GB	3	32843	2406	2495	2789	FR	2821
FR	5	20310	2088	2400	2868	GB	2475
CA	7	12186	1081	1342	1861	CA	2104
CH	6	12453	1265	1233	1260	TWN	1620
IT	9	6684	806	919	1284	KOR	1161
NL	10	6540	654	766	959	IT	1078
SE	8	7572	822	857	768	CH	1056
RU	12	2278	460	147	174	SE	806
BE	11	2631	244	240	313	NL	799
AT	13	2182	267	318	393	AU	459
AU	14	1751	265	340	432	BE	397
DK	15	1477	157	187	158	IL	384
FI	22	553	121	200	304	FI	358
IL	21	604	113	179	299	AT	338
NO	17	761	79	90	112	DK	199
CZ	16	1074	55	54	39	ES	148
ZA	19	683	74	96	115	NO	130
ES	20	663	65	78	130	ZA	123
MX	18	743	41	32	32	HKG	86
HU	23	380	87	108	93	BR	63
NZ	27	197	51	33	52	CHN	62
PL	25	226	37	11	17	IE	52
AR	24	260	18	11	17	HU	50
RO	26	223	14	3	1	NZ	44
BR	28	184	24	30	41	MX	40
LI	29	172	18	13	15	IND	37
TWN	32	1	65	174	732	AR	31
KOR	31	33	8	39	225	VE	29
HKG	30	109	27	25	52	LU	24
CHN			1	1	47	LI	17
SGP						CZ	15
IND			4	10	23	RU	12
						SGP	
						MLS	
						TH	
Total		782352	61819	71661	90364	Total	101419
Source: Industrial Patent Activities in the US(Dept. of Commerce, Various Issues)							

2.2. RANKING BY INDUSTRY AND ITS CHARACTERISTICS

This section analyses the numbers of patents granted by the U.S. to the inventions made by corporations and individuals in Taiwan and Korea during the period from January 1 to December 31, 1994. As shown in Table 2, among the patents granted to Taiwan and Korea this year (1443 and 943, respectively), the industrial sector that represented the largest percentage is electric/electronic equipment (SIC:36,3825) for both the countries, totaling 334 and 536, respectively. The sub-sector that accounted for the largest portion of the electric/electronic equipment sector is communications equipment/electronic parts (SIC:365-367), acquiring 225 and 476 patents each.

Table 2 Numbers of Patents Granted by the U.S. to Taiwan and Korea According to Major Patent Areas

SIC Description	Taiwan	Korea
SIC 28(Chemicals and Allied Products)	**46(3.2)**	**52(5.5)**
(SIC 281,286: Basic Industrial Inorganic & Organic Chemistry)	**21**	**17**
(SIC 283 : Drugs and Medicines)	**5**	**11**
SIC 30(Rubber & Miscellaneous Plastic Products)	**56(3.9)**	**16(1.7)**
SIC 32(Stone, Clay,Glass & Concrete products)	**19(1.3)**	**10(1.1)**
SIC 34(Fabricated Metal Products)	**201(13.9)**	**27(2.9)**
SIC 35(Machinery, Except Electrical)	**258(17.8)**	**158(16.8)**
(SIC 354 : Metal Working Machinery & Equipment)	**38**	**8**
(SIC 357 : Office Computing & Accounting Machines)	**32**	**79**
SIC 36,3825(Electrical & Electronic Machinery)	**334(23.1)**	**536(56.8)**
(SIC 365-367 : Communication Equip & Electronic Components)	**225**	**476**
(SIC 366-367 : Electro Comps & Accessories & Commun. Equip)	**215**	**387**
SIC 37,348(Transportation Equipment)	**77(5.3)**	**21(2.2)**
(SIC 373-375,379 : Other Transportation Equip)	**53**	**2**
SIC 38(Professional & Scientific Instruments)	**110(7.6)**	**84(8.9)**
Others	**342(23.7)**	**39(4.1)**
TOTAL	**1443(100)**	**943(100)**
Source: Patenting Trends in the US (US. Dept. of Commerce)		

This trend is more significant for Korea. 56.8% of the patented inventions is in the sector of electric/electronic equipment (SIC:36,3825, 536 patents), and a majority (50.5%) thereof is accounted for by its central field, i.e., communications equipment/electronic parts (SIC:365-367, 476 patents). It can be presumed from this that the patents granted to Korea are concentrated in the field of electronic parts, more specifically, semiconductors centering around DRAMS. What should be further noted is that Korea is in 5th place in the 1994 ranking for this area, meaning that England, Italy, the Netherlands, Canada, Switzerland and Sweden are below Korea in the number of patents in this field. However, in addition to the sector of electric/electronic equipment (SIC:36,3825), only one sector exceeded 100 in the number of U.S. patents granted, that is, machinery (SIC:35, excluding electric/electronic equipment), which recorded 158 patents and represented 16.8% of the total. Thus, the greatest characteristic of Korea in its U.S. patent acquisition by technological sector may be summarised as follows:

The greatest factor in Korea's climb to 10th place in the ranking by country for the number of U.S. patents granted was the significant contribution of the electronic/electric sector (SIC:36,3825) by count, 546, and ranking 5th in the same sector. The fact that this one sector accounted for 57% of the total number of patents granted to Korea indicates that Korea is still far from achieving a well-balanced R&D capability across all the different technological sectors. This in turn indicates that the infrastructure for technological development in Korean industry is quite unevenly distributed, with a heavy concentration in one specific sector.

On the other hand, while similarly to Korea the sector that brought Taiwan the largest number of U.S. patents is electric/electronic equipment (SIC:36,3825), it only represents 24.9% of the total (1,443 patents). The sector with the second largest number of U.S. patents, 258, is machinery (SIC:35, excluding electric/electronic equipment), accounting for 17.9% of the total. The third is metal products (SIC:34), which represents 19.1% with 276 patents, followed by three other sectors with 100 or more patents each. These are professional & scientific instruments (SIC:38) with 125 patents (8.7%), gum and other plastic products (SIC:30) with 118 patents (8.2%) and transportation equipment (SIC:37) with 100 patents (6.9%). These five sectors together account for 63% of all the patents granted by the U.S., demonstrating a greater presence of technological development infrastructure than Korea. The sector in which both the countries are weak is chemical and related products (SIC:28), with 57 patents for Korea and 49 patents for Taiwan, leaving them down in 13th and 14th places, respectively. Thus, while the two countries have two common characteristics, namely, the recent sharp increase in the number of U.S. patents granted and weakness in the chemical area, a significant difference can be found in how broad their patent-level technology extends. In the next section, differences between these countries as observed in patent applicants will be revealed.

2.3 CHARACTERISTICS OF KOREA AND TAIWAN AS OBSERVED IN PATENT APPLICANTS

This section attempts to identify what kind of groups of corporations or individuals are active in applying for U.S. patents. As shown in Table 3, the greatest characteristic of U.S. patent applicants in Korea is that the Samsung Group (Chaebol: Korean-type plutocracy), including Samsung Electron and many others, have by far the largest number of U.S. patents, 459, demonstrating the Group's prominent position among Korean corporations. In fact, the Group alone accounts for 48.7% of all the U.S. patents granted to Korea, and more specifically, Samsung Electronics alone accounts for 42.2% of the country's total by achieving 398 patents. The Lucky Group, the second to Samsung, has 218 U.S. patents in total (Gold Star Com: 144, Gold Star Electron Co.: 59, others 15), followed by the Hyundai Group with a total of 66 patents (7.0%) granted to Hyundai Electron, Hyundai Motors and Hyundai Heavy Industries, the Daewoo Group with 19 patents granted to Daewoo Electron, and so on. Thus, the number of U.S. patents that are known to be held by the Four Great Groups (Chaebol), excluding those that are suspected, comprise 80.7% of all the patents granted to Korea.

Table 3 Numbers of Patents Granted to Major Korean Companies

	Patents	**%**
Samsung Group	459	48.7
(Samsung Electronics)	398	42.2
(--- Electron Devices)	45	4.8
(--- Aerospace)	6	
(--- Construction)	1	
(--- Corning)	2	
(--- Chemical)	1	
(--- Heavy Indtrs)	6	
Lucky Group	218	23.1
(Gold Star Com.)	144	15.3
(Lucky Electronics)	59	6.3
(Lucky Inf & Comm)	2	
(Lucky Electric Machinery)	1	
(Lucky Ltd.)	12	1.3
Hyundai Group	66	7
(Hyundai Electronics)	52	5.5

(Hyundai Motor)	13	1.4
(Hyundai Electronics USA)	0	
(Hyundai Heavy Industrs)	1	
Daewoo Group	19	2
(Daewoo Electronics)	19	2
KIST(National)	14	
KCTRI(National)	9	
Others	159	16.9
Total	943	100

Next, the characteristics of U.S. patent applicants of Taiwanese nationality will be identified (refer to Table 4). The first characteristic is that the organization holding the largest number of U.S. patents is not a private company but a national body called the Engineering and Technology Research Institute. A national body can further be found in the third place, the Administrative Institute National Science Committee, implying the important role that these public research organs are playing in this country. The second characteristic is that manufacturers of PCs, peripheral devices and semiconductors stand higher in the ranking (second, fourth and fifth, respectively), but the number of U.S. patents they each hold is much smaller compared with Korean companies and groups. The third characteristic is that quite a few patent applications have been made in the names of individuals. Patents held by individuals are 934 in number, accounting for 64.7% of the total, in contrast to the small shares represented by private companies and national research institutes, 18.6% (268) and 10.3% (148), respectively. This perhaps indicates that large companies are not central performers of technological development activities but a great variety of venture-type small companies are an important drive of these activities, leading to many applications for U.S. patents under the names of individuals. The next section will examine to which technological sectors the patents held by these major applicants in Korea and Taiwan belong .

2.4. SECTORS OF PATENTS HELD BY MAJOR APPLICANTS

As mentioned in the previous section, the Taiwanese applicant holding the largest number of U.S. patents in the country is a national organization called the Engineering and Technology Research Institute. The scope of technological sectors its U.S. patents cover is as follows:

Table 4

		No. of Patents	%
1	ITRI*	117	8.1
2	UMC	48	3.3
3	NSC*	23	1.6
4	TSMC	12	-
5	ACER	11	-
6	Green Master Industries	11	-
7	Silitek Corp	8	-
8	U-Lead Industrial Corp	6	-
9	Formosa CFC	5	-
10	C.Richard(Indivisual)	5	-
11	Winbond EC	4	-
12	China Synthetic Rubber	4	-
13	Chunghwa Picture Tubes	4	-
14	L.Leueis(Indivisual)	4	-
15	L.Yuan-Hsin	4	-
16	Y.M.Tung	4	-
	Others(Under 3)	1173	81.3
	Total	1443	100
	(Company)	268	18.6
	(National/Public)	148	10.3
	Indivisual	934	64.7
	N.A	93	6.4

As shown in Table 5, the makeup of the 117 patents granted to the said institute extends across 73 technological sectors according to the International Patent Criteria (IPC). Among these areas, the largest number of U.S. patents was granted in the sector of HO1L (semiconductor devices, solid state devices, manufacturing and processing methods, devices therefore: details of semiconductors or other solid state devices: devices comprising semiconductor or other solid state components, etc.). However, the number of patents classified as HO1L is 30, representing merely 15.9% of the total (189), followed by C07C (organic chemicals: cyclic compounds and carbocyclic compounds) and C08G (organic high molecular compounds: polysaccharides and their derivatives) with 11 patents each, and G11B (information storage based on relative motions between information storage carriers and converters) and H04N (visual communications: including TV) with 9 patents each. Patents in these five sectors total 70, representing 37.0% of all the U.S. patents granted to the institute, and patents in the top ten sectors (totaling 99) account for 50.3% of all (refer to the note below for the titles of the major technological codes other than the above-mentioned five sectors).

Table 5 Sectors of U.S. Patents Held by the Engineering and Technology Research Institute (Taiwan) (in 1994 : Number of US Patents)

Patents	No. of Fields	Total Patents	Share	IPC		
30	1	30	15.9	H01L		
11	2	22	27.5	C07C/C08G		
9	2	18	37	G11B/H04N		
8	1	8	41.3	C09K		
6	1	6	44.4	G06F		
5	3	15	52.4	C08F/C08L/G03F		
4	1	4	54.5	H03K		
3	7	21	54.5	B29C/C12N/G02B		
2	11	22	65.6	B01D/B03C/B05D		
1	43	43	77.2	A01N/A23K/B01J		
	Total : 73	Total : 189	100			
Note :ITRI got 117 US patents in 1994, according to [PATOLIS] searching.						
: As a patent is applied for several patent clasification sectors, total number of patents gets to 189						
Source : [PATOLIS]searching						

In comparison, the sectors to which the patents held by Samsung Electron, the Korean applicant holding the greatest percentage of all the U.S. patents granted to the country, extend will be analyzed below. Table 6 shows the makeup of the 40 patents granted to Samsung Electron by the U.S. across different technological sectors. This table, in other words, may be said to indicate which technological sectors the company has targeted in its strategic development activities. Of all the 410 U.S. patents held by the company, 396 have been identified by their IPC codes, which extend across 78 sectors. Some of these 396 patents were applied for and granted in more than one sectors; including these overlaps, the patent applications made successfully by Samsung total 544. Based on this total, the technological sector in which the largest number of U.S. patents was granted is H04N (visual communications: including TV) with a total of 170 patents (31.3%), followed by H01L (semiconductor devices or solid state devices, manufacturing and processing methods, devices therefore: details of semiconductors or other solid state devices: devices comprising semiconductor or other solid state components, etc.) with 70 patents (12.9%), G11B (information storage based on relative motions between information storage carriers and converters) with 69 patents (12.7%), G11C (static storage of information storage: storage device and mechanism for digital/analog information) with 46 patents (8.5%), and G06F (digital computers which execute at least part of computation electrically) with 23 patents (4.2%). These top five sectors have 387 U.S. patents in total, accounting for 69.5% of the company's total of 544 patents (refer to the note below for the titles of the major technological codes other than the above-mentioned five sectors).

Table 6 Sectors of U.S. Patents Held by Samsung Electron (Korea) (in 1994 : Number of US Patents)

Patents	No. of Fields	Total Patents	Share	IPC
170	1	171	29.7	H04N
70	1	70	41.9	H01L
69	1	69	53.9	G11B
46	1	46	61.9	G11C
23	1	23	65.9	G06F
16	1	16	68.7	H03K
11	3	33	74.4	F25B/H04N/H05B
9	2	18	77.6	G02B/H04B
8	1	8	79	G05B
6	2	12	81	D06F/G05F
5	2	10	82.8	F25D/G02F
4	5	20	86.3	G01R/G03G/H01S, other 3 sectors
3	6	18	89.4	B41J/G03B/G05D, other 2 sectors
2	11	22	93.2	A23B/A47L/B65H, other 8 sectors
1	39	39	100	A47J/A61L/A62B,other36 sectors

Note :Samsung Electronics got 396 US patents in 1994, according to [PATOLIS] searching.As a patent is applied for several patent classification sectors, total number of patents is 575.

Source : [PATOLIS]searching

As shown in Table 7, the number of U.S. patents has increased rapidly and jumped up to 22nd place in the ranking by company, surpassing even SIEMENS, BAYER and BASF in Germany and PHILIPS in the Netherlands . Compared with 16 Japanese companies named in the top 50 in this ranking, Samsung Electron ranks lower than only nine Japanese major companies, i.e., Cannon, Hitachi, Mitsubishi Electric MFG., Toshiba, Matsushita Electric Industrial, Sony, Fujitsu, Fuji Photo Film and Sharp, but higher than Ricoh, Sumitomo Electric Industries, Olympus Optical, Honda Motor, Konica, Asahi Optical, and Pioneer Electronic.

Table 7 Transition in Number of U.S. Patents Held by Samsung Electron

Year	Patents			
pre-1980	0			
1981	0			
1985	0			
1986	1			
1987	7			
1988	8			
1989	26			
1990	61			
1991	148			
1992	250			
1993	347		Total	1259
1994	410	Source: See Table 3		

3. U.S.-PUBLISHED PAPERS AND POSITION OF EASTERN ASIAN COUNTRIES

In the previous sections, we have analyzed the position of two eastern Asian countries, Taiwan and Korea, according to the number of patents granted by the U.S., one of the important indices for estimating technological development strength from the aspect of industrial technology. This section will look at the numbers of scientific papers, which is an index for showing the level of science of a specific country or organization. The data used here are limited to those for scientific papers published in the U.S. retrieved by searching through the JOIS database of the Japan Information Center of Science and Technology (JICST) and INSPEC.

3.1. RANKING BY AUTHOR'S NATIONALITY

First, the data obtained by searching through INSPEC, a database concerning physics, electrical engineering, electronics, control theory, computer, and computer utilization, will be used to analyze the numbers of scientific papers in these fields published in technical magazines issued in the United States for respective fields will be analyzed. As shown in Table 8, the numbers of papers published by authors of Taiwanese and Korean nationalities have increased at a much higher rate than those of major developed countries since the latter half of the1980s. As a result, the total number of papers published by Taiwanese and Korean authors increased to around 4,300 by 1996, which compares favorably with the number of papers published by German authors in the early 1990s.

Table 8 Transition in Number of U.S.-published Papers by Author's Nationality

Year	Korea	Taiwan	Japan	Germany	USA
1980	34	81	3071	2028	37304
1982	72	88	3358	2195	37565
1984	98	139	4456	2576	45635
1986	177	311	5423	2931	47586
1988	271	410	6159	3272	54469
1990	421	752	7487	3941	57166
1992	628	1196	8095	4493	55807
1994	1353	1875	9885	6101	61780
1996	2281	2081	11300	7414	58431

3.2. RANKING BY COUNTRY BASED ON MAJOR KEYWORDS

Here, a ranking of author's nationality will be created by searching through papers published in the U.S. with keywords "solid state components" and "mechanical properties" under the category of "hardware technology" and "computer utilization" under the category of "software technology" and grouping top 1,000 papers in the order of date of issue by respective nationalities. These are the more important among keywords, since they are associated with groups of U.S.-published papers that are placed highest in the keyword-based total number rankings for 1981 and 1994(Hayashi[1][3]).

Table 9-1 shows the ranking by author's nationality of papers published in the U.S. (top 1,000 in the order of date of issue) with the keyword "solid state components". The country in first place in 1981 is naturally the U.S., the publishing country, and second place, Japan, followed by the former U.S.S.R., France, England, Germany, Canada, India, Italy and Switzerland, all forming the top ten countries. In 1981, the nationalities of the contributors of papers keyworded with "solid state components" distributed among 30 countries; of them, among eastern Asian countries other than Japan, Taiwan, China and Korea were ranked No. 12, 14 and 16, respectively. In 1994, the number of authors' nationalities of the top 1,000 papers published in the U.S. increased to 36, with eastern Asian countries other than Japan also increasing to five, including China, Taiwan, Korea, Hong Kong and Singapore with heightened ranks of 11th, 12th, 13th, 18th and 21st, respectively.

Table 9-1 Papers Published in the U.S.: Keyword "Solid State Components"Number of Papers by Author's Nationality (1981, 1994)

1981	Nationality	No. of Papers	**1994**	Nationality	No. of Papers
1	USA	631		USA	532
2	JPN	79		JPN	141
3	RUS	43		DEU	44
4	FRA	23		GBR	40
5	GBR	22		FRA	35
6	DEU	20		RUS	28
7	CAN	17		CAN	24
8	IND	10		ITA	20
9	ITA	7		SWE	17
10	CHE	7		CHE	13
11	NLD	7		CHN	12
12	TWN	6		TWN	11
13	ISR	3		KOR	11
14	CHN	2		NLD	11
15	DNK	2		BEL	10
16	KOR	2		AUS	6
17	HUN	2		ISR	5
18	BEL	2		HKG	4
19	AUS	2		IRL	4
20	ESP	2		BRA	4
21	BRA	1		IND	3
22	SWE	1		DNK	3
23	AUT	1		ESP	3
24	POL	1		SGP	3
25	NOR	1		AUT	2
26	YUG	1		CSK	2
27	SAU	1		PRT	2
28	EGY	1		ARG	2
29	TUR	1		BGR	1
30	RMN	1		POL	1
31				NOR	1
32				FIN	1
33				HUN	1
34				ZAF	1
35				NZL	1
36				JOR	1
	N.A	93		N.A	8
	Total	1000		Total	1000

Next, characteristics of papers published in the U.S. with the keyword "mechanical properties" will be identified (refer to Table 9-2). In 1981, the number of countries from which contributions were sent was 34, with the U.S. and Japan as Nos. one and two, followed by Canada, England, Germany, France, India, Sweden, Australia and Italy, forming the top ten countries. Among the contributing countries, eastern Asian countries other than Japan are China, Korea and Taiwan, with rankings 13th, 20th and 27th, respectively. In 1994, the total number of contributing countries increased to 44, and eastern Asian countries other than Japan increased to five to include China, Taiwan, Korea, Hong Kong and Singapore with their ranking moving up to 11th, 12th, 13th, 18th and 24th, respectively.

Table 9-2 Papers Published in the U.S.: Keyword "Mechanical Properties"

		1981		1994	
	Nationality	No.of Papers	Nationality	No. of Papers	
1	USA	644	USA	584	
2	JPN	55	JPN	55	
3	CAN	42	CAN	34	
4	GBR	31	GBR	32	
5	DEU	30	RUS	28	
6	FRA	25	DEU	25	
7	IND	21	CHN	24	
8	SWE	8	TWN	20	
9	AUS	8	FRA	18	
10	ITA	7	IND	16	
11	RUS	7	AUS	13	
12	NLD	6	ITA	13	
13	CHN	6	NLD	10	
14	CHE	6	KOR	9	
15	FIN	6	ISR	9	
16	ISR	5	POL	7	
17	POL	5	BEL	7	
18	BEL	4	CHE	6	
19	CSK	4	SWE	5	
20	KOR	3	EGY	4	
21	AUT	3	MYS	4	
22	ZAF	3	SAU	4	
23	HUN	2	NOR	3	
24	DNK	2	AUT	3	

25	NZL	2	GRC	3	
26	KWT	2	SGP	2	
27	TWN	1	PAK	2	
28	ESP	1	FIN	2	
29	BRA	1	ESP	2	
30	NOR	1	YUG	2	
31	TUR	1	IRQ	2	
32	RMN	1	TUR	2	
33	EGY	1	COL	2	
34	PRR	1	BGR	1	
35			ALG	1	
36	N.A	55	KWT	1	
37			HKG	1	
38	Total	1000	IRL	1	
39			DNK	1	
40			BRA	1	
41			VEN	1	
42			HUN	1	
43			BHA	1	
44			PRT	1	
			N.A	37	
			Total	1000	

Table 9-3 shows an analysis of 1000 papers published in the U.S. with the keyword "computer utilization" under the category of "software technology". The number of countries contributing U.S.-published papers in 1981 was 29, which increased to 37 for 1994 papers. In addition to Japan, eastern Asian countries appearing in the 1981 ranking by author's nationality include Taiwan and Hong Kong only, in 20th and 28th place, respectively. The 1994 ranking includes six countries from this region, Taiwan, Korea, China, Hong Kong, Singapore and Malaysia, which earned 8th, 12th, 14th, 20th, 29th and 31st places, respectively.

Table 9-3 Paper published in the U.S. Keyword "Computer Utilisation"

1981	Nationality	No.of Papers	1994	Nationality	No. of Paper
1	USA	742	1	USA	620
2	JPN	30	2	JPN	57
3	CAN	21	3	CAN	40
4	DEU	20	4	DEU	39

5	ITA	20	5	ITA	32
6	GBR	18	6	GBR	31
7	NLD	17	7	FRA	25
8	AUT	9	8	TWN	15
9	FRA	8	9	AUT	13
10	BEL	8	10	SWE	12
11	FIN	8	11	RUS	8
12	RUS	6	12	KOR	7
13	IND	5	13	ESP	7
14	NOR	5	14	CHN	6
15	SWE	4	15	NLD	6
16	ISR	3	16	BRA	5
17	ATU	3	17	ISR	6
18	CHE	2	18	POL	4
19	ILE	2	19	YUG	3
20	TWN	1	20	HKG	3
21	ESP	1	21	NOR	3
22	HUN	1	22	STU	2
23	DNK	1	23	IND	2
24	BRA	1	24	BEL	2
25	POL	1	25	CHE	2
26	SAU	1	26	FIN	2
27	ZAF	1	27	SAU	2
28	HKG	1	28	BGR	2
29	NIJ	1	29	SGP	1
			31	HUN	1
			32	MLY	1
			33	DEK	1
			34	ARG	1
			35	ZAF	1
			36	NZL	1
			37	CSK	1
	N.A	27		N.A	36
	Total	1000		Total	1000

The rankings by author's nationality of papers published in the U.S. according to the data obtained by searching the JOIS database based on the three major keywords may be summarized as follows. First, the numbers of contributing countries of papers with the keywords "solid state components", "mechanical properties" and "computer utilization" show obvious upward trends by increasing from 30 to 36, 34 to 44 and 29 to 37, respectively. Similarly to the results of searching the INSPEC database, this indicates that the areas of science and basic research are experiencing a shift from a concentration in a few specific countries, or more specifically, the triple-pole structure of the U.S., Europe and Japan, to a more distributed structure over many countries. Second, as the number of contributors from eastern Asian countries increased, the places of these countries grew into the upper half of the ranking. From these two results, it can be assumed that the technological development infrastructure of eastern Asian countries has increased by international standards in the aspect of science.

3.3. RANKING BY COMPANY

The number of scientific papers of the Samsung Group (Chaebol) with Samsung Electron as its center published in the U.S. has increased in proportion to the number of patents granted to the group by the U.S. (refer to Table 10). In general, the research domain of scientific papers is "science", whose specific categories fall in the domain of basic research. The rapid increase in research papers under the name of the Samsung Group, therefore, reflects a rapid increase of the group in its development strength at the level of basic research. With this in mind, the papers published in the U.S. in 1994 will be analyzed to identify which **technological** areas the group has targeted strategically at the basic research level.

Table 10 Papers of the Samsung Group published in the U.S.

Year	No. of Papers
1982	0
1984	2
1986	2
1988	6
1990	15
1992	40
1994	87
1995	99
Total	378

Fig. 1 shows the papers under the name of the Samsung Group published in the U.S. in 1994 organized according to technological areas. Analysis of the papers of the group in the second-layer technological areas tells us that the area with the largest number of papers, 27, is NC (electronic engineering), followed by ND (communications engineering) with 19 papers. The technological areas with 10 or more papers are all in electronics and communications.

Figure 1
Main Research Fields by Samsung Electro
Number of Published Papers in the US in 1994 by Samsung

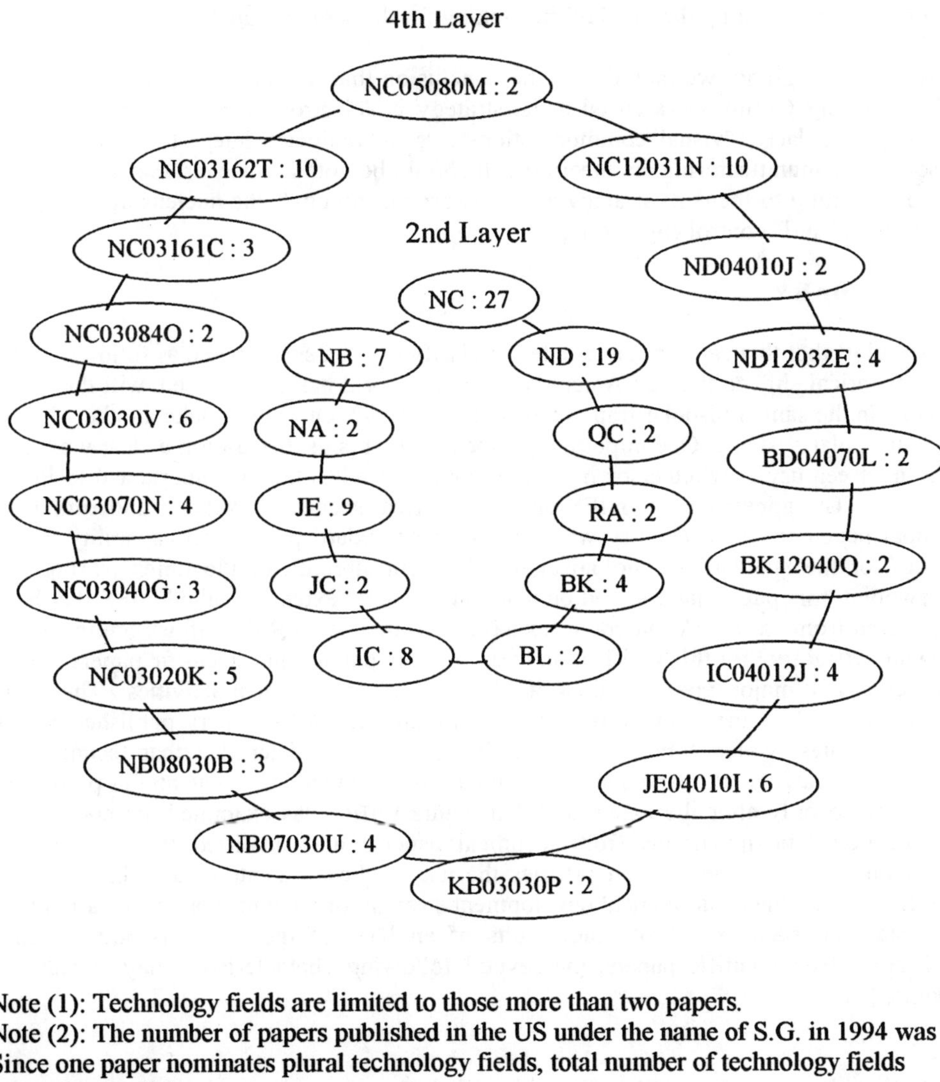

Note (1): Technology fields are limited to those more than two papers.
Note (2): The number of papers published in the US under the name of S.G. in 1994 was 87. Since one paper nominates plural technology fields, total number of technology fields ammounts to 125.
Note (3): Technology code is shown in the paper.

Source: compiled from JOIS

Turning our eyes to the fourth layer, which is the most finely categorized basic research area addressed by the Samsung Group, the technological areas with the largest number of papers, 10, are ND12031N (TV sets, visual communications general) and NC03162T (semiconductors integrated circuits). These areas are followed by NC03030V (solid state device manufacturing engineering general) and JE40101 (figure and image processing general) with 6 papers each, and NC03020K (solid state device materials) with 5 papers, forming the top 5 of the ranking for the same group.

From the foregoing, we may derive the conclusion that the area of basic research that the Samsung Group has focused in its strategy is "electronics and communications", more particularly, "visual communications", "semiconductor integrated circuits" and "solid state manufacturing engineering". It should be noted, however, that the group is also beginning to focus its strategy on "software technology", the nucleus of "computer utilisation" and "control engineering".

4. SUMMARY

The points that this paper has attempted to clarify may be summarised as follows. After the historical shift to the Pax Americana that the world has experienced, will the world maintain the same uni-polar frame in the 21st century? Or, will it make another shift to a multi-polar system? One important perspective that is useful to form a clear view into the next century is whether or not NIES countries will play an important role in the history. The appearance of NIES on the historical scenes is likely to undermine the global hegemony of a country which holds the standard production capability of the times, or the standard technological development strength of the times. From this viewpoint, this paper has focused on verifying to what extent so-called NIES that have appeared from eastern Asian countries have built their own R&D infrastructure. Here, the analytical tool for this verification was sought in patents and scientific papers, which represent two major types of output of research and development activities. The target was restricted to the number of U.S. patents and scientific papers published in the United States, which I believe is a viable measure to determine that technologies covered in patents and papers are of an international level. U.S. patents, in particular, are granted only after the Patent and Trademark Office has examined the contents of applications for inventions from technical aspects and qualified them as newly-developed technologies, which justifies the use of these patents as a viable index to determine that the technological development strength of a country or organization is at international standards. From the results of analysis of these patents and searches through these scientific papers, the several following characteristics may at least be pointed out.

The first characteristic that has been identified through patent analysis is found in the fact that, since the 1990s, the numbers of patents granted to inventions made in Taiwan and Korea have moved upward at the highest pace ever to the upper part of the ranking by country for U.S. patents granted. As of 1994, these countries are counted in the top ten in this ranking by country. This result, although being individual-year-based, implies that the technological development strength of these countries has reached a level that cannot be neglected on the international technological scene.

The second characteristic can be found in the contents of patented inventions. While the U.S. patents held by Taiwan are full of variety, those held by Korea show a heavy concentration in electronic and electric equipment. The third characteristic is that, while an overwhelmingly large share of all the U.S. patents granted to inventions made in Taiwan is represented by U.S. patents held by a broad range of venture-type individuals, most of U.S. patents granted to Korea are held by Chaebol groups, more specifically, the Samsung group alone representing a share near 50%. The first characteristic drawn from searching through scientific papers published in the United States is that the number of U.S.-published papers by either country has begun to increase at a remarkable pace since the latter 1990s, justifying us to point out that both the countries have advanced rapidly in their scientific or basic research levels. Second, from the results of searching for the numbers of papers by author's nationality based on the major keywords "solid state components", "mechanical properties" and "computer utilization", it can be pointed out that eastern Asian countries, including China, have reached a level that cannot be overlooked, both in terms of number and ranking.

Finally, the third point that should be noted is that the results of verification made on the Samsung Group in Korea, which permits a more focused search, show exactly the same increasing trend; that the group has focused its strategic target of basic research on technology, with hardware technology centering around electronic and communications engineering as its nucleus; and that, limited to the research level in this field, the group is very likely to have already reached the world level. These findings may permit us to assume that the rise of eastern Asian countries centering around NIES is now intensifying a global trend toward the international distribution of technological strength in conjunction with a paradigm shift in technology.

References :

1. Asamoto,G.(1995), Technology Accumulation and Development of Taiwanese Computer
2. and Semiconductor Industries, in Chen,B.H.and Hayashi,T.(eds), Technology
3. Development and Tranfer in Asia, Bunshindo, Chap.5.
4. Cantwell,J.(1991), The International Agglomeration of R&D, in Casson,M.(ed), Global
5. Research Strategy and International Competitiveness, Basill Blackwell.
6. Chen,B.H.and Hayashi,T.(eds)(1995), Technology Development and Tranfer in Asia,
7. Bunshindo.
8. Etemad,H. and Dulude,L.S.(1985), R&D and Patenting Patterns in 25 Large MNEs, Centre
9. d'Etudes en Administration Internationale, Ecole des Hautes Etudes Commerciales,
10. No.85-05.
11. Hayashi,T.(1), Changing Mechanism of the American Technologies and the Inter-
12. relationships of the Technologies, Rikkyo Economic Review, Vol.49, No.3,Jan.,1996.
13. ------- (2), Software Technologies and Pradigm-shift of International Competitiveness,
14. in Komoda,F., Nishiyama,K,Hayashi.T and Kaneko,S.(co-authored), IT and Analysis
15. of Interrelationships of Technologies, Chuoukeizaisha,1996.
16. ------- (3),Technology Transfer by Japanese Companies, Industrialization of Eastern
17. Asian Countries and their R&D Capabilities, Asian Management Studies, No.2,
18. March 1996.
19. Hirakawa,H.(1993), Economic Development of NIES and their Technological
20. Accumulation, in Hayashi,T. and Komoda,F(eds), Technological Innovation
21. and the World Economy, Minerva, Cap.7.
22. Kajiwara,H.(1995), Development Strategies of Asia, Toyokeizaisha.
23. Kim,Y.(1988), Industrialization of Eastern Countries and the World Capitalism,
24. Toyokeizaisha.
25. National Institute of Science and technology Policy(1995), Science and Technology
26. Indicators:1994, Science and Technology Agency.
27. OECD(1996), Reviews of national S&T policy : Republic of korea.
28. ----- (1996), Innovation, Patents and Technological Strategies.
29. ----- (1995), Chinese Taipei : The Origins of the Economic Miracle.
30. Ree,J.C.(1996), Product Development System of Korean Firms, Rikkyo Economic Journal,
31. No.51, July.
32. Pak,U.(1989), Technological Development of Korea, Bunshindo.
33. Townsend,J., Henwood,F., Pavitt,K. and Wyatt, S.(1981), Science and Technology
34. Indicators for the UK, Innovations in Britain Since 1945, SPRU.

CHAPTER IV

MONITORING CENTRAL AND EAST EUROPEAN EXPERIENCES OF KNOWLEDGE TRANSFER DURING THE TRANSITION PERIOD

EXCELLENCE AND SOCIAL RELEVANCE IN HUNGARIAN HIGHER EDUCATION

András Patkós
Department of Scientific Affairs, Ministry of Education and Culture
Hungarian Republic

1. Trends in Hungarian Higher Education since 1989

Higher Education in Hungary has gone through essential transformations during the period since the change of the political system has been started. The number of students has almost doubled, while the number of teachers/researchers employed in this sphere has suffered a 15% reduction. This decrease, however, was much milder than in other institutions of the R&D sphere, therefore the part of higher education in the human resources of research and development has increased from 41,2% in 1989 to 59,8% in 1996 (see Table 1). The teaching load of the staff increased substantially. This circumstance might explain, why the number of the teachers having PhD degree grows somewhat slowly relative to the non-educational R&D institutions. In any event, the fact that 69,5% of the researchers with PhD or equivalent degrees has been working in the sphere of higher education in 1996, clearly demonstrates the importance of these institutions for the future of research, development and experimental production in Hungary.

The statistical analysis of the composition of R&D expenditure shows that the private (entrepreneurial) sphere has recovered gradually after the shock of the political changes, and the increase is continuous and relatively fast since 1995. The same is even more true for the foreign sources of R&D coming either through special aid-programs or by collaborative projects. The reason for the diminishing of the part of R&D in the Hungarian GDP (see, Table 2) is caused, no doubt, by the stagnation of the support received from the central budget or through special funds operated under the supervision the Ministry of Finance. It is popular to say, that one needs at least the doubling of science related expenditures until 2003. However, a careful approach is to be chosen, since one should not change the proportions of the support from the private entrepreneurial sphere, the state budget and the international projects, which amounts at

A. Inzelt and J. Hilton (eds.), Technology Transfer: From Invention to Innovation, 257–263.

Table 1

Evolution of the number of R&D units and of personnel between 1989 and 1996

	1989	1990	1991	1992	1993	1994	1995	1996
Number of R&D units								
full number 1)	**1 312**	**1 256**	**1 257**	**1 287**	**1 380**	**1 401**	**1 442**	**1 461**
outside Higher Education	**379**	**316**	**257**	**207**	**302**	**295**	**333**	**341**
in Higher Education	**933**	**940**	**1 000**	**1 071**	**1 078**	**1 106**	**1 109**	**1 120**
Number of Researchers								
full number 1)	**33 836**	**30 256**	**26 763**	**24 110**	**23 012**	**22 401**	**20 859**	**20 485**
outside Higher Education	**19 879**	**16 212**	**12 571**	**9 815**	**9 138**	**8 504**	**8 194**	**8 236**
in Higher Education	**13 957**	**14 044**	**14 192**	**14 295**	**13 874**	**13 897**	**12 665**	**12 249**
in HE as %	41,2	46,4	53,0	59,3	60,3	62,0	60,7	59,8
No. of PhD/DLA holders 2)								
full number	**5 869**	**5 697**	**5 634**	**5 655**	**5 680**	**6 257**	**5 979**	**6 404**
outside HE	**2 023**	**1 895**	**1 791**	**1 660**	**1 587**	**1 690**	**1 681**	**1 954**
in HE	**3 846**	**3 802**	**3 843**	**3 995**	**4 093**	**4 567**	**4 298**	**4 450**
in HE as %	65,5	66,7	68,2	70,6	72,1	73,0	71,9	69,5

Remarks:

1) R&D institutes, HE R&D units, for-profit R&D units, other R&D units. Researchers on pension or employed in other forms of activity are not taken into account.
2) C.Sc., D.Sc., Member or Corresponding Member of HAS. (New style PhD not yet included)

Sources: Research, development and experimental production (HOS) 1989, 1990, 1991, 1992, 1993, 1994, 1995, 1996 (Preliminary)

MEC Department of Scientific Affairs, dr. Gyula Engloner (1997 October)

Table 2

Financial sources of Hungarian R&D

B HUF

Year:	89	90	91	92	93	94	95	96
Entrepreneurial sphere	15,4	13,1	11,0	9,9	10,1	11,5	15,0	17,2
Central state budget	6,8	8,0	9,1	11,0	11,9	14,7	15,4	20,6
Special R&D Funds	10,9	11,8	6,0	8,9	11,0	10,8	7,9	3,0
Others (internal, foreign)	0,7	0,8	1,0	1,8	2,3	3,3	4,0	5,2
Total	33,8	33,7	27,1	31,6	35,3	40,3	42,3	46,0
Total expenditure as in % of GDP	1,96	1,61	1,09	1,08	1,0	0,93	0,80	0,69
Total share of central budget and of special funds as in %	52,4	58,8	55,9	63,0	65,1	63,3	54,1	51,3
Total share of sources independent of central budget as in %	47,6	41,2	44,1	37,0	34,9	36,7	45,9	48,7

Remark: The source of the fund KMÛFA appears from 1994 in the central budget, but for the continuity of the statistics it is shown among the special funds.

Sources: Research, development and experimental production (HOS) 1990, 1991, 1992, 1993, 1994, 1995, 1996 (preliminary)

MEC Departments of Scientific Affairs, dr. Gyula Engloner (1997 October) present to 50%-40%-10%, respectively. Any increase in the support from the central budget, should induce by well-thought motivating measures also increased support from the private sector.

The share of the higher education in the total R&D expenditures seems to be stable for several years (Table 3). It amounts to one quarter, which is similar to the figure characterising industrialised countries. The situation concerning R&D investments is somewhat obscure. Probably, the construction of the big new university campus in South-Buda is excluded from the statistics, though the buildings, laboratories and other facilities will not serve exclusively educational purposes.

The general tendency of increased importance of higher education in Hungarian R&D is not reflected by the technology oriented activities sponsored by the State Committee for Technological Development. This agency is running the Central Fund for Technological Development, which is the only fund serving exclusively the goals of market oriented applied research. This fund lost substantially its real value, and the part taken by institutions of higher education has been shrinking continuously since 1992 as is evident from Table 4. This tendency seems to be extremely dangerous in view of the global importance of the increased weight of the private sector in financing technological research activities and higher education itself.

2. Global trends - Hungarian response

The study of the Group on the Science System of OECD entitled "University Research in Transition" finds a global tendency towards increased role of the industry in financing research at universities. This tendency is accompanied by growing demand for the economic relevance of the research projects. In general, short term, market oriented research, conducted in systemic linkage between universities and companies becomes a dominant feature of academic life.

A shift can be perceived from originality of the research towards contract work providing quick and reliable answers to well-defined questions of immediate impact on production. A certain romantic aura of the research career has disappeared and the systematic, well-organised mass-activity of R&D loses its social reputation. The globalisation is felt stronger in R&D, than on the average: multinational companies develop new products, test the quality of their components in those laboratories all over the world, where the required quality services are ensured to be the most advantageous possible.

Many scientists and external observers see in this tendency a certain danger for academic freedom and the compulsory public nature of the university research. As a consequence in the traditional advanced technology countries the prestige of the research activities is on decline, the group of researchers is ageing, and the income gap relative to the private sphere is increasing.

Table 3

R&D expenditures relative to the GDP and on current prices in Hungarian Higher Education

R&D expenditure	**89**	**90**	**91**	**92**	**93**	**94**	**95**	**96**
Total R&D expenditure relative to GDP in %								
full	1,96	1,61	1,09	1,08	1,00	0,93	0,80	0,69
HE	0,26	0,24	0,22	0,23	0,22	0,24	0,19	0,17
R&D expenditures in B HUF								
total	**33,8**	**33,7**	**27,1**	**31,6**	**35,3**	**40,3**	**42,3**	**46,0**
HE	**4,4**	**5,1**	**5,4**	**6,6**	**7,8**	**10,3**	**10,2**	**11,1**
HE as in %	13,0	15,1	20,0	20,9	22,1	25,6	24,1	24,1
Total R&D investment in B HUF								
total	**4,0**	**3,3**	**2,2**	**3,4**	**3,6**	**4,7**	**4,7**	**5,3**
HE	**0,4**	**0,5**	**0,5**	**0,7**	**1,1**	**1,4**	**1,4**	**1,2**
HE as in %	10,0	15,2	22,7	21,2	30,1	29,8	29,8	22,6

Sources: Research, development and experimental production
(HOS) 1989, 1990, 1991,1992, 1993, 1994, 1995, 1996
(Preliminary)

MEC Department of Scientific Affairs, dr. Gyula Engloner (1997 Oct

Table 4

Evolution of the Central Fund for Technological Development (KMÛFA), and the R&D support directed from it to Higher Educational Institutions between 1990 and 1996 (current prices)

Year:	**90**	**91**	**92**	**93**	**94**	**95**	**96**
KMÛFA M Ft	**6 813**	**7 047**	**3 877**	**5 588**	**4 652**	**4 119**	**3 350**
R&D support to HEI M Ft	**252,1**	**180,8**	**431,9**	**600,0**	**855,6**	**420,1**	**301,5**
as in % of KMÛFA	3,7	2,6	11,1	10,7	18,5	10,2	9,0

Remark: The KMÛFA data are decreased by income subtractions.
Sources: Situation of the technicians and engineers and the R&D, 1988-1993 (OMFB, 1995)
Communication from OMFB on 1st of November 1996 and 28th of October 1997.
MEC Department of Scientific Affairs, dr. Gyula Engloner (1997 October)

In Hungary, the response of higher education was to put more emphasis on the knowledge transfer role of the institutions, rather than to try to strengthen the positions of the originality and the basic nature of the research. The improvement of the methods of training receives more attention also because of the urgent needs caused by the doubling of the number of students in the past 6-7 years. The higher educational institutions work very hard to receive from the state further state-financed student places, since the income from this represents the basic financial source for running the institutions. There is a tendency to emphasize among the performance indicators those related to the educational process.

R&D activity has reappeared only very recently among the performance indicators of the institutions. With the 1996 amendment of the Higher Education Act a separate sum directly supporting R&D has been introduced, which is established annually in proportion to the number of PhD-holders in the higher educational sphere. It represents about 2.5% of the whole Higher Education budget for 1997. The distribution of half of this sum among the institutions is based on a ranking table formed with data obtained on the number of the PhD-holders, the running projects financed by at least of one of the main R&D support agencies, and the number of the PhD-students in a given institution. The bonus given to external contracts prompted several university researchers to disclose existing contracts, hidden until now from the university administration.

3. Strategic university-industry partnership

Research contracts, however, cannot be considered as a stable and reliable income source, upon which university financial plans can be based with confidence for several years. Also in this case companies are quite restrictive concerning the ownership of the new scientific or technological results, since the contracts mostly refer to very specific problems.

Stability and the freedom of research seem to be better supported in a new form advocated by several multinationals settled in Hungary in the past decade (among them Nokia, Eriksson, various pharmaceutical companies, etc.). Strategic partnership means inter-departmental or even inter-university collaboration of research groups and a group of interested companies, enterprises. They should focus their research activity on a wider, but well-specified area, not narrowed down by daily problems of production. In this way there is space for basic research (its successes might contribute beneficially to the public profile of the sponsoring company), and academic freedom is not restricted so strongly. It is better to term this type of activity as oriented basic research than applied research.

The success of these university based research centers depends essentially on their interrelationship with the educational process. A center might provide thesis-themes for large number of MSc or PhD students, working under the direction of the university professors associated with the research program of a center. Additional grants and stipends might be provided to these students by the companies, attracting in this way the best students from an age group. Even some junior postdoctoral positions might be maintained by the companies for keeping young talented experts associated with the center, instead of prompting them to leave for some well-paid position outside the academic world.

The importance of such non-traditional partnerships and services is growing. Its development should harmonize with the Humboldtian values of academic research. Only academically sound and valuable research can be economically and socially relevant.

The Innovation or Industrial Parks have chosen their sites in Hungary giving preference to the neighbourhood of universities. Also in the very recent past two research institutes of the Hungarian Academy of Sciences, emphasizing applied, marketable products in their research strategy and practice were taken over by the universities located on the same campuses. In this way several universities are in a very good position, when the State Committee of Scientific and Technological Development and the Ministry of Education and Culture are co-operating in shaping a new form of research support specifically meant for strategic university-industry partners legally committed to long term collaboration in the field of leading technologies. The evaluation of the results and the renewal of such centers should be done every 3-4 years, taking into consideration the whole spectrum of their activities from basic research to education.

In conclusion, I would give a rather optimistic vision of R&D in the Hungarian Higher Education. Strategic activities contributing to balanced financing of the institutions will be harmonised in the framework of new integrated organisations encouraging multidisciplinary co-operations. Increased interest of international companies, recently settled in the country, in higher education will certainly bring a new era of high quality technological research at universities.

INNOVATION POLICY AND TECHNOLOGY TRANSFER AS A PART OF STRUCTURAL POLICY IN POLAND

T. MARKOWSKI
University of Lodz, Department of City and Regional Management
90-214 Lodz, Poland, 39 Rewolucji 1905 St.

Abstract: The paper presents the recent concept of innovation policy and instruments of technology transfer elaborated within the Task Force for Structural Policy of Poland established by the Polish Government together with European Commission in October 1996. The Task Force for Structural Policy assumed that the crucial elements of harmonisation and integration of Polish economy with the EU are not only restructuring of traditional industries but first of all building a legal and institutional system supporting innovation and technology transfer to the business. An essential aspect of Poland's of the structural policy is the regional approach to innovation. The paper consists of two parts. In the first theoretical aspects of a innovation and innovation systems are discussed. The second part contains the list of major recommendation the Task Force have proposed for the Polish government.

1. Introduction

Amongst the key difficulties Poland will face in the forthcoming years, the issue of restructuring of many essential sectors of the economy is especially important. Structural policy will provide for appropriate conditions and set the right mechanisms of deep restructuring of Poland's economy. This, in turn, will determine the course of its integration in both the European system and, in a wider perspective, in the world economy. Current difficulties and tensions in the area of Poland's co-operation with European Union partly result from the infirm condition of Poland's structural policy and slow pace of the restructuring processes in many essential sectors of the county's economy (e.g. coal industry, fuel industry, transportation, metallurgic industry, power industry, telecommunication and agriculture). Without speeding up and deepening that process, the division of work beneficial to both parties, based on specialisation and co-operation will not be established.

At the same time, future perspective of Poland's integration with the European Union inclines Polish Government to broad usage of experiences of member countries of the Union in the scope of structural policy. In this context it is important to adjust the pace

A. Inzelt and J. Hilton (eds.), Technology Transfer: From Invention to Innovation, 265–284.

and directions of these changes to the parallel structural policy changes which are presently happening in EU member countries.

To this effect the Polish Government, acting in the close co-operation with foreign partners, announced in October 1996 the establishment of a Task Force for Structural Policy of Poland in order to provide a forum for an in-depth analysis of Poland's current structural policy framework, to draw-up future orientations for structural policy in Poland and to consider respective supporting roles of European institutions, Polish local and regional agencies and partners from the non-governmental sector in implementing such orientations. Structural policy oriented in this way included the following main components:

-regional development;
-ownership transformation, deregulation and demonopolisation of economy;
-market and consumer protection;
-technical and technological progress and;
-sectoral restructuring programs.

2. The national innovation system: theory and definition

Thus, the modern approach treats the innovation process as a sequence of interactions from the moment an idea comes into being to its implementation and dissemination, aimed at inducing change in the product, technology, organisation or society. Implementing innovations quickly and efficiently has become one of the key factors enhancing competitiveness in increasingly complex and global markets. In the literature on the subject, all sources point out that innovation is a socially complex, complicated and difficult phenomenon. This is due first of all to the fact that innovation combines the features of three spheres which it links together: science, technology and production. The principal features of innovation, in the modern sense of the term, include the following [Guinet 1995) :

Innovation is localised, which means that its creation and integration into the economy takes place in a specific spatial context. It thus takes the specific form of localised, external information-related economies. This feature is fully consistent with the notion of "club goods" i.e., those intellectual resources which constitute the technological knowledge of the players belonging to a network (club) [Institutions and Instruments of Regional, Competitiveness 1996]. Thus, if no organised innovation system exists in a different spatial location, technology transfer becomes impossible. For if such a system is non-existent, we are dealing simply with one-off technology transfers and insular development models. Therefore, spatial development and other phenomena related to urbanisation processes may serve as important factors in stimulating the development of an innovation system.

Innovation is an integrating process, which means that autonomous units adopt special forms of reciprocal organisation and integration in such fields as manufacturing, research, marketing or financial planning. Guinet calls this "investment in organisation". The main emphasis is laid on controlling the transfer of information, knowledge and services. Such integration is based on a well-developed information infrastructure of the constituent units, which join forces in order to achieve a common goal of an innovative character.

Innovation is a learning process, resulting from the accumulation of specific knowledge and information, useful for the functioning of the firm. It is an interactive process which draws on both internal and external resources. An innovative firm is characterised both by its capacity to accumulate knowledge, and by the efficiency of its production processes. Learning is based on the resources available to the firm.

Innovation is a social phenomenon, as it disturbs established patterns of production and consumption. It is, therefore, important to stimulate favourable attitudes among the public by developing both a good relationship and a pro-innovation atmosphere, particularly in connection with large-scale projects.

Innovation is a process of creative destruction not only in the social, i.e., external, dimension, but also within the firm itself. Any product-related innovation must correlate with changes in organisation and management, which should in turn anticipate the needs dictated by cost calculation.

Innovation is rooted in culture and has links with history and tradition. Such factors should be taken into account when building the foundations of an innovative economy, as they can have both a favourable and a detrimental influence on the development of nations or regions. Exceeding the limits of social approval for innovation may result in innovation being blocked by the public. An enterprise stimulating climate is thus important for both creating and diffusing innovations.

Innovation is both costly and risky, which is especially acutely felt by small enterprises. Hence the special importance of various types of financial and insurance institutions which reduce the risk and costs incurred by the innovative firm.

2.1. The concept of a national innovation system

1. The modern understanding of innovation as an integrated, network process serves as a starting point for any systematic approach to the issue of developing new technological and organisational solutions, as well as their transfer and application in economic practice. These phenomena occur within what is described as an innovation system, i.e., one oriented towards innovation. It involves specific institutional, organisational, economic-financial and information-related arrangements, which influence the role played by its particular components and their interdependencies, the progress of technology transfer, and the level of innovation in an economy. Innovation systems may be discussed both on

> the level of the national economy as a whole as well as the on regional or local levels. In the first case, the term national innovation system (NIS) is used, while in the latter case we are dealing with a regional (or local) innovation system (RIS). Different definitions and characterisations of an NIS can be found in the extensive literature on the subject. Of special interest is the definition put forward by Lundvall, who treats the NIS as a configuration of productive and scientific/technological subsystems, institutional arrangements, and the interdependencies between the two. Elements of this system are influenced by features specific to a given country, such as: historical experience, value systems or culture [Lundvall 1992].

The NIS components are inter-linked by means of diverse relationships in the fields of planning, information, the law, organisation and financial issues, which in turn are a consequence of both the financial and procedural mechanisms adopted, and certain co-operative arrangements. Contemporary national innovation systems in highly developed countries are characterised by: long-term research projects conducted by universities and research laboratories a high level of innovation in enterprises, well-developed links between science and industry. The state's innovation policy functions as a factor facilitating and co-ordinating the smooth progress of innovation processes.

The structure of an NIS, its national specificity, adopted solutions and mechanisms, the links between its components, and its interactions with the environment influence the level of innovation and competitiveness both in the economy as a whole as well as among domestic firms. An efficient NIS may facilitate better utilisation of limited resources and accelerate progress through the more effective management and organisation of resources, a more efficient combination of imported and domestic technology, and its more effective adaptation to and assimilation by the economy as a whole.

2.2. The regional innovation system

A regional innovation system (RIS) is a specific forum for co-operation between various types of organisations and institutions functioning in a region, whose principal objective is to develop entrepreneurship and innovations in that region. Such entities comprise: regional authorities (voivodship-level) and local-government authorities (at the town and gmina level), regional development agencies, institutions of higher education, R&D institutes, technology transfer centres, advice centres, specialist and professional associations, financial institutions, consultancy firms, manufacturers and service providers, their R&D units, etc. All these constitute a structure which functions as a network linking all the agents active in the sphere of innovation and technology transfer, so as to form a Regional Innovation System . The RIS is treated as a fundamental factor of internal and external (from the point of view of the region) integration between the economy and science. It is also an important factor in a region's capacity to participate in European integration and derive benefits from this process.

In the regional dimension, innovation systems are demand-oriented. Regional authorities and institutions are better prepared to take into account the broad spectrum and specific, branch features of firms operating in a region, as well as regional and inter-regional links. Regional systems differ significantly in this respect from an NIS, which has a more supply-side orientation. The government and central institutions should mainly emphasise the importance of research problems, the science and technology sphere, and international co-operation. The functioning - and management - of this kind of system, which consists of entities of diverse organisational, legal and functional characteristics, is a complex and delicate task. The main problems and functions of RIS management include:

-specifying the position and role of a regional innovation and entrepreneurship system with respect to the main problems, needs and capabilities of a given region;
-developing a pro-innovation regional structure;
-co-ordinating the activities of particular units in accordance with an adopted regional development programme;
-inter-regional and international co-operation in the fields of technological progress and technology transfer; the efficiency of RIS activities.

An RIS is oriented in its entirety first and foremost towards supporting the innovation potential of enterprises - particularly small and medium-sized ones, established by representatives from academic circles. Small, innovative firms can also become a driving force of economic development and structural transformations in regions.

An NIS consists of many innovation networks, which are co-ordinated to a greater or lesser degree. It is not, however, a simple mosaic of sub-networks. The component networks may be intimately inter-linked or, conversely, isolated. They can also have connections with international, national and regional (local) networks.

A shift in network elements brings about changes in the networks themselves, which in turn affect the structure and functioning of an NIS or RIS, and thus innovation in the economy and domestic firms. This creates a new challenge for state innovation policy and the need to monitor innovative activities - on the state and regional level on the one hand, and with regard to key innovation networks on the other - as well as to identify their key partners, localities and links with other networks, particularly in the context of supranational networks.

3. Modern innovation policy and instruments for its implementation

3.1. Characteristics and evolution of innovation policy

A broad context for innovation policy is provided by the level of economic and technological development achieved, which is reflected in the competitiveness of the economy and domestic firms on foreign markets and by the development strategy adopted by the country in question. Such an innovation policy must also take into account the specificity of the solutions which characterise particular elements of the NIS, links between these elements and their influence on innovation in an economy etc.

Therefore, innovation policy involves different objectives and different institutional arrangements and instruments in economically and technologically advanced countries than it does in those countries "lagging behind" or trying to "catch up" in the race towards socio-cultural development [The European Report 1994]. Modern innovation policy in highly developed countries (e.g., EU members in the 1990s) is increasingly becoming a technological policy oriented towards stimulating the competitiveness of the economy. Governments support the development of selected technologies and high technology products, as well as international co-operation in the fields of science and technology. Greater emphasis is placed today on research which addresses the needs of industry, on improving links between institutions of higher education and industrial firms, and on the education of scientists and engineers.

Innovation policy in these countries is characterised by the changing position and role of the state. Previously, the state was the main agent of innovative policy and endeavoured to impose (by way of regulation) particular objectives of technological development. Nowadays, the most prominent function of the state is that of a co-ordinator and facilitator, which creates the institutional framework for self-regulation in the progress of technological innovation. Moreover, the organisational, institutional and cultural aspects of innovation are now taken into account on a much wider scale in innovation policy.

Countries trying to catch up with economically and technologically advanced states usually experience problems with using technological change as a vehicle of growth, transformation and modernisation. Such countries often lack institutional structures and, even more so, human capital resources, which would allow them to absorb, reproduce, adapt and improve imported technologies [Technology and Economy 1992]. Therefore, a very important component of innovation policy in these countries consists of developing a technological "capability" (or core competence), comprising the knowledge and skills necessary to obtain, assimilate, utilise, improve and create new technologies.

The concepts mentioned above also include engineering skills and technical know-how, organisational skills, and knowledge of the patterns of behaviour and needs of participants in innovative activities (including workmen, engineers and customers).

Favourable conditions for developing a technological capability can be created by importing foreign technological ideas, by developing (in the early phases of economic growth) relatively low-technology industries, and by importing equipment - on condition that elements of foreign technology are efficiently and effectively combined with the knowledge and experience of local personnel. In due course, it will become possible to use the manufacturing experiences acquired in this way to improve the quality of domestic products, which will involve progressively more and more advanced technologies.

In the early phases of developing a technological capability, the task of innovation policy in countries trying to "catch up" should be to orient public R&D institutions towards helping the economy absorb innovations. As the technological capability of firms increases, public R&D institutions should change their orientation and concentrate their efforts and resources on basic research which supports applied studies and R&D work conducted within firms.

3.2. Approaches to implementing innovation policy

With regard to implementing innovation policy, it is possible to distinguish between a problem-based approach, a regional approach, and, to some extent, a sector approach. The most popular of the three - the problem-based approach - consists of solving problems common to the economy as a whole by working out and applying arrangements related to planning, legal regulations, organisation, finance, information and investment. Such an approach combines issues from the fields of science and technology, general and academic education, policy aspects related to systemic transformation and creating institutional and market conditions appropriate to this end, infrastructural policy (information highways, communication systems, transportation systems), international co-operation, and the institutional infrastructure for supporting technology transfer.

The regional approach lies in integrating state innovation policy with intra-regional policy by differentiating between the tasks allotted to government, regional and local structures. The tasks pursued at particular levels are not substitutes for one another. Rather they mutually complement each another in a non-competitive way. A regional approach is concerned, among other things, with regional innovation, institutional structures, special economic zones, regional infrastructural investments (harbour complexes, airports etc.), and, above all else, implementing innovation-boosting projects.

The sector approach is concerned with innovation policy in areas which are of importance from the point of view of national specificity, the development strategy adopted in a given country, or international links. This can be exemplified by agriculture, which requires a specific approach, objectives and instruments. Other areas where the sector approach is often adopted include the energy industry, national defence, and environmental protection.

Another separate issue is that of the small and medium-sized enterprise sector, mainly because of its growing importance as a source of new inventions and a basis for implementing innovations, its stimulating impact on the economy and job creation, and its special social role (promoting entrepreneurship and activating local initiatives). Of particular importance here are small, innovative firms in the high-technology sector, which are breaking new technological ground in the most advanced branches of science and technology.

3.3. The main instruments for supporting innovation and technology transfer in the context of Poland's innovation policy (The experience of EU countries)

Three types of instruments can be distinguished in innovation policy: direct support, indirect support, and infrastructural development (M. Quevit 1997). Various forms of this kind of support (including intervention) are connected with the internal activities of enterprises, and also encompass financial aid in the form of subsidies, the purchase of property and tax allowances. The ultimate goal of this type of aid is to promote innovation processes in particular enterprises.

Indirect support aims at enhancing the general economic environment in a particular area (a state, region or locality). It applies to all forms of economic activity and is oriented towards increasing the competitiveness of firms functioning in a given area. One of its important goals is to develop services which promote technology transfer, boost the sales of local products and improve the relationship between enterprises and institutions supporting innovation. It includes also publicly-funded measures in the sphere of human resources, and, in particular, those aimed at obtaining the technological know-how and professional skills directly connected with the innovation process.

With regard to the infrastructure, public intervention applies to such public utilities as technology centres, information systems, telecommunications systems (e.g., teleports), etc. Infrastructural needs should be assessed not only in terms of the number of specific installations, but first and foremost in terms of the quality of services offered in connection with the infrastructure. This is particularly important in the case of innovation, which involves by its very nature mainly non-material investment.

With regard to the relationship between scientific and technological potential on the one hand, and the modernisation and innovation needs of enterprises on the other, two categories of instruments may be particularly important for the restructuring of Poland's economy:

-instruments and measurements which promote interaction between research institutes and industry,

-instruments helping establish firms based on new technologies by utilising domestic technological resources and scientific and technological potential.

Technology transfer can be defined as the supply of technologies to the market. It is a specific form of communication, often of an interactive character, which involves various feedback loops linking senders and receivers [van Geenhuizen, Nijkamp 1996]. Technology transfer comprises all forms of innovation diffusion and technical education. In most cases, it is a market process, whereby technology is bought and sold [Malecki 1991, which means that it is introduced, together with the relevant know-how, into economic practice. Such transfers usually take place between the science and research sector, on the one hand, and the manufacturing sector, on the other. The same process also goes on within the economy and in contacts between individual inventors and the economy.

The approach to technology transfer should clearly vary, depending on the type of the enterprise concerned. In actual fact, the forms of technology transfer and the network links between institutions of higher education and industry differ significantly from one another, depending on the type of industrial partner involved. Four types of enterprises can be distinguished:

1. Large enterprises based on high technology. Their research potential is comparable to or even greater than that of institutions of higher education. Contacts of such firms with academic centres are of a long-term character and involve both domestic and international institutions. The crux of the matter is to find suitable partners for such enterprises, in co-operation with whom they could carry out transnational and competitive preliminary research. In the Polish economy, enterprises of this kind are unlikely to gain a dominant position for the time being.

2. Large enterprises based on low technology. Technological change progresses in such entities at a somewhat lower pace, and their market position depends on factors other than technological complexity. In Poland, such enterprises are to be found in the shipbuilding, mining and construction industries.

3. SMEs based on high technology. The main challenge accompanying the formation of such firms is to process research results in close co-operation with the academic community. In some countries, the existing educational and academic culture does not favour undertakings of this kind, which forces their governments to take initiatives promoting such entities.

4. SMEs based on low technology. Such firms are unable to gain direct access to the scientific and technological resources of higher-education institutions, and cannot become involved in comprehensive technology transfer originating at the university level. In the case of such enterprises, technology transfer instruments are based on a global approach, which is better suited to their needs.

4. An assessment of the National Innovation System

The potential for innovation in the Polish economy is still low compared to other OECD countries. Only in some aspects is the level of innovation comparable to that of the new or prospective EU members. Such a situation poses a threat to the international competitiveness of Poland's economy and domestic firms. The low level of innovation in the Polish economy is reflected above all in the low proportion of R&D outlays in the country's GDP: 0.8 % in 1995, as compared to over 1% of GDP in OECD countries. Other relevant indicators include:

-the decreasing number of Polish inventions patented in recent years, which is not compensated for by the increasing number of foreign inventions patented in Poland;
-the persistently low contribution of high-tech industries to overall industrial output; the low (but increasing) proportion of new and modernised products in overall sales.

The insufficient level of innovation in the Polish economy is a sign of the poor state of the national innovation system. The principal deficiencies of this system include:
-no direct links between scientific and economic policy;
-responsibility for formulating the state's scientific policy, financed from the state budget, has been delegated to the Scientific Research Committee (KBN), which has close links with the academic community. This gives rise to egalitarian tendencies and conservatism when determining research priorities and the structure of research;
-the limited influence of the business world on the formulation of scientific policy aims and the allocation of research funds;
-the government organ responsible for promoting innovation in the economy, the Technology Agency, is still in the process of taking shape.

Only scant interest has been shown in industry for the results obtained by domestic R&D facilities. The efficiency of the fiscal and financial means employed by the system to boost demand for R&D work among economic entities is far from satisfactory. There are no sufficient incentives, either, for economic entities to develop and patent new solutions. The existing system of tax relief, exemptions and subsidies is imperfect and is a source of much controversy. The market for financial services (particularly high-risk ones) is underdeveloped. Larger industrial enterprises in Poland, whose interest in innovation is limited as yet, mostly look to foreign suppliers for ready-made technologies.

The mismatch between the offer of the domestic R&D institutions and the needs of the transforming economy, results from faulty structural, organisational and economic/financial arrangements. The efficiency and quality of management in R&D units are low. This poses a threat to domestic R&D facilities, which face competition from foreign enterprises and scientific and technological units.

There is a lack of strong capital and organisational ties between units operating in the sphere of science and technology, on the one hand, and industry on the other, as well as between these units themselves. This concerns R&D institutes, units of the Polish Academy of Sciences and institutions of higher education. The internal facilities of Polish enterprises are inferior to their counterparts in developed countries.

Inadequate mechanisms of technology transfer exist, both between domestic units and between those units and foreign partners (licence agreements, direct foreign investment). The institutional infrastructure facilitating technology transfer is underdeveloped, and the existing financial and capital mechanisms are inadequate too. Weak capital and business-insurance markets are affecting the degree to which Poland can utilise the financial potential offered by foreign financial institutions. This situation also restricts Poland's capacity for advanced-technology transfer and affects its chances of attracting foreign entities active in this field.

Poorly developed regional innovative systems. This mainly concerns the various types of regional engaged in stimulating innovation and technology transfer, such as innovation centres, „incubators", agencies and associations, capital funds, and information, consultancy and training facilities.

Such kinds of institutions are only just beginning to function in Poland. Certainly they are numerous, but rather diffused, and lack sufficient authority, financial and organisational resources, and experienced personnel. The experience of Western countries indicates that the development of regional innovative systems is a long-term process which calls for a mature regional policy, involving financial support from the budget.

The regional links that are emerging around the leading urban centres vary considerably in terms of their efficiency and capacity to stimulate innovation and facilitate its dissemination and transfer. Accordingly, there are growing disproportions appearing in regional development, leading in some cases to the „marginalisation" of certain regions.

More powerful stimulants for small-scale innovation are lacking. With regard to stimulating technology transfer and innovation among SMEs, the state's policy has only reached the stage of implementing a small number of national- and regional-level programmes. Hopefully, this will prepare the ground for a comprehensive pro-innovation policy geared towards SMEs. The mechanism used so far to stimulate innovation and technology transfer among such enterprises has been hardly efficient, whether in terms of institutional solutions or financial and organisational arrangements. Not enough is being done to help establish and develop small firms relying on advanced technologies.

Among the more encouraging factors in favour of innovation and technology transfer are the following: The relatively high level of professional and vocational education in Polish society. The proportion of educated people is growing, as is the number of secondary schools and institutions of higher education (the new ones are mostly privately owned). One the other hand, this rapid growth is partly taking place at the expense of teaching standards. Among other problems in this field we may mention the inadequate financing provided for institutions of higher education, and their conservative organisational structure, which stands in the way of strengthening contacts with industry or modifying the curricula.

The high academic and research status of some scientific and R&D units (the exact sciences, some research institutes and certification units); and the good standing enjoyed by certain disciplines in Poland (mathematics, physics, astrophysics, chemistry).

The increasing volume of direct foreign investment which is transferring new technologies and organisational methods to Poland. One should take note, however, of the uneven distribution of foreign investment in Poland. The main reason behind such a state of affairs is the patchy economic and institutional infrastructure and, in many regions, the absence of an active policy aimed at attracting investors. Therefore, foreign investment has only a marginal influence on local communities and recovery in depressed areas suffering from high unemployment. One should also take note of the increasing number of foreign inventions being patented in Poland, which indicates the growing stature of our country.

Continuing systemic reforms which create favourable conditions for a more enterprising and innovative economy in Poland. In many branches of the economy, fundamental systemic changes have been put into effect (liberalisation of business activities, development of competitive markets, growth of the private sector, restructuring). This favours competition, which in turns stimulates innovation. On the other hand, some solutions adopted in the fields of privatisation, restructuring and finance - as well as the practical side of the transformation - have undermined the favourable effects of systemic changes on enterprise innovation. To give a few examples, the choice of buyers for privatised Treasury-owned enterprises often places other domestic enterprises at risk of losing their markets; industrial R&D units are in some cases excluded from privatisation agreements; capital from privatisation is occasionally used to finance consumption.

Polish legislation governing the protection of intellectual property is considered to be satisfactory and comparable to that of other OECD countries. The problem, however, is that the laws in question are too loosely enforced which affects the security of business transactions in Poland. Also knowledge of legal regulations pertaining to intellectual property is inadequate.

5. Forms of pro-innovation activity

The main goals of innovation policy should be to improve the international competitiveness of the Polish economy and domestic enterprises and increase the share of new, modernised and environmentally friendly products in overall industrial output. To achieve this, a coherent, modern innovation system at both the national and regional level needs to be created. The principal tasks include: establishing a market for innovations, achieving a high absorption capacity in the economy, investing in human capital, and constructing infrastructural and institutional systems stimulating intellectual development and the practical application of knowledge (innovations), which will enable the economy to achieve a sustainable competitive edge.

The fundamental systemic changes in the economy introduced after 1989 need to be continued, and in particular, the following: privatisation and economic restructuring (in the infrastructural, sectoral and regional contexts), liberalisation, introducing competitive markets, developing of capital-market institutions, and opening up the economy to world markets. Continuation of institutional and market reforms combined with constant an enlargement of the private sector and increasing openness to world markets will create incentives to improve economic efficiency through innovation and enterprise.

A necessary condition for increasing innovation in the Polish economy and firms is an increase in spending on science and research to a level typical of developed countries, that is, exceeding 1% of the GDP. It is also essential that industry should make a far more substantial contribution to the financing of applied studies.

Links between scientific and economic policy. Linking scientific policy with the development goals of the economy should be seen as an important task and responsibility of the government. This necessitates a change in the role and position of the Scientific Research Committee (KBN) with regard to formulating and financing the scientific policy of the state. The main proposals in this area are as follows.

The Chairman of the KBN, acting in the capacity of the Minister of Research and Technology, should outline the principles of the government's scientific policy and submit them periodically (e.g., every years) to the Sejm; For the purpose of formulating strategy in the field of research and technology, a Science and Technology Council should be set up, affiliated to the Minister for the Economy. This would be composed not only of academics, but also of representatives from financial and industrial groups, selected and appointed - by virtue of their personal qualifications - by the Minister. The Board should formulate opinions on the needs of the state in the area of innovation, work out general strategies of scientific and economic development based on adopted economic and social goals, and evaluate the implementation of R&D programmes and projects;

Functions connected with the administration of funds and financing of research - in accordance with specific criteria and the policy of the government - should be delegated to a network of agencies responsible for organising tenders, preparing programmes involving institutions of higher education, research institutes and enterprises, monitoring research activities, and analysing the results. Such agencies should be subordinated to the Minister for the Economy as the agent responsible for implementing industrial policy. They should concentrate on organising research primarily in the form of joint programmes involving several units. A vital role could be played here by the newly established Technology Agency.

A certain number of members of the Basic Research Committee and the Applied Research Committee of the KBN should be designated by appointment, and the proportion of such members should gradually increase (e.g., to 50%). The appointments should be made by business circles concerned with science and technology and by the government. It is also necessary to increase the elected representation of industrial institutes on the Applied Research Committee.

The Technology Agency should receive financial backing at the initial stage (3-4 years) so that it will be capable of providing broad-range support for innovative projects involving advanced technologies, which often require considerable financial resources. For the sake of efficiency, the Agency should function through a network of existing regional centres or ones newly created with its help.

Profound reconstruction in the sphere of science of technology is necessary. The aim of reconstructing the R&D base is to adapt it to the structural models adopted in OECD countries, while preserving the proven solutions worked out by R&D institutes and centres in earlier years. Structural and ownership transformation of the R&D base of the economy will require utilising the existing privatisation path. Financial and organisational/ institutional mechanisms will have to be employed together with corresponding legal regulations pertaining to the following:

financing the existing state-run industrial institutes working in areas related
to national security (energy, defence, agriculture);
establishing R&D units within industrial holding companies;
employee-ownership as a possible privatisation path for R&D units;
transfer of shares in R&D units to institutions of higher education, other R&D units and regional organisation.

The target structure of the R&D sector should be achieved by implementing a government programme, which should not last more than three years. With regard to reforming the Polish Academy of Sciences (PAN), it is recommended that changes in this area should lead to the establishment of capital and organisational links between PAN units on the one hand and institutions of higher learning and industrial R&D units on the other. Another desirable development would be the emergence of links between

PAN units and various non-profit and public-service organisations. Such links do not necessarily imply the loss of academic identity and autonomy for the PAN units concerned.

Reinforcing mechanisms for transferring and absorbing modern technologies. One of the most important goals of innovation policy in Poland is to introduce modern mechanisms for transferring and absorbing new technologies. This refers to the transfer of both foreign and domestic technologies and solutions. The principal tasks of innovation policy in this area comprise:

- support for the development of an institutional infrastructure facilitating the adoption of modern technologies, particularly on the regional level;
- strengthening the units functioning on the scientific and technological services market;
- encouraging the establishment of funds and capital institutions investing in high-risk projects;
- an active licensing policy;
- support for foreign direct investment;
- stimulating demand in the economy for applied research and development projects conducted in stimulating demand in the economy for applied
- research and development projects conducted in Poland, as well as their results in the form of innovations constituting the property of entities operating in Poland.

Promoting and properly enforcing the norms pertaining to the protection of intellectual property. The policy on intellectual-property protection is of crucial importance for the security of business transactions and the long-term stability of the economy. It should also attract foreign investment to Poland. The main proposals in this field include the following:

- introducing a Patent Code and Patent Proceedings Code,
- improving the efficiency of rulings in criminal proceedings related to intellectual-property protection, among other things, by providing suitable education and training for judges and other personnel of the courts of justice
- establishing a monitoring system and channels for disseminating information on patents, science and technology, utilising the existing network of scientific/technological and economic information centres;
- measures regarding education and disseminating information on the norms of intellectual-property protection, e.g., through changed curricula at all levels, training for employees of central and local government institutions, including customs offices, agencies and centres for promoting enterprise and technological progress, foreign direct investment;
- partial financing from the budget for several patent and licensing consultancy centres; at the initial stage, such centres would require

subsidies from the Technology Agency, after which they should be self-financing.

Support for developing regional innovation systems. An important task of any effective state innovation policy is to support the development of regional system. This concerns above all institutions promoting entrepreneurship and innovation in a given region, as well as measures supporting regional investments of a technological character (possibly with an inter-regional impact), and, to some extent solutions based on the idea of Spatial Economic Zones (SEZs). The establishment of such zones is a policy instrument which has a selective impact on the economy of a region. It is, therefore, by its very nature, efficient only when applied in rare and exceptional cases. Thus, in view of the mounting pressure from local communities for the establishment of SEZs, government policy must set an unequivocal limit on their number and link their establishment to the development of regional innovation systems. Support for the development of a regional infrastructure for entrepreneurship and innovation systems should take into account the marked disparities between voivodships in this field[xlix]. Regional investments of technological character should be supported, especially those which promote innovation and the dissemination of new technologies, influencing not only the economy of individual regions, but also those of neighbouring regions (as in the case of seaports with an industrial complex in the vicinity). Their overall impact is so great that their adjustment to the new needs of the transforming economy cannot be left to proceed in a spontaneous fashion.

Supporting the development of the communication and transport infrastructure The technological infrastructure (e.g., telecommunication and data-transmission networks) helps disseminate innovations and, at the same time, serves as one of the most powerful stimulants of innovation and technological progress. The Polish data-transmission infrastructure should be in line with world standards and enable the processing and unhindered flow of data in global networks. Experience gained in the course of developing and utilising networks for scientific purposes should allow Poland to become a partner in the global information society, a rapid growth in which predicted for the first decade of the 21st century.

The development of the capital and business-insurance markets is one of the main factors encouraging economic entities to adopts innovations. This may require a contribution from public funds to help establish high-risk funds and a system of guarantees and insurance (representing the founding capital) so as to reduce the risk involved in innovation projects. Legislation currently being prepared on industrial funds, and on guarantees extended by the Treasury and certain corporate bodies should be promulgated, as should the draft regulations governing the functioning of regional investment funds etc.

Enhanced innovation potential will require a substantial increase in the country's educational potential. This should be the main form of investment in human capital, determining the prospects of economic development and the living standards of society.

As regards entrepreneurship and innovation, the education system should aim at preparing society - and especially the younger generation - to meet the challenges posed by new technologies, forms of organisation, means of communication etc. General and specialised forms of education in entrepeneurship and innovation must be treated as an important source of pro-innovation attitudes in the future, creating a favourable atmosphere for entrepreneurship and technological change, but first and foremost as a stimulant for establishing new, innovative firms, including those promoting high technology. Bridging the enormous gap between the level of education in towns and in the country represents an important challenge for education policy makers. One of the principal problems is how to update the curricula. Of special importance in this context are subjects related to modern production technologies and organisational forms, evaluating technologies, as well as entrepreneurship and technology transfer in the broad sense of the term. The state's help will be particularly required in providing primary and secondary schools with access to information networks: this will involve covering the costs of the relevant equipment and the fees paid to service providers. An increase in the level of education of society, measured using formal criteria, has been achieved over the last few years - in some cases, at the expense of declining teaching standards. This situation calls for radical reform that will ensure that the emerging competitive market of educational services will be based on sound principles.

Encouraging the development of small and medium-size innovative firms support for the innovative undertakings of SMEs should be one of the chief priorities of innovation policy on the regional level. In particular, the policy encourage the establishment of small, innovative firms in high-technology industries. To achieve this, a full range of legal, organisational/institutional, planning, fiscal and financial instruments will be required. Projects in this area should be undertaken in co-operation with the Technology Agency and non-governmental institutions of a nation-wide character. Another important task is supporting the formation of institution networks, both nation-wide and regional, facilitating the transfer of technologies to SMEs through the creation of „incubators", innovation centres and technology parks. Public funds may be used to finance R&D work, technology transfer and implementing projects for SMEs through different channels, such as: (1) research grants, (2) implementation grants,(3) public orders, and (4) financing programmes for innovative projects. The operations of R&D units contributing to studies on technologies, consultancy and training organised for SMEs should be financed mainly from public funds (the state budget, as well as regional budgets). The principle should be adopted that support of this kind will be extended first of all to enterprises linked to „incubators", innovation centres and technology parks. Measures should be taken to provide an extensive range of programmes in the areas of education, training, information and consultancy, access to information networks etc., so as to generate demand for innovative solutions among SMEs. Enterprises that take advantage of this offer should have all or a substantial part of their costs reimbursed from public funds. Other means of encouraging SMEs to embark on innovative projects include financial arrangements. The following measures may be mentioned in this context:

-a system of exemptions and subsidies supporting pro-innovation activities helping SMEs provided by institutions promoting enterpreneurship and innovation;
-introducing financial incentives for SMEs to obtain certain types of certificates (e.g. certificates of quality or safety).

In addition, support should be provided for developing a network of regional, inter-regional and international contacts relevant to innovation and technology transfer to SMEs, such as various kinds of agreements and contracts, innovation fairs, exhibitions, exchange programmes, and assistance in the creation and development of networks of firms.

Innovation policy makers should seek genuine (and not just verbal) social and political approval for science and technology. This can be achieved, among other things, by means of nation-wide consultations and discussions involving academics, the clients of research units, parliament, ministries and the general public. Procedures for determining priorities in science and technology should also be introduced, taking into account the results of such consultations, as should procedures for technology assessment (that is, studies to determine the social, cultural, ecological and other effects of the implementing new technologies or extending the range of use of existing ones).

Poland's agricultural policy is a crucial element of national structural policy on the way to European integration. Specificity of this sector, rural character of polish economy and a society (38% of the population lives in the countryside) will require the developing of specific instruments for promotion of innovation in this sector. The existing structural, economic, financial, legal and institutional framework which constitutes the innovation mechanism of agriculture lacks efficiency either as a means of attracting or disseminating technological and organisational information. This system does not match the needs and challenges of the transformation. Polish agriculture, being in a state of fragmentation, is incapable of developing a mechanism for attracting innovation on its own. Hence, for a long time to come the mechanisms for introducing biological and organisational innovations in agriculture will have to be maintained. This task requires the active involvement of the state. In the long run, a gradual shift will be necessary towards regional, demand-oriented models of support for agriculture. Vertical integration will also have to be stimulated by market institutions and processing plants.

With regard to fiscal/financial and organisational solutions, the existing system of financial support for biological improvements in agriculture (allowances, subsidies, tax exemptions) will have to be retained throughout the period of Poland's association with the EU. Moreover, Poland should secure the right to apply these instruments and measures for a certain period of time after joining the EU.

Disseminating biological and technological advances in agriculture requires farmers' associations and branch organisations to be established. These should be capable of integrating marketing activities on a large scale with a system of "market-oriented" subsidies and activities promoting new solutions. Without support from the state and regional budgets, and in the future the European Commission budget, this task will be difficult, if not impossible.

Polish associations and organisations of food producers and processors should joint the European system of farmers' branch associations and organisations, which perform most of the functions connected with technology and economy in the EU agriculture and food-processing industry. Their activities are financed in full or in part from EU funds. Partial privatisation of institutions and agencies promoting new advances is advisable, although the state should still play a major role in many areas (such as, e.g., providing extensive training, information and promotional programmes, and partially financing, consultancy costs related to agriculture). The same is true of those privatised entities offering consultancy services and disseminating new advances in the field of animal and seed production.

6. Conclusions: fundamental challenges and problems

The openness of an economy to innovation is conditioned first of all by the efficiency of its markets, construed in terms of product diversity and competitive mechanisms. This general assumption underlies the entire reasoning behind the suggestions formulated above, concerning the creation and reinforcement of an infrastructural and institutional system promoting innovative processes, their transfer and dissemination in the national economy. It is an issue which gains particular importance in the present conditions of systemic transformation, when elements of such a pro-market innovative system are still weak or, in many cases, non-existent. Moreover, many elements of the innovation system are still oriented towards the former type of economic structure. Many of the institutions that make up this system still focus on vertical and intra-sectoral links. They function on the periphery of the market mechanism or even outside it and seek survival in permanent financing from the public sector, and especially from the state budget. It is our belief, therefore, that one strategic goal of structural policy should be to stimulate the formation of an efficient domestic innovative system, consistent with a market economy.

Through methodical co-ordination of the activities of various innovation-promoting institutions (at the central, regional and local levels), such a system will shorten the path to an internationally competitive economy and will guarantee the development of an economic structure generating an ever increasing added value. It is in this market-oriented sense that we treat innovative policy as a strategic, supra-sectoral component of structural policy. The general assumption is that efficient markets and human intellectual development constitute today the main factors of development.

The history of civilisation indicates that intellectual resources - and no longer cheap and readily available raw materials (the consumption levels of which tend to stabilise) - represent the main development factor determining a sustainable competitive edge.

Poland has largely exhausted its raw-material, which means taking into account the globalisation of production as well as the globalisation and stabilisation of raw-material markets - that a competitiveness-enhancing strategy should capitalise on the increasingly efficient utilisation of resources, which can be achieved through investment in human capital. The main idea behind the selection of instruments for innovation policy is based on the assumption that structural policy in the course of adjusting to EU principles and standards will focus on constructing efficient systems for the absorption of innovation that stems from the most advanced economies. Such an assumption is a consequence of Poland's political aspiration to join the economic structures of the European Union as soon as possible.

Bibliography:

1. Geenhuizen, M.van and Nijkamp, P. (1996) Technology Transfer: How to Remove Obstacles in Advancing Employment Growth. in: Production of Knowledge and Dignity of Science. Warsaw: EUROREG, p. 81.

2. Green Paper on Innovation.(1996) European Commission, Brussels-Luxembourg.

3. Guinet, J. (1995). National System for Financing Innovation. OECD, Paris p. 21.

4. Institutions and Instruments of Regional Competitiveness -Supplementary Report. (June 1996), Task Force for Regional Development in Poland. Warsaw.

5. Instruments of Innovation Stimulation and Technology Transfer -Supplementary Report. (June 1997) Task Force for Structural Policy. Warsaw.

6. Kotler, F. (1994) Marketing: Analysis, Planning, Implementation and Control. Warsaw: Gebretner & S-ka, p.322 (in Polish).

7. Lundvall, B-A (ed). (1992) National System of Innovation. Towards a Theory of Innovation and Interactive Learning. London.

8. Malecki, J. (1991) Technology and Economic Development: The Dynamics of Local, Regional and National Change. London: Longman, p. 283.

9. Outline of Regional Development Policy for Poland - Final Report. (June 1996) Task Force for Regional Development in Poland, Warsaw.

10. Poland's Structural Policy on the Way to European Integration, -Final Report. (June 1997) Task Force for Structural Policy, Warsaw.

11. Quevit, M. (1997) Methodology of Evaluation of Regional Innovation Potential. Louvain, (non published material).

12. Technology and Economy. The Key Relationship. (1992) OECD, Paris.

TECHNOLOGICAL TRANSFORMATION AND ORGANIZATIONAL CHANGES IN UKRAINIAN AVIATION AND BELORUSSIAN ELECTRONIC INDUSTRIES: TWO DIFFERENT STRATEGIES

I.EGOROV
Centre for S&T Potential Studies
National Academy of Sciences of Ukraine
60 Shevcheriko Blwd.. 251-032, Kiev,
Ukraine
A. SLONIMSKI
Economic Research Institute
Ministry of Economics of Belarus
Slavinski st. 1. bld. 1, 220045 Minsk, Belarus

1. INTRODUCTION: WHY THESE TWO EXAMPLES?

These two sectors were chosen for analysis of transformation processes because they both used to play important roles in the economic of these countries in the past and they continue to do so at present. Computers and electronic components of Belorussian 'Integral' company. and Ukrainian 'Antonov' (AN) cargo planes have represented relatively high technological capabilities of these Soviet republics. After the collapse of the Soviet Union in 1991 both industries were suddenly cut from their main partners and consumers by the newly-established borders and customs regulations. but they still try to keep themselves afloat and even to penetrate into new markets.

As the other Soviet enterprises. factories in these sectors were part and parcel of the Soviet economic system where decision-making, procedure were highly centralised and enterprise itself was responsible almost exclusively for internal issues ranging from meeting production quotas to social conditions of the workforce.

In these industries. as well as in other military related sectors, the director of the enterprise had complete authority. This authority extended not only downwards through the production shops, departments and laboratories, but upward to the government as well. These relations between the government officials and top industrial managers have been preserved with some cosmetic changes in these sectors, as well as in other key industries that make them important for understanding difficulties of transformation processes in Ukraine and Belarus as a whole.

At the same time, these industries have started to implement different strategies of survival and development that have determined different dynamics of the main indicators and the means of technological and organisational adjustment to the new environment. It is natural to assume that various industries will respond differentially to the crisis of the economy but it is important to clarify the reasons shy and to what extent these strategies could be different.

A. Inzelt and J. Hilton (eds.), Technology Transfer: From Invention to Innovation, 285–298.

So. in first part of this article. changes in Ukrainian aviation industry will be considered. and in the second part.. Belorussian electronic industry dynamics will be analysed. The third part will be devoted to comparisons of the two industries and to general conclusions about their comparative advantages and future development. There are very few special publications and statistics about the situation in these industries in recent years which is explained by the need not to reveal commercial secrets. That is why our study is mainly based on interviews made between 1994 &1996 with managers and other specialists of the enterprises and scarce information from periodicals.

2. THE UKRAINIAN AVIATION INDUSTRY: ATTEMPTS TO PRESERVE EXISTING POSITIONS

The Ukrainian aviation industry was created as an integral part of the of the Soviet military industrial complex. Just before the collapse of the USSR in 1991, it comprised more enterprise that were subordinated directly to Moscow. Two of the these enterprises produced airplanes (in Kiev and Kharkov) and one - helicopters (in Tyachiv, Western Ukraine). The others participated in production process as subcontractors. There is no reason to consider helicopter enterprises in this article because it actually stopped all operations in 1992 . Even optimists do not hope it would be possible to restore production without substantial external support for several years [1]

Aviation (plane's) sub-sector was luckier. Probably, because three enterprises that were really lame and critically important for the further development of the industry's future,, were situated in the. Ukrainian territory: The "Antonov aviation complex "the motor Such engine factory. Ukraine has also several repair factories that could be used for servicing of aviation industry production and eight different design bureau that belonged to the Soviet aviation industry. Missile design bureau were also involved in aviation industry's R&D.

Fragmentation of the Ukrainian aviation industry, could be illustrated by, the fact that the Kharkov factory had much closer ties with Russian design bureau and in Soviet times it produced TU - type planes only.

The Ukrainian aviation Industry did not have enough capabilities to start airplane production independently. The main reason for this situation lies in the f act that this industry has not been formed as a separate one from the very beginning. but only as a part of the aviation industry of the whole Soviet Union. It means that level of co-operation in technology and production spheres with enterprises in other republics of the Soviet Union was exceptionally high. About 85 % the aviation enterprises were deployed in Russia. Other Soviet Republics had to rely heavily on supplies from them to assemble airplanes.

The main design and technology bureau of the aviation industry (TU. SU. MIG and others) were also. in Russia. Only Ukraine among former Soviet republics had its own Antonov design bureau but it was not enough to develop all planes components at the highest level. Ukraine had no metallurgy to produce special alloys for the aviation industry and even up to 80% of machine tools and other equipment for Ukrainian aviation enterprises have been supplied from the enterprises that were situated in Russia.

So at the beginning of 1990's the first main feature of the Ukrainian aviation industry was it high dependence in co-operation with Russian partners despite the fact that Ukrainian enterprises have possessed special place in the Soviet aviation industry first of all due to the engine production (in Motor Sich factory) for several types of Soviet planes and to the Antonov aviation complex. It could be possible to mention mutual but non symmetric dependence between Russia and Ukraine in the aviation industry because AN has provided R&D and technological support for a family of cargo planes in Russian aviation factories, such as AN-38 medium sized plane (different modifications) in Novosibirsk factories: AN-124. AN-74 and AN-32 –in Ornsk and Ullanovsk factories. AN- 74 Arsenyev (Far East) and Samara factories: and AN-27 that was produced by Polish aviation factory.. But this dependence was not critical for the majority of Russian enterprises. After the collapse of the Soviet Union, Russia has started to substitute factories from the other parts of the Soviet union in technological and production chains. Co-operation between Ukrainian and Russian enterprises became a complicated process also for political reasons tee.

But the aviation industry belongs to so-called 'complex production systems'. that include a lot of elements and where changes could not be done immediately. At the same time. Russia had no resources and technological capabilities to start to build new cargo planes from zero stage. Some factories in Russia were interested in technological supervision from the side of Antonov bureau and they preferred to use engines from 'Motor Sich' for a number of reasons (plane's design, price level and so on), despite leading airplane producers such as Illyshin and Typolev complexes started up to equip their planes with American or British engines to meet Western standards for noise reduction level and environmental protection.

On the other hand. Ukrainian enterprises had no direct contacts with foreign companies. and no experience of working in market conditions. Ukrainian managers had stable personal and business relations with their Russian colleagues only. These reasons become crucial in making decisions to prolong co-operation despite all political and economic barriers

To facilitate the process of co-operation and to avoid political difficulties. industrialists in both countries pushed the Governments to conclusion of the Agreement about production and S&T co-operation in the defence industry on November 18, 1993. Actually, this agreement comprised a number of areas that were not directly connected with the military sphere, such as .civil aviation. In accordance with this agreement, no quotas or other restrictions could be established on export and import of military production. This production is proclaimed duty. free in both countries. Aviation enterprises became participants of the agreement because the Russian Defence Ministry has announced a preliminary order for several cargo planes for it's air forces [2].

In 1995 the Russian and Ukrainian governments decided to make the next real steps to restore co-operation between enterprises and to create an institutional framework for it. In accordance with the agreement between the two countries about the creation of financial and industrial Groups (FIG) in March 1995, joint activity in the aviation industry had to be stimulated. In 1995 about 50 enterprises (nine from Ukraine. and approximately forty from Russia) decided to establish a new multinational company 'International Engines" with about 215,000 employees (75,000 of them from Ukraine). Special agreement between the two countries about unification of aviation Industries and co-operation in this sector was also signed this time..

FIG members had not to pay custom taxes. internal value added taxes .and could also receive loans from the banks of' both countries on preferential conditions. But Ukrainian Parliament has suspended the creation of FIGs with foreign companies. The main objection to the deal lies in possible danger for Ukrainian sovereignty over its key industries.

But even without implementation of this FIG agreement. Ukrainian aviation enterprise's managers have not seen other variants of the survival and further development of their enterprises except co-operations with the Russian aviation industry

Creation of the new AN-70 multi-purpose cargo plane became the core of co-operational process. It is still considered as one of the most successful examples of existing technological co-operation and technology transfer between two countries.

The project was started in the Soviet Union in 1987. In accordance with initial plans, it had to be finished in 1991 to change the relatively obsolete AN-74 and to enter the world market with a new plane of this type. A lot of latest ideas of Soviet specialists and results of reverse engineering were used at the design stage of AN-70 creation.. The new plane could be used both as a cargo plane or as. passenger plane for 52 persons), and it could be converted in one or another modification in a matter of hours. Its speed is about 700 km per hour and flight range is 4,500 km. Engines were made in Motor Sich factory [3]

.

Political disputes and economic crisis in the Former Soviet Union has led to numerous delays, despite the first plane having already been sold to one gold mining company in Siberia (Russia), even before a test flight. programme started.

The Russian Defence Ministry has provided 80 % of all the budget for creation of this new plane and , Ukrainian Ministry for Machine Building and Conversion and the other 20%. 75% of all components were received from Russia. This project can initialise orders for at least 500 planes. There are plans to start its mass production in two enterprises (in Kiev, Ukraine and in Samara. Russia). It means that it could give jobs to

27000 highly- qualified specialists in Ukraine and 50000 in Russia. There is also a theoretical possibility to compete with new European cargo plane FLA. That is still on the stage of design

Attempts to enter the market as soon as possible led to the catastrophe (crash) of the first assembled plane in spring, 1995. The second plane was assembled the next year. and it started test flights in summer 1996. The plane needs 250 checking flights and approximately two years of other tests to be commercialised. But AN-70 enjoyed some success in the international air show MAX-97 in Zhukovskv (Russia) in August. 1997 [4].

It is important to mention that Antonov company is trying to develop several types of new planes. So. in spring 1997, the company produced first middle- distance plane AN-140. that has started test flights. This plane could be equipped both by Motor Sich TB3 117 BMA-CB engines by Pratt & Whitney PW127A engines [5].

Unfortunately, first tests were not very promising. The very first flight in September..1997 had to be postponed for several days. despite the special arrival of the Ukrainian President to the airfield to stress the importance of the event. Specialists explain this situation by the scarcity of resources and low technological discipline among subcontractors: Motor Sich engines were assembled at the last moment. and they were installed without appropriate tests. In the Soviet times, Antonov factory usually used to build three airplanes to be sure that at least one of them could meet standards completely. and be ready for flight testing programme in time.

On the other hand. after the first successful test flight on September 17. 1997 The . Ukrainian aviation industry received a contract for 40 planes. from national air carrier "Ukrainian Airlines ". The plane's characteristics are better than similar to Saab or Fokker planes have. and Ukrainian planes will be much cheaper.. Russia has stopped work on it's 11- 112 and TU-130 planes of this class. That means Ukrainian aviation industry has some chances to occupy this segment of the market at least in the Former Soviet Union.

Ukraine has also won a tender to build an aviation factory in Iran to assemble AN-141 airplanes.

Another Ukrainian hope was connected with the new AN-180 plane for 150-180 passengers, flight distance 3000 km with two very efficient low noise D-27 engines from Motor Sich to meet Western standards. Antonov bureau started to design planes at the beginning of the 1993. but the project is proceeding slowly

Ukrainian enterprises could also produce cargo planes Russian (AN-124) or Mrya (AN-225) that were created the end of 1980's and relatively new and relatively new AN-34 short-distance multi purpose plane.

Due to high capabilities of the Ukrainian specialists and their relatively low salaries, Russian designers started even to build some new Russian planes in Ukrainian factories as it was with the new middle-distance TU-334 passenger airplane Engines 436TI for TU-334 will be produced in 'Motor Sich factory in Ukraine. The plane itself' could be produced in Ufa. Taganrog (Russia) and Kiev. But in recent years Tupolev has received less than 10% of finance promised by the Russian government to start mass production [6].

Viability of all these projects connected with plane production are not evident. The industry is in a deep crisis, despite some promising technological advances. It kept itself afloat because of direct state support and the special scheme of airplane leasing. Ukrainian companies that were created with an active Antonov's participation, have arranged commercial flights of cargo planes throughout the world. and they pay part of the profit to support main Antonov enterprise and design bureau. But after several scandals with smuggling, the Ukrainian government had to reconsider this practice. New regulations became tougher, and the share of taxes to the state budget became much higher

At the same time the state could not provide enough money to support the Industry directly. Main problems, as the head of the Antonov company, Mr. Pyotr Balabyev mentioned, are connected with wrong financial regulations and lack of state orders [7] The Ministry for Machine building and Conversion programme provides only 20-30% of the money that the company needs to survive.

Despite creation of the new and potentially competitive airplanes, Ukrainian aviation industry follows old-styled strategy in economic and technological spheres.

So Ukrainian specialists tried to build airplanes based on the best results of the Soviet aviation technology. Actually. they underestimate the changes that emerged in the aviation equipment throughout the world in recent years. first of all in aviation electronics. This establishes serious barriers for Ukrainian aviation production in the world market. Capacity of the post- Soviet markets is not enough to make such highly sophisticated aviation industry efficient.

There were some promising propositions for co-operation from the other countries. that have been rejected because of these reasons

So the American company Federal Express has expressed Intention to test AN-38 for correspondence delivery, but some characteristics of the plane have to be improved significantly. US Air Force officials also expressed an interest in heavy Russian planes but with Pratt and Whitney engines and General Electric electronics. Ukrainian side could not guarantee that these changes could be made in time

It has become more and more evident that the Ukrainian aviation industry needs to transfer new know-how from outside to preserve or improve its competitive positions in the market. Operations in the world market of high-tech products require also higher level of' co-operation with different partners. introduction of new standards and long-term organisational efforts. Ukrainian enterprises have just started to use some new (electronic) equipment from the leading, world producers, but four main obstacles prevent this being done more effectively.

First, lack of own capital and underdevelopment of national financial sector that could not provide money for long-term projects.

The second obstacle is the existing practice of intellectual property' protection. In Soviet times this problem was not important. Even at the beginning of 1990's a lot of information was transferred from one partner to another without relevant documents and protection. As was mentioned above a key role has been played by already existing long-term personal relations between officials in Russian and Ukrainian aviation industries [8]. It means that Ukrainian specialists have succeeded in the tacit knowledge transfer process, but only from traditional Russian sources. Other official ways of technology transfer were used on a very limited scale from the other hand reverse engineering was very popular in the Soviet aviation industry that made it easy to use some new results from abroad. Property rights are still not determined in accordance with international practice. Such a situation is not acceptable in relations with Western partners and suppliers.

The third obstacle is connected with the old-style management of the enterprises and relations that exist between them and the state.

In accordance with the existing Ukrainian laws, there could be no privatisation in the main enterprises of the aviation industry (despite the strong feelings among the personnel of the enterprises that all the best parts of the factories and design bureaux have been sold by the top managers.

There were no serious changes in the processes of production and organisation in recent recent years. despite the shift in economic and institutional environment. Such an institution is not a stimulating one for potential Western partners.

Ministries have the authority to control main processes in the enterprises through the system of subsidies, tax exemptions and state orders. Usually the state keeps a majority stake in the aviation industry enterprises, and state bureaucrats have no intention to share their power with anyone else. From the first glance, it looks strange to be interested in keeping non-profit enterprises, but bureaucrats use them as channels that help to convert state subsidies Into personal income through "friendly" banks that arrange numerous delays with payments to make short-term loans for trading companies.

The fourth obstacle is connected with the outflow of highly qualified personnel from the industry. In 1989 the Ukrainian aviation and space industry had more than 150 thousand workers and engineers. By mid 1996, the overall number of employees has been reduced by more than 40%. In 1996. the level of wages in Ukrainian aviation industry was at least 3 times lower than in gas or energy sectors. Design bureaux have preserved less than half of their R&D manpower : the brain drain was particularly high in 1991-1996. This process was accompanied by the further ageing of the personnel that makes the future even more problematic. Thanks to the massive outflow of specialists. the Ukrainian aviation enterprises have lost their leadership in composite materials and some other fields. they had among similar enterprises of the former Soviet Union. But 'simplification' of production that was observed in the engineering industry of the Former Soviet Union [9] has still not became a dominant tendency for the main enterprises of the industry. At the same time, some subcontractors of the main aviation enterprises were compelled to move to the lower segments of the market to improve their competitive position.

3. THE BELORUSSIAN ELECTRONICS INDUSTRY: MARKET INCENTIVES AND STATE PRESSURE

In the Soviet era, the bulk of Belorussian electronic enterprises were part of the integral company in the all Union Ministry of electronic industry. At the end of the 1980's the main specialists of six Belorussian factories and four design bureaux were connected with production of different chips (400 ml. per year) and semiconductor devices, (1500 ml per year). Belorussian enterprises also produced computers. but they used mainly reverse engineering methods for that. These methods were more or less effective in the 1970's and the beginning of the 1980's when Soviet specialists could re-produce IBM-360 370 type computers. But at the end of the 1980's microprocessor technology became more sophisticated and Belorussian engineers could not even copy the Intel 286 processor.

After the collapse of the Soviet Union. the traditional market for electronic enterprises started to shrink because of the general decline in production and fierce competition from foreign companies. On the other hand. it is evident that electronic companies could produce a wide range of goods without much modification to either staff or capital equipment, especially if they had specialisation in purely electronic components. Other studies of the post-Soviet economics also show an unambiguous relationship between industrial sectors and re-organisation which is consistent with the proposition that Industries which are less complex and asset specific will be more likely to re-organise successfully [10]. That is why Belorussian electronics producers have suffered less from the general crisis than their Ukrainian or even Russian colleagues. Some data about main indicator's dynamics are in table 1.

Table 1

Some indicators of the development of the Belorussian electronics industry in 1990-1996 (based on the Report of the Ministry of electronic industry, of Republic of Belarus 1997.

Indicator year	1990-1996
Integral chip production	373 164 (mil.)
Semiconductor devices (ml.)	1114 865
Share of export outside CIS	0.1 27. 1(%)
Number of types of integral	1235 1185 chips
Number of types of	268 461 semiconductor devices
Number of types of liquid 0	40 crystal indicators
New types of consumer goods 0	150 launched

Data in the table demonstrates the general decline in the volume of production but positive changes too. First of all, diversification of production and export growth. During the years of independence, export has grown from $0.22 ml. to 58.175 ml. in 1996, and in 1997 it could reach $12 ml. Only in 1996, Belorussian electronic enterprises started to produce 130 new types of semi-conductor devices and integral chips. In 1996 electronic factories from Belarus had 45% of the East Asian chip market for watches and 7% of the market for calculators. The state has provided Integral with special $5 ml. for equipment, that gave the opportunity to buy new equipment from the Korean LG company. This in turn has helped to conclude new contracts with Motorola and LG on supplies of chips and other components from Belorussian factories.

Some Belorussian electronic enterprises have created joint ventures with companies from CIS states that could not find their place after losing military contracts. So two such enterprises that produce electronic watches started their work in Ukraine [11]

Belorussian enterprises have also started to deliver some new products for the internal market. first of all, magnetic cards for banks, public telephones and so on. Special state S&T programme 'Belelectronika' was started in 1994. It includes such sub-programmes. as 'Belorussian TV set', import-substitution in consumer electronics and some others [12]. But the Industry really became export-driven and foreign orders determine the main directions of technological transformation. At the same time it is evident that adjustment to the world market is accompanied by the process of simplification of production. Belorussian enterprises stopped to produce a number of relatively sophisticated types of final products. Despite level of renewal of production reaching 23% per year in 1995. 1996 Belorussian factories were moved to the lower segments of the market where they compete with Chinese companies. Low value-added products prevent revenue increase and starting to compete in more lucrative and fast-growing sectors.

One of the main reasons for this situation lies in fact that problems of technology transfer have not played an adequate role in the process of transformation. Enterprises have spent a lot of money on support of the so-called social sphere keeping the level of unemployment extremely low but not on purchasing licences to produce hi-tech products. Usually technology transfer has the form of instalment of new machines to produce relatively simple electronic devices. or in training courses for the limited number of specialists abroad in famous foreign electronic companies.

Intellectual property rights protection is not in the focus of Belorussian specialists. They received only 46 patents in 1995-1996 but all in Belarus and Russia not in the world leaders in electronic industry.

The general tough economic situation in the country also contributed to the erosion of technological capabilities of the electronic industry in Belarus. The country is lagging behind most Eastern European countries in reforming it's economy. There is no large-scale privatisation. no real control over inflation and other processes in the financial sphere. The official exchange rate of the national currency is artificially low which leads to the direct loss of electronic enterprises because they have to buy raw materials abroad at relatively high prices and to sell their products cheaply. Managers of electronic enterprises mostly rely on state regulation and special credits, that as they hope would help them to survive and probably to improve their competitive positions. Integral officials have even prepared a special draft of the law that gives electronics many advantages should the abolition of value-added taxes occur. In this case they will be able to raise the volume of production by 20% and export by more than 2.1 times by the year 2000. This growth will be accompanied by the stabilisation of the number of employees that declined by more than 40% to less than 20 thousand in 1990-1996 Cheap loans from the state could help the enterprises to restore their financial strength. which was undermined by high inflation in 1991-1996. But it is not evident that such a step will be the best remedy for the efficiency improvement in state-owned enterprises.

4. COMPARATIVE ADVANTAGES AND COMMON PROBLEMS

Despite serious differences in their functioning and strategies, it would be possible to mention common problems which the Belorussian electronic and Ukrainian aviation industries experienced in the mid 1990's.

First of all it is a lack of transformation efforts in the economic sphere. Enterprises of both industries are suffering from Soviet-type methods of internal regulations, reliance on state support and unwillingness to start privatisation. Privatisation is not a universal solution., but it is evident that the state could not provide adequate financial support for both industries in the near future. The political climate in both countries especially in Belarus is not favourable for foreign companies. If large-scale privatisation in Belarus was stopped after the election of Mr. Lukashenko as the president of the country in Ukraine privatisation procedures are not transparent and they have been tailored for representatives or local elite (state bureaucrats first of all) and so-called insiders (usually factory managers).

Second, unlike some other industries managers of these enterprises and state officials are not ready to create joint ventures with potential investors as minor partners. Understanding of the need to improve quality of their products is not accompanied by the corresponding efforts in the sphere of technology transfer.

Third, technology transfer itself is mainly limited to simple forms such as use of foreign-made equipment (through import of equipment) or use of standards to enter the external market. Enterprises in both countries are partly learning, through exporting their production but effectiveness of this learning, has to be much higher. As Radosevic showed, it is usually not enough for building appropriate technological capabilities to be competitive in the world market [13]. At the same time. it is evident that Belorussian electronic enterprise unlike the Ukrainian aviation industry became deeply involved in the international division of labour and they have much more diversified contacts with foreign countries.

At the same time the position in the lowest segments of the electronic devices market is very unstable because of potential competition from numerous East Asian and Latin American companies that could easily obtain access to these types of technologies. Potential competitors could use the same advantages - low-paid work force as Belorussian factories.

Subcontracting and joint venture creation are the best ways for the bulk of electronic industry in the former Soviet Union to survive. but it has to accompanied by measures aimed at development of technological capabilities and expansion to the new fast-growing segments of the market.

The Ukrainian aviation industry has another strategy. It's enterprises are trying to enter relatively sophisticated and already divided markets where competition is especially fierce among suppliers. not customers. The prizes for winners are also very high. So for instance the market price of one Russian (AN-124 or Mrya (AN-225) class airplane is approximately equal to all economic help per year, that has been provided by the USA to Ukraine in recent years [14].

It is also clear that this industry is very capital and technologically intensive and it creates a number of obstacles for Ukrainian plane makers. Co-operation with only Russian partners will not be enough for breakthroughs in the world market although co-operation in production and technological spheres in aviation industry is crucial for both countries. The Ukrainian aviation industry is still mainly oriented on CIS markets where demand has fell in several recent years but there are forecasts that in 2000-2010. the growth rate for new passengers will reach 8% per year. But aviation companies in these countries actively buy American and European airplane that undermines positions of Russian and Ukrainian producers in their own markets.

The strategy of the Ukrainian enterprises needs to be much more diversified and flexible. Different types of co-operation with foreign partners are to be developed. Ukrainian aviation industry still has not received any investment from abroad (except for some from Russia). Large plane-makers are not very interested in the emergence of a new competitor in the market. At the same time, small companies like Fokker and aerospace divisions of large Korean companies expressed their interest in joint production. For the Ukrainian enterprises it could be a good chance to raise their technological and marketing capabilities, but psychological and institutional barriers prevent the development of this type of co-operation.

Another opportunity for Ukrainian aviation enterprises lies in the possibility of participating in international co-operation as parts suppliers (subcontractors). In 1996. some experience was received by the Motor Sich factory that fulfilled an order for 5000 engines from the famous Italian car maker lveco. This type of co-operation could improve technical standards and will open new opportunities for transfer of technological know-how.

Special divisions of the Ukrainian aviation enterprises could focus on cargo flights and technical assistance for the existing fleet of Ukrainian made planes in the former Soviet Union, India, Poland and other countries. It could create jobs for Antonov for at least 20-25 years. because about 25,000 AN planes were made during recent decades.

As to the Belorussian electronic industry the experience of other post-Soviet countries provides a clear indication between Industrial sector and organisational reform. So electronic industries are most likely to engage in organisational reform while the heavy industries are least likely. But in the Belorussian case. political power has established strong limits on transformation processes which prevent enterprises from developing some positive tendencies that took place in 1993-1935.

A lot of problems for both industries are connected with uncertainty in. the legal sphere. Processes of innovation and technology transfer in Ukraine and Belarus are mainly regulated by the old Soviet laws that create a lot of problems although some changes were made in recent years.

Several laws about R&D and innovation activity were passed through parliaments in both countries But usually they contain only some basic principles that have to be clarified. At the same time, these laws have opened the way for some other legal acts that clarify different aspects of R&D activity. So, several special laws were approved by the Ukrainian Parliament in 1993-1995 about protection of rights on inventions and useful models, about protection of rights on prototypes, about protection of brand names and trade marks, about copyright protection, about "Special Economic Zone Sivash" and some others. These steps slightly improved the situation with property rights regulation in Ukraine. Only in 1994-1995 the Ukrainian budget received more that $3.5 ml. in revenues for the foreign inventor's rights protection.

In Belarus the situation is different. A lot of rules in this sphere are determined by numerous presidential decrees that could be changed by the President without preliminary announcement and approval by the Parliament. It makes both economic and innovation activities more complicated. For instance announcement of the new set of national priorities could mean changes in the level of taxation for the rest of industries to collect money for special priority supporting funds.

A very complicated situation exists in the patenting sphere. After the collapse of the Soviet Union, Ukrainian and Belorussian scientific institutes, design bureaux and enterprises have not received patents, especially Western ones. It is important to note that the absolute majority of patents were inherited by Russian institutes and Ministries as the leading or main Institutes in particular sectors. Ukrainian and Belorussian enterprises and institutes usually have no foreign patents at all. This is a tremendous liability when trying to negotiate with foreign companies and attempting to market high-tech products abroad As a result of economic crisis and devaluation of assets institutes and enterprises have usually nothing to pay for patents and licenses. So the Belorussian electronic 'Integral' company had to ask for a special $5 mil. loan from the state to buy new chip production technology in 1996

Both the Ukrainian and Belorussian Patent Offices provide information on how to file foreign patents and how to choose a patent attorney. But very often this information is useless because inventors lack the funds to do this operation. This includes not only funds to file patents, but also to pay fees and to maintain them abroad. There is a significant interest in patenting in the Former Soviet Union. For instance, only in 1991-1994 more than 50 000 applications have been filed for Ukrainian patents more than 90% of them from Ukrainian Inventors. Unfortunately, the Ukrainian State Patent Office does not have the resources needed to handle all those applications in an expeditious manner. So far, 2.500 patents have been granted and about 3000 trademarks registered.

But at the same time the number of patents registered abroad continued to decline. As Dyker showed overall number of patents from the former Soviet Union fell by more than 3 times from 1990 and now nearly equals to the corresponding figures for relatively small country Hungary [15]

On the other hand. in the Ukrainian aviation industry, some managers are suspicious of all inventions that were made outside their company or at least the Former Soviet Union.

It creates another obstacle for the technological and organisational changes in Ukrainian aviation industry. It is clear now that the best solution for this sector lies in the combination of different strategies of technological and economic development with assertion on the more active institutional changes and technological transformation including more active foreign technology acquisition.

Despite all difficulties it seems that the Belorussian electronics industry has better perspectives to find it's niche in the world economy than the Ukrainian aviation industry. Of course there is no guarantee that negative influence of the political environment could not destroy already created capabilities, but initial conditions it has are definitely better. The market for electronic components is not sophisticated and access to customers is easier than in aviation.

At the same time, signs of degradation in the technological capabilities in both countries are evident and the process could be irreversible in the near future. Without intensive technological exchange, it would not be possible to compete even in the domestic market of hi-tech products.

References:

1. (1995) Novosty . June 3
2. (1993) Golos Ukraine. December 5
3. (1994) Novosty. December 22
4. (1997) Zerkalo Nedely. August 23
5. (1997) Finansova Ukraina. January 21
6. (1995) Moskovskie Novosty. January 8,
7. (1995) Zerkalo Nedely, August 28
8. (1994) Zerkalc, Nedely, December 24
9. Kuznetsov E. (1997) Learning. to Learn: Emerging Patterns of Enterprise Behaviour in the Russian Defence Sector, 1992-1995, in J. Sedaitis (ed.) Commercialising High Technology : East and West, Rowman and Littlefield Publishers Lanham. pp. 181- 204
10. Aldrin A. (1997). Defence Enterprise Adaptation in St. Petersburg in J. Sedaitis (ed.) Commercialising High Technology: East and West, Rowman and Littlefield Publishers, Lanham, pp. 205- 238
11. (1997) Belorrusskaya Gazeta. June 30
12. Belarus A. L. Ponomar' (1997), "Belmlkrosistemy"-Algoritm Uspcha. Tehmika Ekonornika. Organizatsia 2. pp. 12-15
13. Radosevich S. (1993). Strategic Technology Policy for Eastern Europe, Miimeo. SPRL. University of Sussex. 45p.
14. (1994) Kievskay Pravda. April 4
15. Dyker D. (1995), Technology and Economic Transformation, STEEP Discussion Paper No. 21 . SPRU. the University of Sussex.

BARRIERS TO INNOVATION IN BULGARIA

Some Results of a Survey (July 1997)

Dr . ROSSITSA CHOBANOVA
Institute of Economics, Bulgarian Academy of Sciences
3 Aksakov St., 1040 Sofia, Bulgaria

1. Introduction

One of the key factors for reviving Bulgaria's economy is successful development of an active technological culture and, in particular, of understanding, knowledge and implementation of innovative activities at various levels. The speed of transition towards market economy and European integration depends to a great extent on the activation of the national innovation potential as well as of the process of transmutation of invention into innovation. Some contemporary information about these processes is available from a survey on barriers to innovation in Bulgaria, conducted in July 1997 with a support of Ministry of Education and Science in Bulgaria and EC Phare

Programme.

This paper has presented some results of this survey, defining main sources of innovative ideas, factors assisting and factors hampering innovation in Bulgarian enterprises and has formulated conclusions and recommendations for policy makers.

2. Methodology of the survey

The methodology of the survey carried out is based on the proposed OECD guideline for collecting and interpreting technological innovation data [1] and on the approaches and experience developed in the Institute of Economics at the Bulgarian Academy of Sciences. 112 persons were interviewed in July 1997 in the form of filling in a questionnaire. The data refers to enterprises with innovative projects during the last two years.

The significance of the survey should not be overestimated, in spite of the large number of units covered; yet the number of indices allows for most of the indices to be of an orientation, yet not representative character. The indices included and surveyed

A. Inzelt and J. Hilton (eds.), Technology Transfer: From Invention to Innovation, 299–308.

comprise, in the first place, most general characteristics of the organisations with respect to their branch and regional affiliation, legal status (independent, part of a conglomerate,. division of an enterprise or another unit, availability of direct relations with other enterprise/s), size (according to the number of employees), main economic activity (production organisation, main type /according to the economic specialisation of production, intensity of R&D and exports). Along with this, the characterisation of the professional, educational and employment status of the person interviewed has allowed to identify the units of the survey and to seek possibilities to determine the influence of the environment which forms the sources of innovative ideas and specifies the impact of the factors assisting or hampering innovation.

Judging from the composition and the structure of the respondents, the conclusion could be drawn that the subjects of the present survey are well-known and respected specialists, with considerable knowledge and experience, which could be considered representative in view of the responses to the questionnaire.

3. Some results of the survey

The results of the survey have referred to the sources of innovative ideas, factors assisting and factors hampering innovation.

3.1. Sources of innovation ideas

The sources of innovative ideas in the questionnaire have been classified into two groups: internal and external ones. The answers are ranked as follows: “often”, “sometimes” and “no”. There are seven specified internal sources of innovative ideas:

- top management
- in-house incentive schemes
- marketing
- production
- in-house incentive schemes
- monitoring to technological development
- personnel with specific qualifications.

Further analysis is concentrated on the answers ranked “often”.

Among the **internal sources of innovation**, of greatest significance for Bulgarian innovative enterprises is the initiative of the top management - in general, and particularly for the enterprises with 100 to 500 employees. Second rank specialists and in-house R&D which applies mostly to larger firms (with over 2000 employees). Of a relatively limited significance are the in-house incentive schemes for monitoring of technological development (insofar they exist in larger firms and less so in small enterprises).

The **sources of innovative ideas which are external** to the enterprises have been distinguished into two groups: from domestic and from foreign sources.

The domestic external sources of innovative ideas are:

- public support programmes of innovation (funds, branch unions, regional associations etc.)
- fairs, exhibitions, meetings
- competition situation
- acquisition of embodied technology
- acquisition of disembodied technology
- influenced by different forms of training
- co-operation and proposals from customers
- co-operation with consultants
- co-operation with sub-contractors
- co-operation with other (related, complementary production) enterprises
- co-operation with research institutes and universities
- science and technical literature
- patents
- commercial literature
- legislation, norms, regulations
- standards
- taxation and other specific economic regulations.

Within the group of questions concerning domestic external sources of innovative ideas of smallest significance is the competitive situation, and within the second one - the scientific and technical literature available.

The influence of foreign competition or the various forms of co-operation with sub-contractors is of a relatively limited significance. The variations based on the size of the enterprises are insignificant.

In Bulgarian enterprises the importance of internal sources of innovative ideas is greater than that of external ones. Of special significance is the interest on the part of the top management which generates innovative ideas, as mentioned above. In terms of external sources of innovative ideas, the co-operation with customers/suppliers and with other firms is of importance.

3.2. Factors assisting innovation

In the period of development when the innovation activity is in a very low level [3] it is very important for policy makers to know which are the key factors assisting innovation projects in the country.

The strongest factors assisting innovation in Bulgarian enterprises indicated by the survey are:

- interest on the part of the top management
- co-operation of R&D with marketing and production
- personnel with specific qualifications
- co-operation with customers/suppliers and with their firms.

The percentage of answers to the questions about all factors assisting innovation ranked "often" is shown in Fig 1.

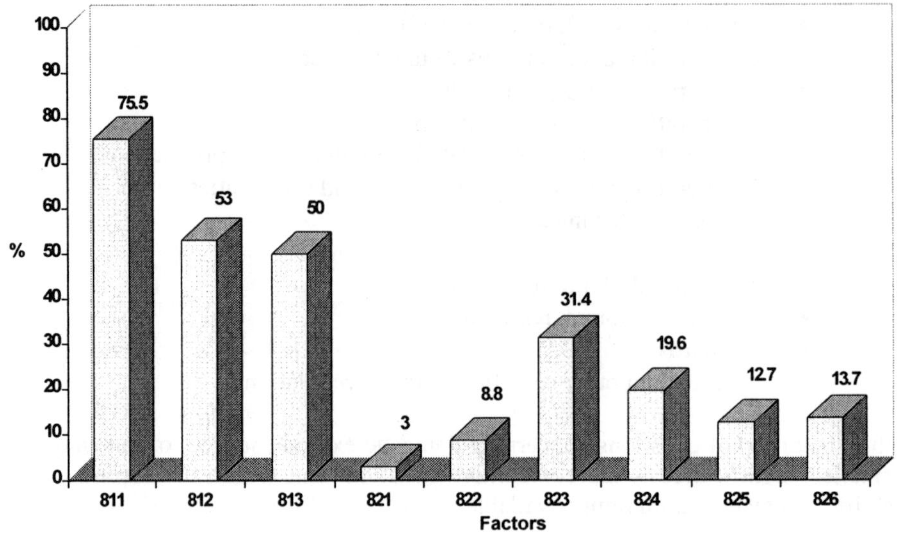

Figure 1.

F

actors assisting innovation ranked "often" (%)

Legend:
811 contribution of top management
812 co-operation of R&D with marketing and production
813 personnel with specific qualifications
821 public support programmes for innovation
822 use of advisory services
823 co-operation with customers/suppliers
824 co-operation with other firms
825 co-operation with research institutes
826 co-operation with universities

Factors which are supposedly assisting but, in practice, do not assist innovation in Bulgaria are: public support programmes for innovation, co-operation with research institutes and universities, use of advisory services.

3.3. Factors hampering innovation

Among the factors hampering the success of innovative projects the questionnaire has listed economic factors, lack of innovative potential and other reasons.

The factors which have hampered innovation most often are:

- the lack of appropriate sources of finance (according to over 75% of the respondents
- too small expenditure on R&D (according to over 60% of the respondents)
- limitations imposed by the corporate tax, the imports regime and value added tax.

Two other factors which hamper innovation are the lack of a place for innovation in the firm's strategy and the lack of necessity to innovate due to earlier innovation. Other factors ranked as frequently hampering innovation are: the excessively long pay-off period on innovation, high innovation expenditure, excessive perceived risks, qualitative shortcomings of the own R&D, lack of skilled personnel, lack of information on contemporary technologies and markets, difficulties in controlling innovation costs, deficiencies in the ability for external servicing, lack of opportunities for co-operation. The percentage of answers to the questions about economic factors and those of innovation potential hampering innovation ranked “often” is shown in Fig. 2.

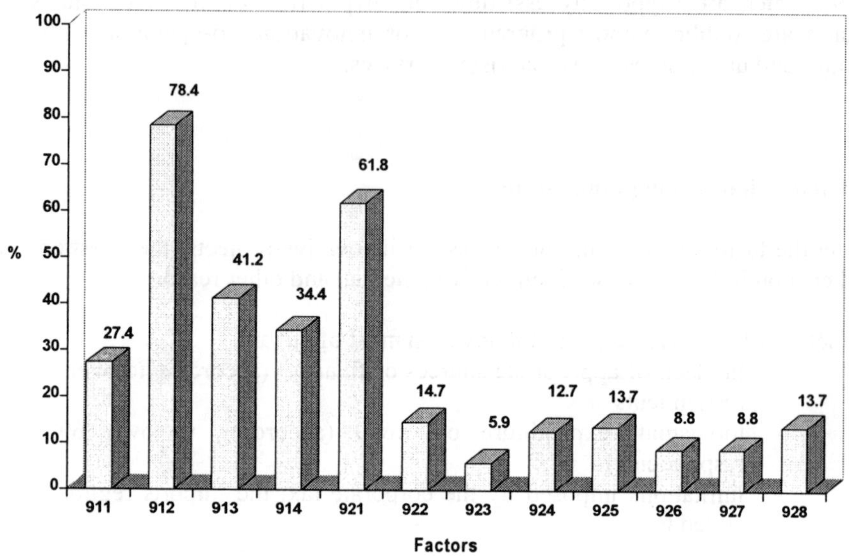

Figure 2. Factors hampering innovation activities and ranked "often"

Legend:
911 excessive perceived risks
912 lack of appropriate sources of finance
913 innovation expenditure too high
914 pay-off period of innovation too long
921 R&D expenditure too small
922 qualitative shortcomings of own R&D
923 lack of skilled personnel
924 lack of information on technology
925 lack of information on markets
926 innovation costs hard to control
927 deficiencies in the ability of external services
928 lack of opportunities for co-operation

Less than 20% of the respondents (an average of 15% for all factors) mention as frequent barriers to the implementation of innovation projects the following factors: qualitative shortcomings of own R&D, lack of skilled personnel, lack of information on contemporary technologies and markets, innovation costs being too hard to control, deficiencies in the ability of external services, lack of opportunities for co-operation. Only 15% of large enterprises have mentioned excessive perceived risks in the carrying through of innovative projects as an "often" hampering factor. About 30% of the specialists define the lack of sufficient information on markets as an "often" hampering factor.

The analysis of the factors presenting the opinion of the respondents on the rest of factors hampering innovative processes indicates that most important are considered: the lack of technological opportunity; the shortcomings in legislation, norms, regulations, standards, certificates; the lack of customer responsiveness to new products and processes; the uncertainty in the timing of innovation.

4. Conclusions and recommendations

The process of transmutation of invention into innovation in Bulgaria is influenced by a number of factors. The main barriers to innovation in Bulgaria consist, above all, in the poor financial situation of the firms, the lack of modern equipment in most of them, insufficiencies in banking and taxation, as well as in economic legislation, which hampers the process of obtaining credit for such a risky activity. The grey and black economies and Mafia structures in the country also play a negative role. Those factors exert strong negative influences over the R&D departments which appear to be the main generators and promoters of innovative ideas. As a result the state of innovative activities in Bulgarian enterprises does not correspond to the country's potential, especially in terms of the human resources quality.

A number of factors have a significant influence over the innovative behaviour of the enterprises in Bulgaria.

Innovative activities depend on the size of the enterprises. Larger firms have greater expenditure on innovation, especially when they are profitable and export-oriented. Legislation has an indirect influence in that respect. The spending on R&D is included in the total expenditure of the enterprises and in this way the tax basis is decreased. Due to the relatively constant nature of innovation expenditure and the smaller turnover in SMEs, the latter are more restricted in their opportunities for innovation in comparison with larger firms. In this respect, there is a need for creation of supportive public and state structures.

Innovative activity in Bulgarian enterprises is closely related to the markets for their production. On the whole, the "narrow" and "shallow" Bulgarian market, as well as the level of domestic competition, do not stimulate innovation. That is the reason for the higher percentage of export-oriented Bulgarian enterprises among those spending on innovation. In this respect, international competition is of special importance for innovation in Bulgarian enterprises.

One of the indicators characterising the level of innovation activity in enterprises is the ratio between the value of their R&D expenditures and their turnover (R&D intensity). In order to make some conclusions about size of enterprises and their R&D intensity the surveyed ones have been put in three groups. In the first group are enterprises with a value of this ratio bellow 1 per cent, in the second one - between 1 and 4 per cent and in the third - above 4 per cent. Most enterprises belong to the first group - 41 per cent of the cases. It has shown that the level of innovation activity in the most surveyed innovative enterprises is low. The variations for the separate enterprises are not significant which means that both small and large units allot almost equal percentage of finance for R&D activities. If the marginal value is 100 employees, the number of enterprises for the first and the third groups is almost equal; only the second group - between 1 and 4% shows a more significant variation.

For small scaled and opened economies as Bulgarian one a very important indicator of potential for development is the relation between the level of innovation intensity and export intensity (ratio between export and turnover). This relation has characterised technology transfer as a process of transmutation of invention into innovation in those kind of economies as well. The survey has shown that the growth of the R&D expenditures in the surveyed innovative enterprises is accompanied by an increase of their export intensity, although for small enterprises the raised R&D intensity does not directly influence the export intensity.

There is an evident tendency towards decreasing the state's direct intervention in innovative activities at the expense of the market forces. The necessity to modify the influence of state policy in innovative activity from direct investment in high technologies to creating a competitive environment and diffusion of new technologies will be the basis of the policy in this area.

At present, the main factors assisting innovation, regardless of the size of the enterprises, are internal for the enterprise: the contribution of the top management and the availability of personnel with specific qualifications. The great significance attributed to the co-operation of R&D with marketing and production as a factor assisting innovation indicates that R&D activities should be subjected to the requirements of the market, and those of international markets in particular, instead of being carried out as an end in itself.

It is a concern that the co-operation with research institutes and universities (the public research base) is not perceived by the representatives of the enterprises as a factor assisting innovation, although the specialists from the institutes are of a different opinion. It is a general stand that, at present, public support programmes for innovation and the use of advisory services are very rarely factors assisting innovation. Bearing in mind the generally limited resources for innovation in enterprises, the concentration of scientific potential in the public sector, as well as the experience of certain countries (e. g. Italy, where public research is of a greater significance for innovation than in-house R&D, especially with regard to renovating capital goods), state policy should be directed towards the development of scientific research in public organisations aimed at the achievement of competitiveness of Bulgarian enterprises on the international markets. Special attention is to be paid to the role of the state in supporting the establishment of an infrastructure which would connect the R&D public sector and SMEs.

Consumers, in their function as innovative sources, are also a factor for innovation in this country. The survey has rated it as acting in the area of building. Yet the general slump in consumer demand at the present stage hampers the influence of that factor on a national scale. On the other hand, external demand is an especially important and stimulating factor for innovation in Bulgarian enterprises.

At present, both Bulgarian and foreign patents are of little significance for innovation in Bulgarian enterprises; this is a barrier to enforce innovation in the country.

In terms of diffusion of technologies, the greatest expectations lie with the public sector (an opinion shared also by other countries of Western and Central Europe). The highly skilled workforce, as well as the material assets in some sectors, are considered factors which contribute, on the one hand, to their own active involvement in innovative activity (especially in SMEs) and, on the other hand, to the adapting of newest foreign technologies to Bulgarian enterprises.

In larger firms, an important factor for the adoption of foreign technologies and technological transfer is the availability of in-house R&D. The results of a previous survey carried out by the Institute of Economics with the Bulgarian Academy of Sciences [2], as well as the experts' discussion provided after the survey, indicate that the process of an accelerated closing down of the respective R&D departments during the past seven years receives a negative assessment from the viewpoint of the opportunities for orientation in choosing new technologies to be embodied in the enterprises.

The elaboration of the idea to implement newest western technologies in Bulgarian enterprises and to use them as a basis for production of goods both for the supplier-countries of the technology and for the markets on the Balkans, the Black Sea region, etc., emerges as a possibly positive (and perhaps the only realistic one) alternative for the country's development.

Emphasis should be laid on the opportunities for foreign investments in customer-oriented R&D. This would be favoured by the low price of the highly skilled workforce in Bulgaria, as well as by the conditions for its reproduction determined by the national traditions in awe for education and professional training [4].

Of special significance for the public support innovation programmes in this country are the international innovative flows. However, it is very important to distinguish between technological-informational flows, equipment flows and the contribution of foreign research and technological knowledge in order to elaborate efficient strategy and policy for their attraction and use.

International programmes for innovation and co-operation could also be of importance. Yet, attention should be paid to the strongly negative assessment of the experts involved in the discussion of the results of the survey on the co-operation within some European programmes between Bulgarian research teams and Greek co-ordinators.

The protection of the interests of Bulgarian scientific potential, as well as the providing of conditions for its reproduction, requires the establishment of an institutional structure which would define the priorities and would co-ordinate the co-operation of Bulgarian teams with European and other foreign programmes.

References

1. OECD (1992) Proposed Guideline for Collecting and Interpreting Technological Innovation Data, Oslo.

2. Chobanova R. (1996) Technology Change in Bulgarian Economy. National Fund for Scientific Investigations in Bulgaria, OHN N430.

3. Chobanova R. (1998) Main Barriers to Innovation in Bulgaria and Evaluation of Possibilities for Overcoming Them, in Chobanova R., A. Dimitrov (eds.), Barriers to Innovation. CIELA Member of the Wolters Kluwer Publishing Group, Sofia, pp. 53-95.

4. Chobanova R. (1996) Technology Transfer to Bulgaria, in J. Djarova (ed.), Foreign Direct Investment in Central and Eastern Europe. The case of Bulgaria, Rotterdam, pp. 141-154.

PROTECTION OF INTELLECTUAL PROPERTY IN ROMANIA; EXPLOITATION OF INDUSTRIAL PROPERTY AND TECHNOLOGICAL TRANSFER

MIRCEA VALER PUSCA; RUXANDRA RÎMNICEANU
Ministry of Research and Technology
21-25 Mendeleev Str., 70168 - Romania

STEFAN COCOS
State Office for Inventions and Marcs
5 Ion Ghica Str., 70018 - Romania

Abstract.

After a brief review of intellectual property in Romania, the paper describes the objects of intellectual property which are or can be protected in this country by surveying the copyright and related rights, including software protection, and protection of industrial property (trademarks, inventions, design, restriction of unfair competition) and integrated circuit topographies.

In this respect, the legal and institutional environment for intellectual property is presented, by briefly surveying the provisions of Romanian regulations in force as compared to specific European legislation and the TRIPs provisions.

Finally, the environment for industrial property and technology transfer exploitation is presented together with the actions taken for stirring innovation, exploitation of inventions and technology transfer.

A. Inzelt and J. Hilton (eds.), Technology Transfer: From Invention to Innovation, 309–324.

1. Brief Review of Intellectual Property Protection in Romania

In Romania, protection of intellectual property has a long tradition; thus, in 1862 the "law of Romanian mass-media" was enacted which, in a separate section, made provisions concerning the copyright of authors of literary and artistic productions.

The issue of specific regulations in the area continued with the enacting of the first Romanian law on trademarks (in 1879) and the first Romanian draft-law on patents (in 1880); the latter draft-law was not passed.

In 1906, following the development of industry and trade, the first law on Romanian patents was enforced.

Chronologically speaking, in 1923 the first specific Romanian law on copyright - "The Law on Literary and Artistic Property" - was enforced.

As is already known, in 1883 the Paris Convention on Industrial Property Protection" was enacted, followed in 1886 by the establishment of the "Paris Union for Industrial Property Protection"; since 1920 Romania has been a member of the Paris Union.

It is also known that in 1886 the "Berne Convention on Copyright Regulation" was passed and in 1927 Romania also signed the Convention.

Finally, it is worth mentioning that Romania is a member of the World Organization for Intellectual Property Protection and an active participant in most of the actions initiated by the "unions" Romania is a member of.

2. The Objects of Industrial Property Which Can Be Protected in Romania

According to the provisions of the Convention for the establishment of the World Organisation for Intellectual Property Protection and the Paris Convention for intellectual property protection, intellectual property includes:

- copyright and related rights;
- industrial property;
- scientific discoveries.

For registration, monitoring and control of enforcement of copyright and related rights regulations, the Romanian Office for Copyright (ROC) has been established, as a specialised body answerable to the Romanian Government. The Romanian Office for Copyright has been founded in 1996 following the enforcement of the Law N° 8 on copyright and related rights; in previous years the Romanian Agency for Copyright

Protection, subordinated to the Ministry of Culture, was authorised to monitor the enforcement of the specific regulations.

The responsibility for the protection of industrial property, in compliance with the specific regulations and the provisions of international conventions to which Romania is a signatory, reverts to the State Office for Inventions and Trademarks, as a specialised governmental body.

2.1. COPYRIGHT AND RELATED RIGHTS

Law N° 8/1996 makes provisions for copyright and related rights grouped in four sections the Law includes both provisions for royalties and criminal procedure as well as other related provisions.

In the first section, the Law makes general provisions concerning the subject, object, content, period, exercise, transfer of copyright ownership as well as special provisions regarding certain productions (film, audio-visual, software, artistic, architectural and photographic productions, etc.).

To comply with the above provisions, the subject of the copyright must be the physical entity or entities who produce a work, and the capacity of subject of a copyright is likely to be ceded (according to the provisions of the law).

The copyright object is the original literary, artistic or scientific production, regardless of the manner of creation, the actual mode or form of expression and irrespective of their value or purpose; the copyright object also extends to derived productions: translations, versions, revisions, compilations of literary, artistic or scientific productions, etc. which embody creativity.

As regards the copyright content, it is sufficient it to list the sole legal rights of authors, which were not provided by the previous legislation - e.g. sole distribution right, sole right of public communication, the right to import, adapt, translate or the suite right recognized to artists - to emphasize the substantial changes produced after the enforcement of the new law on copyright and related rights.

For the first time in Romanian legislation, the second section of the Law makes provisions for royalties to persons who mediate the access of the public to original productions:

- the royalties of renderers and performers,
- the royalties of producers of sound and audio-visual recordings,
- the royalties of broadcasting and television corporations.

The third section of the Law regulates, for the first time in Romania, the collective management of copyright and related rights as a form of managing those productions for which rights cannot be solely exercised, because of either the nature of the production, or the way it is exploited.

The bodies for collective copyright or related rights management are private legal entities established as a result of free association of copyright holders who carry on a non-profit activity in compliance with the provisions of Law N° 8/1996.

The above section also makes provisions for the duties of the Romanian Office for Copyright and the penalties applicable in case of infringements of the Law.

The final section of the Law - final and transient provisions - contains provisions for the scope of the Law, i.e. criteria for granting protection to authors, artists, broadcasting and television producers and corporations, who are not Romanian citizens.

According to the provisions of the Law, foreigners as holders of copyright or related rights are entitled to the protection provided by international conventions, treaties and agreements to which Romania is a signatory, and if no such document exists, they are entitled to be treated on equal terms as Romanian citizens, on condition that, in their turn, Romanian citizens are similarly treated in the respective countries.

2.2. INDUSTRIAL PROPERTY PROTECTION

Starting with the provisions of the Paris Convention for Industrial Property Protection and combined with the provisions of the Treaty on Intellectual Property of Integrated Circuits - concluded in Washington in 1989 - the objects of industrial property include:

- technical productions, including inventions and utility models, design and topographies of integrated circuits;
- signs associated with products or services, including trademarks or commercial and service brands, business name and geographical indications;
- avoidance of unfair competition.

It is necessary to specify here that, although integrated circuits topographies are not mentioned in the provisions of the Paris Convention and are, generally, included in the wider range of intellectual property due to their protection features - similar to the other objects of industrial property, and that they are protected as a result of being registered by the State Office for Inventions and Trademarks - we have included integrated circuits topographies into the area of industrial property.

Below we briefly describe the manner in which the objects of the above industrial property are protected in Romania in compliance with the provisions of the Law.

2.2.1. Protection of Technical Productions

At present technical productions are protected by the following laws in force:

* Law N° 64/1991 on patents protection;
* Law N° 120/1992 on the protection of industrial designs and patterns;
* Law N° 16/1995 on the protection of integrated circuits topographies.

All the above-mentioned laws have been enforced after 1990 and are adjusted to the specific European legislation.

As a general rule, the subject (holder) of the industrial property right is the author of the invention, industrial design, pattern or topography; he/she can cede his/her right to his/her invention before or after registration for protection, but before taking a decision about the provision of protection or after the protection certificate is granted; moreover, after the protection certificate is granted, he/she can cede total or partial right to which the protection certificate entitles him/her.

In Romania, the objects of industrial property right are as follows:

- inventions - which are protected according to the provisions of Law N° 64/1991 on patents protection;
- industrial design - which is protected in compliance with the provisions of Law N° 120/1992 on the protection of industrial designs and patterns;
- integrated circuits topographies - which are protected according to the provisions of Law N° 16/1995 on the protection of integrated circuits topographies.

Unfortunately, at present there is no Romanian Law providing protection for utility models; interested foreign persons can request a patent for protection of utility models in Romania in compliance with the provisions of the Paris Convention.

The holder of an industrial property certificate, issued by the State Office for Inventions and Trademarks, is granted sole right of utilization in Romania of the protected objects; the holder is entitled to forbid issue of the following documents to third parties without his/her consent:

- for products: production, marketing, tendering, utilization, import or storage for marketing, tendering or utilization;
- for processes - their application.

It is worth noticing here that for protection of inventions and integrated circuits topographies an attributive system is available in Romania (the protection certificate is granted following confirmation of compliance with the legal provisions concerning the novelty, inventive character and applicability of the protected solution), whereas for design protection a declarative system is applied. All possible claims are to be settled in court.

We emphasize the fact that the State Office for Inventions and Trademarks intends to amend the Law N° 129/1992 so as to make possible for the attributive system be applied to design protection as well.

Foreign physical or legal entities, resident or based beyond Romania's frontiers, are subject to the provisions of specific Romanian protection legislation on industrial property, in compliance with the international conventions to which Romania is party, or on a reciprocal basis.

The law explicitly provides the liability of foreign legal or physical entities to be represented before the State Office for Inventions and Trademarks by an authorised representative.

2.2.2. Protection of the Signs Associated to Products or Services

As regards marks, at present they are regulated by Law N° 28/1967 on trademarks, commercial or service brands, but a new "law on marks and geographical indications" will soon be enforced; therefore, below we briefly review the present law on marks, emphasising the novel provisions of the Law on Marks and Geographical Indications which will soon come into force.

The Law defines marks as "distinct signs used by enterprises to distinguish their products, productions and services from identical or similar products or services of other enterprises and stimulate quality improvement of products.

In compliance with the provisions of the Law, manufacturing enterprises are liable to register the mark of the products for domestic consumption, whereas the use of marks by trading companies and service providers is optional; the present Law on marks provides sole right of legal entities to register marks.

As the title indicates, the protection object is the mark (manu-facturer's brand, trade or service mark), defined as a distinctive sign associated to a product or service which allows the consumer to distinguish among similar products or services.

Mark registration provides its holder sole right of use for 10 years from the deposit formation, as the registration terms provide; upon request of the mark holder registration can be renewed every 10 years.

The right to the mark use can be transferred (totally or partially); the Law makes no provisions for the case in which the entire assets of a company are transferred.

To comply with the provisions of the Paris Convention and the Madrid Arrangement, to which Romania is party, the Law makes provisions regarding the national treatment applied to foreigners and for recognition of convention - derived and exhibition priorities.

The new Law on marks and geographical indications preserves the previous definition of the mark, but makes provisions for registration of marks by both legal and physical entities.

As regards the content of the protection provided to a registered mark holder, the new Law contains a more comprehensive definition; "the mark holder is entitled to:

a) use an identical sign for products or services similar to those whose marks are registered;
b) use a sign which, for reason of identity or resemblance with the mark, or for reason of identity or resemblance with the products or services for which the mark is registered, would create confusion among consumers or be associated with a former mark;
c) use a sign identical or similar to the mark of products or services different from those for which the mark is registered, at a time when the latter is re-nowned in Romania and if, as a result of sign use, somebody takes advantage, without just reason, of the distinctive character or renown of the mark, or use of the sign would be detrimental to the mark holder".

In relation to the above provisions, the trademark holder can forbid the following:

- laying trademarks on products or packages;
- supply, marketing or storage for marketing of marked products;
- import or export of products with the respective trademark laid on;
- use of the trademark for documents or advertising.

The new Law also provides that the trademark can totally or partially be transferred, but if all assets of an enterprise are transferred, except for the trademark, the right to the trademark is cancelled.

The new Law makes a very important provision concerning the right of persons interested in requesting the deprivation of the trademark holder of his/her right granted by the respective trademark if, with no just reason, the trademark has not actually been used in Romania for an uninterrupted interval of 5 years with the products or services which it was registered for.

In accordance with the specific legislation, the Law provides the possibility to register certification marks and geographical indications; however, the right to them cannot be transferred.

The Law has an attributive character, and the trademark is registered following an assessment by the State Office for Inventions and Trademarks.

The Law maintains the trademark validity (for 10 years and every 10 years a renewal is possible) and the national treatment of foreigners.

2.3. AVOIDANCE OF UNFAIR COMPETITION

In compliance with the provisions of Law N° 11/1991 on the control of unfair competition, "the acts or facts contrary to fair marketing or manufacturing usages are qualified as unfair competition"; the Law divides the acts of unfair competition into contraventions and infringements of the law.

Among the acts of unfair competition qualified as infringements of the law mention can be made of:

- usage of a company name, a corporate image or packages capable of being taken for those of another producer;
- production, import, export or storage, marketing or selling of goods bearing false endorsements about patents, the nature and characteristics of goods and the name of the producer or dealer in order to mislead other dealers or consumers.

Among the facts qualified as contraventions we mention the disclosure by an employee of classified information about the activity of the company where he/she discloses to a competitor; we have to notice here that the Law is not very explicit in defining the nature of the classified information had in view and the conditions under which the respective data are considered classified.

The actions of unfair competition are brought to court for trial and if facts of unfair competition result in damage or moral prejudice, the injured person is entitled to appeal to a competent court for a civil trial.

3. The Legal Environment for Industrial Property Protection in Romania and Its Adjustment to Specific European Legislation and TRIPs

As already shown above, the laws providing protection of inventions, industrial designs and patterns as well as integrated circuits topographies have been enforced after 1990 and they are adjusted to the specific legislation in the area.

However, several amendments are being prepared to adjust the respective laws with the TRIPs provisions.

3.1. AMENDMENTS TO LAW N° 64/1991 ON PATENTS

The amendments are aimed, on the one hand, at providing protection to patented processes and methods, and, on the other hand, at taking actions against possible infringements of the rights due from a patent or when the import of products is prejudicial to the rights granted by a patent.

For instance, we mention that:

- the protection content of processes and methods is extended, as the holder is entitled to forbid third parties "to apply and market, store for marketing, sell or import to this end at least the product obtained as a result of a patented process";
- if a prejudice is caused to the right of a holder of a patented process, the responsibility to demonstrate that the process applied in the manufacture of an identical product is different from the patented process reverts to the person assumed to have violated the respective right;
- the possibility to order suspension of duty payment for the import or export of the goods produced by violating process or method patents; the responsibility falls to the Customs officials who can order suspension of duty-payment either ex officio, or upon request of the patent holder;
- order by a law court, and upon request by interested persons, soon after duty payment, of a series of actions for ceasing violation of the rights to which the patent entitles, caused by a third person when putting up for sale imported goods which imply a prejudice to the respective rights.

4. The Institutional Environment for Intellectual Property Protection

We have already mentioned that for copyright protection, the Romanian Office for Copyright, as a specialised body answerable to the Government, has been established to comply with the provisions of the Law N° 8/1996.

The responsibility for protection of intellectual property (inventions, trademarks, design) and topographies of integrated circuits reverts to the State Office for Inventions and Trademarks as a specialised governmental body.

It is necessary to mention that claims concerning industrial property protection are settled in court. The Supreme Court of the City of Bucharest is assigned to grant mandatory and ex officio licences.

We also emphasise that amendments are to be made to the specific regulations, such as the decisions of the Supreme Court of the City of Bucharest can be contested in the Court of Appeal of the City of Bucharest.

As regards commercial secrets, they are protected by the provisions of Law N° 11/1991, which is subject to amendments in order to comply with the TRIPs provisions.

Foreign physical or legal entities, resident or based beyond Romania's frontiers, benefit from the provisions of the specific regulations for protection of industrial property under the terms of international conventions to which Romania is party.

As at present the law on the extension of the European patent (Law N° 32/1997 on the approval of the Ordinance of the Government N° 32/ 1996 for ratifying the Agreement between the Romanian Government and the European Organization of Patents on patent co-operation, signed in Bucharest on September 9, 1994) has been enforced we shall briefly refer below to the protection of foreign patents in Romania.

Therefore, foreign patents are protected at present by:

- the provisions of the Paris Convention;
- the provisions of the Patent Co-operation Treaty (PCT);
- the provisions of the agreement of co-operation with the European Patent Organization.

As in our opinion the provisions of the first two agreements are well-known by the interested persons, below we make a brief presentation of patent protection in Romania under the terms of the European Patent Extension.

In this respect:

- the request for a patent is registered by the European Patent Organisation, to comply with either the European Patent Convention, or the Patent Co-operation Treaty, for which the European Patent Organisation operates as assigned or selected office - Romania is present as a designated country;
- after transmission by the European Patent Organisation, but not before 18 months since the date of the request for a patent or invoked and recognised priority, the State Office for Inventions and Trademarks publishes the request for patent extension;
- the request for a patent is provisionally protected since the date of publication by the State Office for Inventions and Trademarks of the Romanian version of claims;
- after issue of the decision for patent award by the European Patent Organization, the extended European patent grants its holder rights similar to a national patent, on condition that:

* within three months of the issue of the decision for patent award, the patent holder transmits the Romanian version of the extended European patent to the State Office for Inventions and Trademarks;
* the patent holder pays the publication fees provided for by Romanian Law.

Non-compliance with the above terms within the interval legally decided makes the extended European patent null and void "ab initio".

The State Office for Inventions and Trademarks issues the Romanian versions of the respective patents.

5. Exploitation of Industrial Property Rights and Technology Transfer

5.1. TRANSMISSION OF RIGHTS

We have already shown that in Romania the person who can request protection for any object of industrial property is that entity who creates the invention, trademark, design, a.s.o. He/She can request protection for himself/herself or can transfer his/her rights to his/her production to a third entity.

It is worth noticing that transmission of rights is possible:

- previous to the creation of an invention, design, etc., in which case the right to request protection is held by the person who takes over the rights; it is the case of research and invention-oriented contracts, in which case the right to request protection is held by the entity who finances the action. and the creator is due additional royalties;
- after the creation of the invention, but before its registration for protection; in this case for creations that are related to the job of the inventor or that are financed by the enterprise with which the creator works, the respective enterprise has priority in transmission of the right to the respective creation. If the enterprise manifests no interest within three months, the right to the protection certificate is owned by the creator;
- in all other cases, the right to the protection certificate is held by the creator of the respective invention; he/she is entitled to cede his/her right before or after registration by the State Office for Inventions and Trademarks, but before or after a decision is made.

The holder of any form of protection can transmit total or partial right to his/her creation, for which a protection certificate exists, by exclusive or non-exclusive licensing.

We also notice that transmission of rights is possible by legal succession as well.
Restricting ourselves to inventions, we find it necessary to say a few words about the treatment of compulsory or ex officio licences.

Ex officio licences are granted upon request of ministries (the Ministry of Health, specialised ministry - after a previous delay - for no or unsatisfactory application, the Ministry of National Defence) under well-defined provisions of the law, whereas compulsory licence is granted upon request of interested persons, upon termination of a minimum interval of four years since registration or three years since patent issue, if no or unsatisfactory application is reported.

Both types of licence are non-exclusive, are limited in terms of extension and duration, and the body authorised to grant them upon the request of the due entities is the Supreme Court of the City of Bucharest.

5.2. PROTECTION AGAINST FORGERY

All laws concerning protection of industrial property (inventions, designs, topographies, marks) provide sanctions for forgery. The sanction is penal and the injured persons are indemnified according to common laws; at the same time, public inventions are subject to provisional protection until a decision is taken.

As already mentioned, amendments to legislation are drafted in order to comply with the legal provisions with the TRIPs provisions aimed at avoiding forgery.

To avoid abuses, the amendments to the Law provide that "guarantees should be ordered by a court of law consisting of payment by the claimant of a liability decided by the law court".

5.3. NATIONAL AUTHORIZED BODIES FOR TECHNOLOGICAL TRANSFER

According to Governmental Decision N° 413/1996, the National Agency for Technological Transfer and Innovation, subordinated to the Ministry of Research and Technology, has been established as a pilot centre for technological transfer and innovation, which facilitates convention of all national producers and consumers of research results.

Among the multiple duties of the National Agency for Technological Transfer and Innovation we mention:

- creation of appropriate mechanisms for transfer of knowledge, information, technologies, know-how, services and products;
- intensified dissemination of information and technological know-ledge, elimination of existing deficiencies of the system;
- focus of the transfer to small- and medium-sized enterprises;
- attraction, to this end, of financial sources from private funds and international bodies;

- establishment of technology transfer and innovation centres, with multiple duties, focused on innovation companies (such centres are operational, at present, in Jassy, Târgu-Mure°, Bucharest and Craiova).

As an innovative structure, the National Agency for Technological Transfer and Innovation focuses its efforts to:

* provide a flexible financing system and attraction of extra-budgetary financial resources for intensification of technology transfer and innovative applied research;
* support and develop a market for new services and technologies;
* promote co-operation among the operational networks of small- and medium-sized enterprises.

Taking into account the fact that in the transition period most enterprises are undercapitalised and with a dramatic need of funds for investments and development, the state covers up to 50% of the total costs of technology transfer under unredeemable co-financing terms.

To get the above percentage, the enterprise should submit a business plan, accompanied by appropriate documents, and the selection is based on:

- the forwardness of the selected technology;
- its anticipated effects on the development of small- and medium-sized enterprises;
- the technical qualification of the requester;
- the existence of financial resources from its own resources or attracted funds to cover the difference;
- the priorities (e.g. economic, social, etc.) following from the reform programme.

At present, the activity of the National Agency for Technological Transfer and Innovation is focused on financing the transfer of patented research results and extension to the entire area of the research results that make the object of innovation creativity.

5.4. INTENSIFICATION OF INNOVATION, INVENTION APPLICATION AND TECHNOLOGY TRANSFER

As a body of central public administration the Ministry of Research and Technology is entitled and empowered to define, correlate and provide: the strategic planning and objectives in the field of research - development - innovation; the policies for fulfilling the objectives of strategic planning; legal-methodological functional, operational, and financial environment where such policies can be fulfilled; assignments for policies' application; monitoring, assessment and control of policies' application.

Consequently, on January 31st, 1997, on the basis of a proposal by the Ministry of Research and Technology, the Romanian Government issued the "Ordinance (N° 8/1997) concerning Stimulation of R & D and Innovation".

The Ordinance considers technological development, innovation (as an outcome and process), technology transfer, innovation assimilation and application as national priorities.

As another means to stimulate application of inventions and technological transfer we mention the provisions of Article 68 of Law N° 64/1991 concerning patents, according to which "the profit resulting from the application of an invention is exempted from taxes in the first five years of application, calculated from the patent date and is, as a result, entirely available to the enterprise that applies the invention".

Exemption from the tax on profit, following the application of an invention, is meant to stimulate its application. Calculation of the interval of 5 years for tax exemption from the patent date is intended to stimulate technological transfer and assimilation of inventions and restriction of the exemption interval to 5 years is aimed as determining enterprises to create and maintain an innovative environment.

As in a rather high number of situations, innovation as a process is unfortunately prone to be confused with innovation as a product its area being in this way restricted, I would like to direct my attention in the following to the innovation as an activity the Ministry wishes to stimulate at least in the field it is empowered to, as an authority of public administration.

The impact of science on the social life especially during the last generation is a unanimously accepted fact. The idea of the future being strongly subject to the "assault" of renewals all generated in the scientific process, is also generally accepted. On the other side, this end of millenium and beginning of a new century is shadowed by Malraux's warning "the XXIst century will be either religious or won't exist".

During the last years we witnessed the outburst of electronics that will probably become "classicised" during the next 10-12 years, at least from the viewpoint of the interest in microminiaturisation. This was propagated by continuity in the related fields such as computing technique (that enhanced and subsequently included the "outburst" of electronics) and telecommunications (that made use of the progress made in several fields). Future "detonations" can be foreseen such as superconductivity, biotechnologies, new materials, genetic engineering, etc., etc.

With a questionable success, sociologists tried to classify these trends from a theoretical point of view. After a long period the whole world accepted to term these "successive outbursts" as **postindustrial society**, a name that imposed a net differentiation from the previous epoch, without underlining the essence of this revolution. The fact that "nicknames" of a great success at one time such as "the microelectronics era", "computerized society", "information society", "technocratic era" did not last in time seems significant to me. Obviously they covered only certain transient, or stage-connected characters and not the essence of changes.

Let us turn back to the roots of the movement: the scientific discovery. This is in essence a creation, it creates information, generally new information. Associated with other knowledge (either old or rather recent) it can further generate a new structure and/or process that can be a structure and/or process either social, biological and so on. All these will, in their turn, be new, at least in part. Their implementation will call for new information resulting in new points of view, new strategies, new structures (organisational, teaching, new value system, etc.).

This situation, although it might seem elementary, defines a social behaviour and consequently the general characterisation of the society we make our way towards, although at its very beginning, could be reduced to a stable (defining) name i.e.: **innovation society**.

The social promotion of its innovative character is the social subsystem named as innovation mode. The expression of innovative effect of this structural element, the external appearance of movement regarding innovation mode exists in the scientific and technical revolution itself !

The innovation mode includes, first of all, the professional segments, that are properly institutionalized, i.e. the research institutes, the nonrecurring activity of design institutes, prototype workshops, experimental plants, industry of scientific equipment, etc. The movement of inventors as well as the more spontaneous forms of the change (nonstructured or nonpermanent, such as study groups, quality teams, creativity clubs of the young, etc.) operate in the same manner and so are the persons acting individually for the progress of knowledge or implementation of the new scientific knowledge.

All these elements of the movement for renewal are more or less active in a nation's life corresponding to an image that shows a more or less creative society. This society-innovating energy is preferentially focused on either social changes (revolutionary or at least reformative periods), economy renewals (i.e. industrial revolution) or on the techno-scientific achievements (post-industrial society). Although it is possible for one or another side to prevail during a certain period or in a certain geographical area the process seems to be unique, indivisible, with clear effects of each side on all the others. All these processes develop in the framework of social life and cannot generally be separated from it.

The innovation mode becomes in this way a "pilot" of the mode of production as well as of the socio-cultural life mode.

The above mentioned facts demonstrate that innovation acquires a permanent character being also institutionalised on various levels. The revolution "controlled" through scientific knowledge changes into a **perpetual reform**. Yet, only a democratic, pluralistic society really allows the development of a perpetual techno-scientific revolution also hampering the set up of a new category of leaders, albeit "scientific" leaders. It is the education, more than propaganda that allows the society to understand and accept the necessity of the profound, sometimes painful, changes taking the necessary steps forward. And only a democratic state that protects people's rights wouldn't allow any reform/revolution - even scientifically motivated - developed under compulsion, to generate unreasonable sufferings.

In essence this is one of the reasons the Ministry of Research and Technology wishes to promote reform through stimulating research - development - innovation, providing a competitive, completely transparent environment, in order to use the products generated in innovative mode for the socio-economic benefit, such as to increase the efficiency of the Romanian economy and provide for the socio-cultural needs of the citizen (particularly) and (generally) of the Romanian nation.

6. Conclusions

To comply with the above we consider that the Romanian legal and institutional environment is capable of securing a sound protection to the main objects of industrial property.

Romanian legislation, adjusted to the specific legislation of the developed countries and complying, to a large extent, with the TRIPs provisions ensures an appropriate balance between the interests of inventors and patent owners, on the one hand, and patent owners and society, on the other hand.

We also consider that Romanian legal provisions are able to stimulate innovation and technologiy transfer, and if the necessary funds will be available we are convinced that they will demonstrate their efficiency.

TECHNOLOGY TRANSFER IN RUSSIAN INDUSTRY

IRINA KUZNETSOVA
Center for Science Research and Statistics,
Moscow Russia

INTRODUCTION

This report considers technology transfer as a part of the innovation process in industry. Innovation statistics is a new branch of the Russian national statistics. It has been run by the Center for Science Research and Statistics (CSRS) since 1994 in conjunction with the revision of the model of science and technology (S&T) policy in the Russian Federation in transition and respective transformation of the national R&D statistics on the base of both domestic and international experience (1, 2, 3).

The paper discusses methodological problems of introduction of international standards into Russian Innovation Statistics. The results of the first full-scale innovation survey are described. An analysis is presented of innovation activity in Russian industry including innovation costs and some results of innovation. The paper concludes with a detailed study of technology transfer performed by industrial enterprises.

1. Introduction of International Standards in the Russian Innovation Statistics

The CSRS developed and implemented an innovation survey in industry based on the recommendations of the Oslo Manual and compatible with the EU Community Innovation Survey (CIS). The definition of innovation was used in conjunction with the above - mentioned rules. Technological innovations comprise new products and processes and, significant technological changes of products and processes. Innovation occurs when a new or changed product is introduced to the market, or when a new or changed process is used in commercial production. The innovation process is the combination of activities - such as design, research, market investigation, obtaining of equipment and so on - which are necessary to develop an innovative product or production process.

A. Inzelt and J. Hilton (eds.), Technology Transfer: From Invention to Innovation, 325–337.

However, the application of the CIS experience in Russia encountered problems with a number of national specific features which were barriers to adoption. In this connection, economic and organizational peculiarities can be identified, as well as those of the data collection system traditional for Russia, each of them leaving its mark on the program of the innovation survey [4].

In the first place, this is the procedure of arranging the survey in two_stages. This was conditioned by the novelty of this direction for national statistics, unfamiliarity with the concepts and indicators for enterprises accustomed to operating under conditions of centralized planning.

As the first stage, the so called introductory survey was implemented in 1995 to solve simultaneously both analytical and organizational problems for the first time in the field of Russian statistics and within a relatively short time it was necessary to obtain generalized information on the state of innovation activity in industry, comparable with the international data;

- to enable enterprises to adjust to the new methodology and indicators in order to facilitate the transition to the more complicated second stage;
- to identify, by the criterion of innovation activity, a sampled population of enterprises which will provide the base for further statistical survey with an enlarged program.

The introductory survey highlighted a set of methodological problems which should be taken into consideration when designing a full-scale innovation survey in Russia.

Innovation statistics, beginning from methodological issues to practical implementation, is in principle a new and complicated direction for Russian statistics. As shown by the introductory survey, the greatest difficulties for respondents were caused by the understanding of the terms, especially the criteria of subdividing innovations into product and process one.

There is another problem which complicates understanding of the content of innovations. The contemporary initial period of market reforms is characterized by such economic phenomena as intensive diversification of production. In dealing with the demand for a solution, enterprises exceed the framework of their main economic activity in the endeavor to gain sufficient incomes. At the same time, goods manufactured in this way are new both for the enterprise and the industry sector.

Additional methodological problems are caused by the contemporary economic conditions of Russian enterprises. At present, they are characterized by a decrease in innovation activity not so much in quantitative terms as in qualitative terms. The "primitivism" in innovation is visible. Enterprises prefer manufacturing of traditional, already known products in current demand to high-technology products. At the same time, for specific enterprises these products can be new.

Swift economic transformations in Russia have required more detailed_information on innovation expenditure. In this connection we use a detailed classification of expenditure by type of innovation activity, aimed at determining the inequalities and bottlenecks in the promotion of innovation.

As shown above, the first stage of the innovation survey was primarily aimed at studying the level of the innovation activity within enterprises. The second stage of the survey covered a detailed study of innovation strategy of enterprises, trends in innovation and determining factors, mechanisms of technology transfer, and impact of innovation [5].

A questionnaire for the survey of innovative enterprises consists of the following major sections:

— expenditure on technological innovations by type of activity and type of costs;
— expenditure on product and process-innovations (e.g. on R&D) by source of funding;
— sales of innovative products (e.g., exports);
— distribution of product and process innovations by objective of innovation activity;
— acquisition and transfer of new technologies.

The other part of indicators obtained by questionnaire polling is moved outside the framework of comprehensive survey of innovative enterprises. The study of such issues as sources of information on innovations, factors hindering innovations, the evaluation of methods of maintaining and increasing competitiveness of product and process-innovations will be accomplished on the base of a sample survey of 1.000 enterprises.

In the framework of the implementation of the second and following stages of the innovation survey the above - mentioned difficulties of the first stage were taken into account.

2. Innovation activity of industrial enterprises in contemporary Russia

Contrary to the regularities of market economy that should stimulate innovative activity of non-competitive productions, in Russia one can observe quite a different situation. The period of economic transformations in the country, characterized by an abrupt setback in production and effective demand was also accompanied by a steady decline in innovative activity. In the late 1980s the industrial enterprises' indicators in the former USSR hovered between 60 to 70 percent, while with the beginning of the market reforms their value lowered by 3 to 4 times and is continuing to fall. In 1995 the share of industrial enterprises that developed and introduced new or improved products and technological processes did not exceed 6 percent. Forecast estimations for 1996-1998 allow the definite identification of only about 3.7 percent of such enterprises.

The most important characteristics in assessing the scope of innovative activity at all levels of management is its quantitative measurement. The main indicator here is expressed by the cost index figures, namely, the volume of innovation costs. At present (according to the data of the overall statistical survey of industrial enterprises for 1995) it makes up 7.3 trillion rubles, including long-term investments and current expenditure almost in equal shares. It is obvious, that this value is negligible. For reference, the total volume of capital investments into the economy of Russia for the same period of time amounted to 250.2 trillion rubles. The share of capital investment into innovative activity did not exceed 2.6 percent of the total volume of investment for production purposes.

The minimal means spent for innovation in industry are realized predominantly at the expense of internal reserves of enterprises. Taking into consideration high interest rates, they limited the use of credits and loans, only 8 percent of internal funds being spent on innovations and largely owing to the fact, that interest rates were partly received on favorable terms.

The state gave them practically no support in their innovative activity. This is testified by data on the share of budgets at all levels, that is a little higher than 4 percent of the total innovation costs.

In such circumstance the problem of attraction of foreign investments becomes extremely pressing. However, the analysis has shown, that in 1995 their share was only 5 percent of the total investments into innovations. This gives us grounds to assume, that no appreciable changes are expected to occur in the near-time outlook, since today there are no adequate laws that regulate investors' activities and guarantee the infrastructure of investment projects.

Inter-sectoral distribution of innovation resources is characterized by a high degree of concentration. More than two-thirds of all innovation costs are spent on certain types of economic activity connected with the chemical and petrochemical industry and production of machines, equipment, instruments and transport equipment. The other four branches linked with the manufacture of food products, the metal-working industry, metallurgy and the mining industry make up 25.5 percent. As a matter of fact, all the other types of activity come only to 5.4 percent.

The above-mentioned absolute indicators, shown in isolation from their relationships with the sales volumes, give no possibility to adequately realize interbranch and other comparisons. In order to get more correct assessments, the innovation cost intensity indicators were used, the value of which in industry as a whole, is equivalent to 2.9

percent. With the help of that indicator it is possible to reveal the peculiarities of innovation processes in Russia not being taken into consideration earlier, which are indicative of some progressive tendency to use innovations as an instrument in the competitive struggle. The highest indicators of intensity are characteristic of the productions which are generally called depressive, namely, of metal-working industry (9.8 percent), food industry (5.5 percent), machine-building (4.8 percent). And on the contrary, the branches that managed somehow to occupy export market have the lowest values of the indicator under consideration. This means metallurgy (0.6 percent) and the mining industry (1.4 percent).

Innovations are generated as a result of realizing a wide range of different types of activity including R&D, technology exchange, technological preproduction, training of personnel, market research. The priorities of innovative activity change at different stage of economic development of the country. Today they are being given to the processes of introduction in order to provide most rapid reimbursement of outlays. The largest share of innovation costs is oriented towards the acquisition of machines, equipment, plants, and other fixed assets necessary for implementing product- and process-innovations. Considering the additional costs for technological preproduction, the sum total of innovation costs, amounting to 60 percent refer directly to the processes of introduction of innovation. Some types of industrial production, such as recycling (99.1 percent), manufacture of food products (83.3 percent) and non-metallic mineral products (81.3 percent) were almost completely oriented towards similar types of activity.

Research and development, as a most widespread type of innovative activity in industrial enterprises amount to relatively small share (26.9 percent) in total innovation costs. This is determined not only by the principal orientation towards the practical realization of innovations, but also by rather low capital intensity of scientific activity, explained by low remuneration of labour of research personnel. On the whole, research is mainly of pure applied character and is carried out with the objective of satisfying the routine technical or implementation demands of enterprises.

A new characteristic of innovation process complying with the intenational standards accepted in this field, but unusual in Russian practice, revealed dramatic disproportions between different types of innovative activity. Thus, the enterprises, accustomed to the directive-planned economy of the past years with its chiefly centralized methods of distribution of output and sales volumes, underestimate the effectiveness of marketing research. Their share in the total volume of innovation costs formed only 0.4 percent. Even less (0.6 percent) is the share of the costs directed to education and training of the personnel for working with new technologies. It seems, that this is first and foremost due to inadequate novelty of the innovations they carry out. It is confirmed by a small share of the costs (2.4 percent) for purchasing new technologies in the form of patent and non-patent licenses. The difference between the maximum and minimum values of the share of the costs by separate types of activity reaches dozens at certain times.

3. Technology Transfer of industrial enterprises

The processes of technology exchange, necessary for progressive development of innovations have not yet occupied the proper place in present-day reality in Russia (Table 1).

TABLE 1. Indicators of Technological Exchange in the Industry of Russia by Types of Economic Activity: 1995

	Share of enterprises in the total number of innovation-active ones	
	purchasing new technologies	transferring new technologies
Total	41.8	3.2
Mining	69.0	2.4
Food, beverages and tobacco	57.7	2.1
Textiles, clothing, fur and leather	35.6	3.0
Wood, paper, printing, publishing	46.0	2.0
Coke, petroleum, nuclear fuel, chemicals and products, rubber and plastics	49.7	2.8
Non-metallic mineral products ("Stone, clay and glass")	43.1	3.4
Basic metals	43.9	3.5
Fabricated metal products (except machinery and equipment)	33.3	1.2
Machinery equipment, instruments and transport equipment	32.8	4.6
Furniture, other manufacturing n.e.c.	25.9	1.8
Recycling	0.0	0.0

In 1995 only 570 industrial enterprises (41.8 percent of innovation-active ones) took part in the acquisition of new technologies, while 44 enterprises (3.2 percent) - in their transfer. It happened due to a long isolation of the domestic market from foreign achievements in science and technology. But on the other hand, objects of intellectual property remained unclaimed by industry. High cost makes them unpurchasable for many enterprises experiencing financial difficulties.

Such a pronounced contrast observed in the flows of technology exchanges can be chiefly explained by inadequate development of industrial sector research potential. At the same time, it should be noted, that in most branches of industry the intensity level of technology exchange is adequate for the activity of enterprises in carrying out R&D.

Taking into consideration only quantitative indicators when qualifying the number of domestic enterprises that imported new technologies, we can observe the situation bearing in mind the respective foreign characteristics. However, it should be taken into account, that, first, we are dealing with quite a small number of innovation-active enterprises. It can be illustrated by the fact that the share of enterprises purchasing new technologies in the total number of industrial enterprises constituted only 2.4 percent. Second, in contrast to our intra-factory research, the level of intra-firm research in the West is very high. This determines the greater demand of domestic enterprises for borrowing science and technology achievements, including those from the sectoral research institutions. One cannot disregard the dramatic deterioration of the economic situation in the country, that dictates the necessity for taking immediate actions in raising the competitiveness of domestic products.

In many branches of industry the activity of enterprises in purchasing new technologies exceeds the average level, the mining industry branches being most active. Among the branches of the manufacturing industry the food and chemical industries appear to be particularly active. Somewhat higher than average are the indicators of the branches connected with wood manufacture, pulp and paper production, metallurgy and non-metallic mineral production.

High rates of intensity in purchasing new technologies implies high levels of innovative activity of the corresponding branches of industry in perspective. A similar interrelationship can be observed most vividly in the chemical industry and non-metallic mineral production. However, directly proportional relationship cannot be extended to all cases. For instance, only one-third of the enterprises in the branches of the machine building industry used to purchase new technologies, while two-thirds intend to implement innovations in the short term.

The bases of borrowed science and technology achievements was formed by domestic developments that were used by 70 percent of enterprises purchasing new technologies.

The impact of foreign experience on technology innovation in Russia is not great: imported technologies were registered only in 30 percent of enterprises that acquired new technologies. Preference was given to the farthest distant countries: over 20 percent of enterprises had contacts with them and about 9 percent with the countries of the CIS (Table 2).

It is evident from the above table that the greatest activity in importing new technologies from highly industrialized countries was shown by export-oriented branches that have adequate contacts and possess sufficient pecuniary resources for the purpose. This is most vividly demonstrated by metallurgy, where more than half the enterprises that purchased technologies, used to acquire them in the farthest distant countries. On the other hand, the branches manufacturing consumer goods, more sensitive about market demands, that are forced to expand the assortment of products and improve their quality are characterized by high indicators. For example, this refers to textiles and clothing, where 42 percent of enterprises made use of foreign know-how in their innovative activity. Machine-building enterprises displayed comparatively low rate of intensity in their relations with foreign countries, some of their preference being given to the contacts with the countries of the CIS.

TABLE 2. Share of Enterprises-importers of New Technologies in the Total Number of Enterprises that Purchased Them, by Type of Economic Activity: 1995

	Total Number of Enterprises that Purchased New Technologies	of which share of enterprises-importers of new technologies	Including	
			in the countries of the CIS	in the non-ex USSR countries
Total	100	30.4	8.6	21.8
Mining	100	37.9	13.8	24.1
Food, beverages and tobacco	100	29.0	3.6	25.4
Textiles, clothing, fur and leather	100	41.7	2.8	38.9
Wood, paper, printing, publishing	100	34.8	13.0	21.8
Coke, petroleum, nuclear fuel, chemicals and products, rubber and plastics	100	28.4	5.7	22.7
Non-metallic mineral products ("Stone, clay and glass")	100	32.0	4.0	28.0
Basic metals	100	68.0	16.0	52.0
Fabricated metal products (except machinery and equipment)	100	28.6	14.3	14.3
Machinery equipment, instruments and transport equipment	100	23.8	13.4	10.4
Furniture, other manufacturing n.e.c.	100	14.3	0.0	14.3
Recycling	100	0.0	0.0	0.0

The forms of acquiring new technologies reflect the common characteristic features of innovative activity at the present stage of economic development, namely, orientation towards maximizing the rate of practical application of innovation. This is expressed by the intensive purchase of equipment for introducing product- and process-innovations. Such a form of acquiring the achievements of science and technology was used by 53.3 percent of the total number of the enterprises which acquired new technologies.

Another widespread form of acquiring new technologies, like that mentioned above, characteristic of most types of economic activity, is the use of the results of R&D (47.7 percent of enterprises that purchased new technologies). Implementation of most progressive, though expensive, objects of intellectual property in the form of purchasing licenses on inventions, industrial prototypes and valuable models, as well as know-how and contracts on technologies' transfer was realized by 13.7 and 12.1 percent of enterprises, respectively. In a number of branches of industry, such as chemical production and metallurgy, every fifth enterprise used to purchase patent licenses. The share of enterprises practicing other forms of acquiring new technologies varied between 0.5 (purchase of enterprises) and 5 percent (employment of specialists) of the total number of enterprises engaged in purchasing technologies.

On the whole, in 1995 24 patent licences were acquired in farthest distant countries, 19 of them being purchased in the far foreign countries. Most of them were acquired by the enterprises of machine-building (10 licences) and light industry (5 licences).

In 1995 the transfer of new technologies (technology achievements) was carried out only by 44 industrial enterprises. Against the general low rate of intensity of transfer of new technologies most high indicators were registered in machine-building (23 enterprises or about 5 percent of enterprises active in innovations), that is explained by a relatively higher level of intra-factory research in the given branch of industry. The number of such enterprises by the other types of economic activity varies between 1 and 5 that makes from 1 to 4 percent of the corresponding number of innovation-active enterprises.

New technologies developed at industrial enterprises are purchased mainly by domestic consumers. Only 12 enterprises — less than one-third of the total number of enterprises that transferred new technologies — exported them (Table 3).

TABLE 3. Share of Enterprises — Exporters of New Technologies in the Total Number of Enterprises that Transferred Them, by Type of Economic Activity: 1995

	Total Number of	of which share	Including	
	Enterprises that Transferred New Technologies	of enterprises-exporters of new technologies	to the countries of the CIS	in the non-exusse countries
Total	100	27.3	13.6	13.7
Mining	100	100.0	0.0	100.0
Food, beverages and tobacco	100	40.0	40.0	0.0
Textiles, clothing, fur and leather	100	0.0	0.0	0.0
Wood, paper, printing, publishing	100	0.0	0.0	0.0
Coke, petroleum, nuclear fuel, chemicals and products, rubber and plastics	100	40.0	20.0	20.0
Non-metallic mineral products ("Stone, clay and glass")	100	50.0	50.0	0.0
Basic metals	100	50.0	0.0	0.0
Fabricated metal products (except machinery and equipment)	100	0.0	0.0	0.0
Machinery equipment, instruments and transport equipment	100	21.7	8.7	13.0
Furniture, other manufacturing n.e.c.	100	0.0	0.0	0.0
Recycling	100	0.0	0.0	0.0

It should be noted that exports of technologies have not been registered in the following five branches of industry: textiles, clothing, wood manufacture, pulp and paper products; metal-working industry, furniture and recycling. What is more, only 6 industrial enterprises, three of which representing machine-building, exported technologies to the farthest distant countries.

As in the cases of purchasing new technologies, the sales of equipment and R&D results were the most popular forms of transfer (36 and 32 percent of enterprises that transferred technologies, respectively). The sale of patent licences was carried out by 6 enterprises. They have sold 12 licences, half of which have gone to distant countries and include the mining and metallurgical industries.

4. Some Results of innovation activity

Activity of enterprises in the sphere of development and introduction of innovations inadequate as to the above-mentioned parameters of innovation potential is aggravated by the low effect from innovations. Consequently, even those small innovation costs spent today by enterprises are reduced. Unfortunately, we must admit, that this was caused not only by inadequate innovative activity, but, largely, by general unfavorable economic situation in the country (investment crisis, inflation, deficiency of effective demand) which gives no possibility to fully realize new products at the domestic market. The above-mentioned factors do not at all encourage to carry out any technological innovations.

Innovation outputs make up a small value, not exceeding 16 percent of the total volume of shipped products of enterprises which implemented innovations. On the whole, these indicator-values by industry vary within 4 to 5 percent, the major share of sales being presented by non-innovative products. Such a percentage ratio is characteristic of almost all types of economic activity. The greatest relative innovation output (for branches engaged in manufacturing non-metallic mineral products) does not exceed one-third of the total volume of shipped products.

The low efficiency of innovative activity shown by innovation output indicators leads to a low novelty level of industrial products. On the whole, the share of principally new products, together with those considerably improved, amounted to 14 percent of the total volume of sales realized by innovation enterprises. It should be noted, that the highest indicators of principally new products are characteristic of high-technology productions in the machine-building industry where more progressive forms of technological exchange were used.

The phenomena of low efficiency of innovation activity in Russia, mentioned above, are in full measure illustrated also by the situation taking place in the sphere of innovation product exports. Most vividly this is expressed by the indicators of exports intensity. The share of innovation product exports in the total volume of shipped products reached only 4.1 percent, that is not more (in 1.4 times) than the corresponding of innovation costs for the same period.

The analysis of structure of the exports shows that innovations had no appreciable impact on the growth of exports. The principal share of exports (over 80 percent) consisted of the products not subjected to technological modifications. Almost half the exported innovation products refers to the two branches of industry — mining and metallurgy. Both of them are characterized by small science intensity, innovations not being the main factor of exports activity.

Not all the innovation-active enterprises displayed proper activity in exports, only 60 percent of them did. The worst of this is that such a situation is typical for innovation strategy of enterprises for which the purposeful struggle for the markets has not yet become a matter of priority. According to the data given by the enterprises themselves, only 33 percent of them set the task to create new marketing outlets, while 23 percent preferred to preserve the traditional ones. One of the principal guidelines in innovation activity of industrial enterprises for the past three years was the expansion of product range, this being declared by more than two-thirds of innovation enterprises. Such priorities are characteristic of almost all types of economic activity in the manufacturing industry and are determined by intensive diversification of production and products under conditions of sweeping and not always well-elaborated market transformations in the Russian economy.

Reference

1. Centre for Science Research and Statistics (1996a) Innovations in Industry Sectors, Moscow (in Russian).
2. Centre for Science Research and Statistics (1996b) Russian Science and Technology at a Glance: 1995, Moscow.
3. Gokhberg, L. (1995) R&D and Innovation Statistics in the Russian Federation: New National Surveys. In: Science and Technology Statistics in the Partners in Transition Countries and the Russian Federation, OECD/GD(96)56, Paris.
4. Gokhberg, L., and Kuznetsova, I. (1995) Innovation Statistics in the Russian Federation. National Experience and Consequences for the Revision of the Oslo Manual, Group of National Experts on Science and Technology Indicators, DSTI/EAS/STP/NESTI(95)6, OECD, Paris.
5. Gokhberg, L., and Kuznetsova, I. (1996) Innovation Statistics: First Results and Nearest Prospects, Issues of Statistics, 1996:3, Moscow (in Russian).
6. OECD (1992) OECD Proposed Guidelines for Collecting and Interpreting Technological Innovation Data – Oslo Manual, Paris.

CHAPTER V
IDEAS FOR FURTHER CO-OPERATION

INTRODUCTION

Scientific and innovation systems are quite different in various regions of the world. The national systems of innovation in the various Western countries are depending on many specific - historic, cultural, institutional - elements and no one of those countries can claim to have been able to design some kind of optimal set of policy measures in these complex domains.

Transition countries tend to adapt innovation systems used in western economies; however, there is no best formula for innovation. Some science and technology (S&T) systems can perform better than others, but the real challenge is to induce a change in people's minds and behaviour. Unfortunately this is a long and costly process.

A common character of diversified national innovation system is that many actors and factors interact in this complex domains. In socialist countries many players were missing or their activities were restricted. During the socialism the thoughts on S&T policy making were oversimplified. It was a strong assumption that if the authorities said something would happen, it would. Since the beginning of the transition period these economies have seemed to focus on traditional pillars of market economies : monetary system, ownership issues, etc. Much less attention was paid to S&T policy issues, changes or building up innovation systems. There was also much attention to allocating responsibility for directing or managing research activities and allocating fundamental research funding between rival state organisations but little attention to getting developments into the marketplace. While people argued over who would manage change, change was not happening.

There was a general feeling from speakers from the Western developed economies that training and awareness, access to information, networks and more advice on practical marketing skills would be more effective and lower cost than ambitious discussions about government-led structural changes. The intervention of public authorities in planning change seemed to hark back to theories of planned economies which had been left behind by the political changes of the past ten years. Even then, for example, subsidised access to information was of little use without training on how to make use of it.

The business community, which was supposed to be leading the economic development of the transition economies, had not yet done so.

When, and to what extent, the Central and Eastern European countries have made their own choices concerning their objectives and priorities, and have decided the ways and means of introducing or building up their national system of innovation, then common action plans can be envisaged for contributing to the development of particular elements

A. Inzelt and J. Hilton (eds.), Technology Transfer: From Invention to Innovation, 341–346.

or subsystems within their national systems. But this may be premature for some time to come.

Taking account of this, and in order to be both ambitious and modest, comprehensive and operational, it would seem interesting to follow the regional innovation approach, helping particular regions to design and implement a specific local innovation strategy.

However, regardless of the specific features of national development, international economic and scientific integration may give us the opportunity for a common methodological approach.

The discussions so far tend to suggest that some type of co-operation of a generator, pedagogic type, may be needed for Central and Eastern Europe as a first pre-condition or be useful for explaining to the main players (government, scientists, business enterprises) in the different countries, what the stakes are and what technology creation and transfer and innovation are all about. It would more particularly seem to be necessary to explain and convince them about the organisational requirements of such processes and to indicate some possible ways and means for meeting them.

RECOMMENDATIONS

Participants recommended a couple of joint projects (studies, education and training, development of transfer mechanisms) and workshops for their own and for their countries' future co-operation.

Studies

The aim of all studies is to learn from international comparisons. Carefully formulated studies that provide information about what has worked in other nations can stimulate creative thinking and perhaps guide government policy makers in transitioning countries as they build S&T infrastructures. It was felt that speakers from the Western advanced economies underestimate the technical sophistication of the transition states and place them improperly in the same category as newly industrialising economies of the third world. Assessment studies on transition economies therefore may serve not only for self-use but for information for others, helping partnership, network building, investment decisions

- S&T policy supporting studies may concentrate on the following issues with regard to international experiences :
  - How can transition economies change from the current model of innovation to a chain-link model?
  - How can post-modern S&T policy formulation tools for priority setting be applied? (Technology audits, technology assessments, evaluation, foresight.)
  - How can regionalisation of S&T and innovation policy be applied in the age of globalisation?

- How can late-comers join the existing networks and strategic alliances?
- How can countries benefit from the new phenomena in technology transfer relating to high-tech industries?
- Which financial measurements are appropriate to encourage S&T innovation?
- How can transition economies preserve and develop their accumulated personal capabilities in R&D and innovation?

- Some policy issues are recommended to study and evaluate in detail:

- the scientific potential of regions selected, especially in fundamental and applied sciences concerned with the military - industrial complex, which may become the basis of effective technology transfer.
- regional institutions such as scientific schools which may be considered as incubators of innovation activities, identifying similar innovation development problems of regions selected (comparative analysis).
- harmonisation of intellectual property rights and its impact on the economies and their intellectual properties in the transition economies.

This project may exploit the accumulated knowledge of WIPO and social sciences faculties such as Science Policy Research Unit in the University of Sussex. UK; PREST at the University of Manchester, UK; the Max Planck Institute in Munich, Germany; and parallel programmes elsewhere.

Development of knowledge and technology transfer mechanisms

By the conclusion of the workshop, there were many useful mechanisms which would be helpful for transition economies. The workshop concentrated on only two of these : bridging institutions and the development of an adequate legal framework (IPR). Both are important for diffusion networks for technical knowledge, technologies and innovation.

Nowadays the transition states' enterprises, domestic owned private firms do, not have enough expertise to manage knowledge and technology transfer and marketing.

Formation of new competencies (facilitators) in technology transfer is important. This organisation can comprise the university/industry relationship and the development of local (or sectoral) competence poles or innovation systems.

<u>Education and Training</u>
Knowledge based economy builds on a learning society. This type of life-long learning at all levels also includes policy makers. Besides global education challenging transition economies, they have to build new capabilities which include new professions, such as knowledge and technology transfer managers, modern thinking

among S&T policy makers and national and regional developers, higher education managers, S&T policy analysers and researchers.

To fulfil these objectives, it is important

- to strengthen the S&T and innovation policy-making capabilities in education, to introduce curricula into graduate courses and launch post-graduate courses,
- to coach and train young staff members in transition economies.
- to strengthen the advisor capabilities for enterprises in dealing with the background/foreground technology issues in licensing and collaboration agreements, due to diligence and technology audits, and selecting appropriate licensing-in opportunities,
- to exchange experience of foreign and international training programmes

WORKSHOPS

The participants considered proposals for further workshops in the NATO scheme in the light of lessons learned at the Budapest workshop. These NATO programmes need to address specific practical problems, which could be solved without the need for major international reforms. Scientists and business people should be encouraged to form partnerships and otherwise co-operate in studies of national innovation practices, on industry use of new technologies, on national policies that stimulate scientific discovery and technology development.

- NATO should encourage and sponsor workshops directed at increasing the dialogue between Central and Eastern European countries and advanced economies in order to identify common concerns and objectives.
- A workshop devoted to the problem of public policy in the field of conversion of S&T defence capabilities. The workshop should study different strategies: e.g. adjustment to NATO standards; creation of subsidiaries of Western defence enterprises in CEEs. (This workshop may be organised in Kiev or St. Petersburg.)

PRE-CONDITIONS

The transition economies need up-to-date, relevant information on world S&T trends and their own performance. On the one hand, they have to build proper indicators, evaluations, etc., systems which support policy making, international co-operation. On the other hand, they need up-to-date information on S&T and related issues.

- Access to information is critical in today's global economy. Today via the internet, news of current events and databases providing information on the latest technologies and scientific breakthroughs are virtually keystrokes away. Innovation would be difficult without the knowledge about current trends and about what scientific discoveries and technological developments are taking place globally. Information on partner capabilities, their environment, etc. is also important. Access to western European networks, both people and IT networks, is essential.

- Beside partnership building, access may be via IT networks. Among the transitioning nations, access to the internet appears to be very uneven. And in those countries with good internet access and availability of databases, database fees may hamper diffusion of the latest information. Grants to transitioning countries that can be used to purchase computer equipment, communications equipment, and as reimbursement for database "user fees", is one possible mechanism.
- For the promotion of technology transfer and innovation in Central and Eastern European countries, the potential Western partner countries need more information concerning R&D and innovation activities. However, the development of general indicators following OECD practices is well under way. What is needed is deeper knowledge and evaluation of existing potential (strengths and weaknesses) in particular regions or particular sectors or organisations. The right people need to be selected for co-operations. Both sides need a critical assessment of potential Western/Eastern partners and their track record to avoid further bad practice in building partnerships.
- There is a general feeling that Western partners would gain a better understanding of the transition economies through networking/business opportunities. S&T evaluation, technology audit are important tools, not only for policy making in CEECs but to support partnership building at all levels.
- The West needs to build up a critical mass of 'experts'/researchers to deal with sub-regional groups which could simplify dealings.

CONCLUSION

The workshop participants recommended many issues which are important for further network building, strengthening S&T policy making capabilities, innovativeness of CEECs.

There are many other issues; although mention was made, for example, of science and public acceptance, financing aspects, etc., it was not the objective of this workshop to handle all these issues.

There was a tremendous willingness to share experience, capability and knowledge and a feeling that the opportunities provided by the workshop should not be lost. Rather, it was felt that the workshop was a starting point for exploring networks, improving dialogue among NATO member states and partner countries.

Mechanisms now need to be provided/supported for that to happen to assist CEEs in the formulation and establishment of appropriate, S&I policies, technology transfer mechanisms, indicators, organisations, legal framework for improving and diffusing university-industry co-operation.

International organisations and advanced countries have an important role to play in the transformation process and they should not forget their responsibilities to network these countries instead of keeping them marginalised.

NATO Member Countries

Belgium

Mr. COX Roger
Licensing Director
Janssen Pharmaceutica N.V.
Turnhoutseweg 30.
2340 Beerse
Belgium
Phone: 32-1 460 3809; Fax: 32-1-460 3749
E-mail: rcox@janbe.jnj.com

Dr. HUYLEBROECK Guy
Bergbosstraat 217
B-9820 Merelbeke
Belgium
Phone: 32-9-230 6886; Fax: 32-9-230 6886
E-mail: feys@belspo.be

Canada

Mr. DESLANDES Yves
Group Leader
National Research Council Canada, Building M-12
Montreal Road Campus
K1A 0R6 Ottawa
Canada
Phone: 1-613-991 6347; Fax: 1-613-991 2384
E-mail: yves.deslandes@nrc.ca

France

Mr. DE BANDT Jacques
Directeur de Recuercue
CNRS, LATAPSES/IDEFI
250 Av. A. Einstein
F-06560 Valbonne
France
Phone: 33-4-9395 4230; Fax: 33-4-9365 3798
E-mail: debandt@unice.fr

Mr. WESTHOLM Gunnar
International Consultant
Formerly OECD Directorate for Science, Techn. and Industry
Private: 191 Boulevard Murat
F-75016 Paris
France
Phone: 33-1-4224 7112
E-mail: westholmg@aol.com

Germany

T. Gering new address:
70 a rue Principale
Siegen 67160
France

Dr. GERING Thomas
Director, Dipl-Ing.
Technology Licensing Bureau (TBL)
Rintheimer Str. 48.
D-76131 Karlsruhe
Germany
Phone: 49-721-623 214; Fax: 49-721-612 170
E-mail: sb81@rz.uni-karlsruhe.de

Dr. NEUMANN Helge
Director, International Office for Technology Cooperation
WISTA-Management GmbH
Rudower Chaussee 5
D-12489 Berlin
Germany
Phone: 49-30-6392 2230; Fax: 49-30-6392 2204
E-mail: helge@wista.de

Greece

Prof. KIRIAKIDIS George
MITOS S.A. Science & Technology Park of Crete
Box 1447, Voutes
711 10 Heraklion, Crete
Greece
Phone: 30-81-391 271/2; Fax: 30-81-391 295
E-mail: kiriakidis@iesl.forth.gr

Portugal	**Mr. OLIVEIRA Pedro** Office Coordinator Planning and Studies Office GEP-IST Av. Rovisco Pais, 1. 1096 Lisboa Codex Portugal Phone: 351-1-841 7751; Fax: 351-1-841 7614 E-mail: poliveira@gep.ist.utl.pt
The Netherlands	**Mr. ARUNDEL Anthony** Research Fellow MERIT P.O. Box 616 MD-6200 Maastricht The Netherlands Phone: 31-43-388 3702; Fax: 31-43-321 6518 E-mail: a.arundel@merit.inumaas.nl **Mr. DE VRIES Marc J.** Assistant Professor Eindhoven University of Technology, Faculty of Technology Management P.O.Box 513 MB-5600 Eindhoven The Netherlands Phone: 31-40-247 4629; Fax: 31-40-244 4602 E-mail: m.j.d.vries@tm.tue.nl
United Kingdom	**Mr. CAMERON Hugh** Director, PREST University of Manchester Oxford Road M13 9PL Manchester United Kingdom Phone: 44-161-275 5921; Fax: 44-161-273 1123 E-mail: h.cameron@man.ac.uk

Dr. DUFF Edward
Head of IPR and Legal Services
VUMAN Ltd.
Manchester Science Park
M15 6SH Manchester
United Kingdom
Phone: 44-161-226 8746; Fax: 44-161-226 5855

Dr. FEARS Robin
Director, Science Policy
SmithKline Beecham Pharmaceuticals
New Frontiers Science Park, Third Avenue
CM19 5AW Harlow, Essex
United Kingdom
Phone: 44-1279-644 648; Fax: 44-1279 644 926
E-mail: robin_b_fears@sbphrd.com

Ms. HILTON Janet
Technology Transfer Manager
VUMAN Ltd.
Manchester Science Park
M15 6SH Manchester
United Kingdom
Phone: 44-161-226 8746; Fax: 44-161-226 5855
E-mail: jan.hilton@man.ac.uk

Mr. HODKINSON Keith
European Patent Attorney
Marks & Clark
Sussex House, 83-85 Mosley Street
M2 3L9 Manchester
United Kingdom
Phone: 44-161-236 2275; Fax: 44-!61-236 5846

USA

Ms. BOND Jennifer Sue
Program Director, Science & Engineering Indicators Program
National Science Foundation
4201 Wilson Boulevard, Room 965
VA 22230 Arlington
USA
Phone: 1-703-306 1777/Ext. 6925; Fax: 1-703-306 0508
E-mail: jbond@nsf.gov

Mr. GEE Sherman
Communications R&D
Office of Naval Research (Code 313)
800 N. Quincy St.
VA 22217 Arlington
USA
Phone: 1-703-696 4844; Fax: 1-703-696 1331
E-mail: gees@onr.navy.mil

Mr. RAUSCH Larry
Senior Analyst
National Science Foundation
4201 Wilson Boulevard
VA 22003 Arlington
USA
Phone: 1-703-306 1775; Fax: 1-703-306 0508
E-mail: lrausch@nsf.gov

Partner Countries

Bulgaria

Dr. CHOBANOVA Rossitsa
Project Leader
Institute of Economics, Bulgarian Academy of Sciences
3 Aksakov Str.
1040 Sofia
Bulgaria
Phone: 359-2-761 227; Fax: 359-2-882 108
E-mail: ineco@bg.earn.acad.bg

Hungary

Dr. DINYA László
State Secretary
Ministry of Education and Culture, Republic of Hungary
Szalay u. 10-14.
H-1055 Budapest
Hungary

Dr. INZELT Annamária
Innovation Research Centre, IKU
Múzeum utca 17.
H-1088 Budapest
Hungary
Phone: 36-1-118 5674; Fax: 36-1-138 2438
E-mail: ainzelt@omk.omikk.hu

Ms. HAJDÚ Judit
Head of Patent Department for Mechanics and Electricity
Hungarian Patent Office
Garibaldi u. 2.
H-1054 Budapest
Hungary
Phone: 36-1-311 0883, -124 400/516; Fax: 36-1-311 0883

Dr. PATKÓS András
Head of Department
Ministry of Education and Culture, Republic of Hungary
Szalay u. 10-14.
H-1054 Budapest
Hungary
Phone: 36-1-331 4133; Fax: 36-1-331 4133
E-mail: patkos@ludens.elte.hu

Dr. SIEGLER András
Vice-president
National Committee for Technological Development
Szervita tér 8.
H-1052 Budapest
Hungary
Phone: 36-1-117 5900; Fax: 36-1-118 7998

Dr. SWAAN Wim
Institute of Economics, Hungarian Academy of Sciences
POB 262
H-1052 Budapest
Hungary
Phone: 36-1-319 2618; Fax: 36-1-319 3136
E-mail: swaan@econ.core.hu

Lithuania

Ms. MURAUSKAITÉ Danuté
Director
Science and Innovation Research Centre of Lithuanian Information Institute
Kalvariju G.3.
LT-2600 Vilnius
Lithuania
Phone: 370-2-726 900; Fax: 370-2-723 017
E-mail: lii@ktc.lii.lt

Poland

Prof. MARKOWSKI Tadeusz
University of Lodz
Rewolucji 1905 r 41
90-214 Lodz
Poland
Phone: 42-325 307; Fax: 42-325 307
E-mail: zerm@krysia.uni.lodz.pl

Romania

Mr. COCOS Stefan
Expert
Office for Standards, Inventions and Maros
5 Ion Ghica Str.
70018 Bucharest
Romania
Phone: 40-1-615 9066; Fax: 40-1-312 3819

Ms. PREDESCU Rolanda
Expert
Ministry of Research and Technology
21-25 Mendeleev Str.
70168 Bucharest
Romania
Phone: 40-1-650 2080/298; Fax: 40-1-210 9275
E-mail: rpredes@scou1.mct.ro

Russia

Prof. GOUSSAKOV Mikhail
Doctor of Economics
Institute for Socio-Economic Problems RAS
S. Kovalevskaya Str. 14-2-35.
195256 St. Petersburg
Russia
Phone: 7-812-533 7601; Fax: 7-812-316 0521

Ms. KUZNETSOVA Irina
Head of Division
Centre for Science Research and Statistics
11, Tverskaya Str.
103905 Moscow
Russia
Phone: 7-095-229 7818; Fax: 7-095-924 2828

Slovakia	**Mr. BERNADIC Frantisek** Eng/Principal Nesti Delegate of Slovakia INFOSTAT Dubravska cesta 3, 84221 Bratislava Slovakia Phone: 42-1-7370 9240; Fax: 42-1-7537 1463 E-mail: bernadic@infostat.sk
Ukraine	**Dr. EGOROV Igor** Steps Centre, National Academy of Sciences of Ukraine 60 Shevchenko Bwd. 252032 Kiev Ukraine Phone: 380-44-219 1486; Fax: 380-44-216 9591 E-mail: igor@yan.kiev.ua
	Prof. NESTEROV Vladislav President, Academician, Dr. Sci (Tech) Ukrainian Technological Academy Kutuzov Str. 18/7 252133 Kiev Ukraine Phone: 380-44-295 7289; Fax: 380-44-295 7289. -290 1002
	Prof. SKRIPNIK Urij Vice-president, Academician, Dr. Sci (Tech) Ukrainian Technological Academy Kutuzov Str. 18/7 252133 Kiev Ukraine Phone: 380-44-295 7289; Fax: 380-44-295 7289. -290 1002

Other Countries

Japan	**Prof. HAYASHI Takabumi** Rikkyo University 3-34 Nishi-Ikebukuro, Toshima-ku 171 Tokyo Japan Phone: 81-3-3985 2279; Fax: 81-3-3985 4096 E-mail: akabumi@rikkyo.ac.jp
Yugoslavia	**Mr. KUTLACA Djuro** Head of Centre "Mihajlo Pupin" Institute Science and Technology Policy Research Centre Volgina 15 11000 Belgrade Yugoslavia Phone: 381-11-774 452; Fax: 381-11-775 870 E-mail: dkutlaca@labtel.imp.bg.ac.yu

International Organisations

EC	**Mr. SCHMIEMANN Manfred** Principal Administrator European Commission, DG XIII/D/1, EUFO 2165 Rue Alcide de Gasperi L-2920 Luxemburg Luxemburg Phone: 352-4301 33353; Fax: 352-4301 32073 E-mail: manfred.schmiemann@lux.dg13.cec.be
EPO	**Mr. CLARKE Nigel S.** European Patent Office Schottenfeldgasse 29 A-1092 Vienna Austria Phone: 43-1-5212 6338; Fax: 43-1-5212 63292 E-mail: clarke@epo.e-mail.com
NATO	**Mr. JUBIER Alain H.** Programme Director, Science and Technology Policy NATO Scientific Affair Division B-1110 Brussels Belgium Phone: 32-2-707 4668; Fax: 32-2-707 4232 E-mail:
OECD	**Mr. FERNÉ Georges** Principal Administrator OECD 2 Rue André Pascal 75775 Paris Cedex 16 France Phone: 33-1-4524 9346; Fax: 33-1-4524 9767 E-mail: georges.ferne@oecd.org

Hungarian Patent Demonstrators

Hungary	**Ms. BAKOS Éva** Hungarian Patent Office Garibaldi u. 2. H-1054 Budapest Hungary
	Ms. TAKÁCS Zsuzsa Hungarian Patent Office Garibaldi u. 2. H-1054 Budapest Hungary

Friends of Rapporteaurs

Hungary	**Ms. GÁL Noémi** Friend of Rapporteaurs Szilvás u. 11/5 H-2400 Dunaújváros Hungary Phone: 36-25-285 144; E-mail:
	Mr. CSERNA Péter Friend of Rapporteaurs Szilvás u. 11/5 H-2400 Dunaújváros Hungary Phone: 36-25-285 144; E-mail: